Thermal Physics
Entropy and Free Energies

Second Edition

Joon Chang Lee

Department of Physics and Astronomy
University of Southern Mississippi

 **World Scientific**

NEW JERSEY · LONDON · SINGAPORE · BEIJING · SHANGHAI · HONG KONG · TAIPEI · CHENNAI

Published by

World Scientific Publishing Co. Pte. Ltd.

5 Toh Tuck Link, Singapore 596224

USA office: 27 Warren Street, Suite 401-402, Hackensack, NJ 07601

UK office: 57 Shelton Street, Covent Garden, London WC2H 9HE

British Library Cataloguing-in-Publication Data

A catalogue record for this book is available from the British Library.

ISBN-13 978-981-4340-76-2
ISBN-10 981-4340-76-6

Printed in Singapore by World Scientific Printers.

I dedicate this book to my wife Hye-Sook Lee

Preface

The purpose of thermal physics is to study thermal properties of matter, and thermal matter consists of astronomically large numbers of particles. As a result of this huge number, these molecules can exist in an astronomically large number of different degenerate states. Each of these microstates shows what the particles are doing in their configuration space. They constantly move around in that space. Since each microstate represents what the particles do, the large number of microstates means a large number multiplicity of what the system does. This large number of multiplicity is the heart and soul of thermal physics. A simple postulate is introduced to deal with this situation. It says that, given enough time, thermal systems will visit all these microstates with equal frequency and they never get stuck in one; rather they are incessantly moving around from one to another. We then group the microstates into macrostates, and begin to study their macroscopic behavior. The postulate has an immediate — and almost obvious — consequence that the size of each macrostate is the only thing that matters, which introduces the idea of entropy and the second law of thermodynamics, i.e. the maximum entropy principle. For systems in contact with various types of reservoirs, the entropy principle is then transformed into various minimum free energy principles. This is the essential and underlying theme of thermal physics.

We will sing this theme in several different variations. Molecules do X to maximize their entropy. Molecules do XX to minimize their free energy. Spins do Y to maximize their entropy. Spins do YY to minimize their free energy. It was the main theme of the first edition of this title.

Entropy may be regarded as a measure of disorder. I will attempt to clarify what is meant precisely by order and disorder with a large number of examples drawn from our daily life. The present second edition is an expansion of the first edition under the theme of order and disorder. Looking at entropy from this perspective is not only helpful, but also allows us to cover many more topics, particularly the well-ordered behavior of quantum systems near the absolute zero temperature. Thermal behavior tends to be disorderly at high temperatures and more orderly at low temperatures. The topics of quantum ideal gases and quantum fluids are now covered under the theme of order. In spite of the expanded themes, several important topics of thermodynamics could not find convenient spots and therefore I have added a new chapter exclusively for thermodynamics to ensure that it gets its fair share of coverage. Several chapters have been vastly expanded so as to cover more topics.

Two most basic thermal systems are liquid-gas systems and magnetic systems. Among the two the former are more ubiquitous, and among the liquid-gas systems the most basic are the ideal gases. Deriving the entropy of the ideal gases is then the most basic fundamental task, and one may therefore regard the Sackur-Tetrode equation as the centerpiece. In the first edition, I attempted to derive the Sackur-Tetrode equation at the basic level of microcanonical ensemble formalism, but unfortunately there was an error. The error has been corrected.

Let me now explain what led me to write this book and how it has been written. When I went to my first thermodynamics class as an undergraduate, I was puzzled and even angered. The material did not seem like physics and I could not understand why it was a required course. I sat at the back and read something else most of the time. By a miracle, I barely passed the course and graduated. As a graduate student, I felt the same about Statistical Mechanics but it was more tolerable. Then I made a fatal 'mistake'. I chose the subject 'Bose fluid at $T = 0$' for my thesis. Taking a glance at the references given to me by my soon-to-be thesis adviser, it looked like a problem of quantum mechanics which was what I wanted. It was only later when I tried to extend my thesis work to non-zero temperatures that I realized that I had trapped myself in the very field that I had resented during all my student days. I could still navigate around for awhile to practice the many-body theory based on the non-relativistic quantum field theory, but my weakness in the basic thermal physics put a serious limit to what I could do. The ugly reality ultimately closed in on me. I grabbed several introductory books which gave me a crash course. I have written this book in the hope that it may keep students from following the same misguided and costly path that I followed. I feel that I am well qualified for this task because I know how they often get themselves into this pitfall. They think that they already know everything there is to be learnt about things like temperature, pressure and heat, and most importantly, they tend to think that these quantities are all mechanical in nature. They fail to recognize the essential difference between thermal physics and mechanics. In fact, they do not recognize the importance of the fundamental postulate. How do I handle this issue? I have no magic trick. I shall just expound on them in the way I describe below.

In writing this book, I did not try to be succinct. My style is just the opposite. I will expound on the basic ideas again and again, this way and that way, on this page and again on the next page. I joke one way and then another. I bring in this or that analogy from our daily life or something we are familiar with. The book is written in such a casual and unorthodox style. To paraphrase a reviewer, this is not everybody's cup of tea. If you have a burning question such as: (1) Just what is entropy and can it be put in a form I can feel comfortable with? Or (2) what is free energy and what do I do with it? ... then the book may serve you well.

I envision two groups of readers. The first comprises beginners who wish to learn the basic concepts of thermal physics and get a sense of how it works. The second group has already taken a formal course, but wish to revisit the subject for the reasons explained above at length. As such, it should also serve those who have taken a course but who wish to take a crash course before attempting a more advanced course.

The first edition was very readable, thanks to my two daughters, Mi-Kyoung Lee and Sue-Kyoung Lee, who graciously undertook the arduous task of perfecting my English. Mr. Stanley Liu, then the acquisitions editor of World Scientific went out of his way and contributed similarly. All the parts that I did not change for the second edition should remain readable, but many chapters have been expanded while several new chapters have been added. I fear that I would not be able to have anyone correct and edit my English for these chapters. The best I can do under this circumstance is to warn the reader about my difficulty in English. I have never learned when to use the words 'the' and 'a' appropriately, and more importantly, when not to use them. I hope that you will be able to bear with me on this shortcoming.

A good portion of this edition was written while the author was receiving grueling medical therapies. I was still able to work on the book although at a much reduced pace. That was only possible thanks to the dedicated care given to me by my wife and the constant support from my two daughters.

I thank all my family members and all our friends for their support and prayers.

The author also wishes to express his gratitude to World Scientific senior editor Alvin Chong. He meticulously read through the manuscript to uncover many faulty spots and diligently went out of his way to deliver this book project.

Joon Chang Lee
October 2010
Email: joon.lee@usm.edu

Contents

Chapter 1

Introduction to Thermal Physics

A theory is the more impressive the greater the simplicity of its
premises, the more different kinds of things it relates, and the
more extended is its area of applicability. Therefore the deep
impression that classical thermodynamics made upon me. It is
the only physical theory of universal content concerning which
I am convinced that, within the framework of applicability of
its basic concepts, it will never be overthrown.

Albert Einstein: Philosopher-Scientist, p. 33, The Library of
Living Philosophers, Vol. VII (edited and translated by Paul
Arthur Schilpp)

1.1 Bird's-Eye View of Thermal Physics

This book is a story of thermal physics. In order to tell it appropriately, we
have to introduce several concepts and the fundamental postulate. Among
all the concepts the most important one is entropy which, I fear, is quite
subtle and can potentially make some people feel mystified. Just so that the
reader can get a sense of what entropy is and how thermal physics works
before we get into the details of the narrative, I have made up several stories
from our daily life.

Consider a man coming to the greater Hattiesburg area for the first
time. The area consists of Oak Grove, Hattiesburg, and Petal. All three
sections are connected and it is difficult for a newcomer to tell which is
which. This man arrives in the area from I-59 entering into Oak Grove and
thinks that if he drives around a few blocks he will find his destination in no
time. That is not so. He has no map and soon becomes lost. Hattiesburg
has about 500 blocks, Petal and Oak Grove each about 50 blocks. He does
not know where he is. He tries this block and that, turns left and then

right, goes under this bridge and over another one. Twelve hours go by, but he is still wandering around. He is so overwhelmed by the charm of this city (it is true!) that he has forgotten why he came here in the first place. He wants to remain lost for another twelve hours, and more. Question (1): what are his chances of being in Hattiesburg after this many hours of wandering around? ANS: $500/(500 + 50 + 50) = 0.83$. Question (2): If he is in Hattiesburg right now, what are his chances of getting back to Oak Grove after another 12 hours or, say, 24 hours? ANS: $50/(500 + 50 + 50) = 0.083$. The odds are in favor of "yes" for the first question and "no" for the second question, but not overwhelmingly.

Let us now exaggerate and suppose that Hattiesburg has a total of 10^{10} blocks while Oak Grove and Petal still have 50 blocks each. The odds are then near a hundred percent for the first question and near zero for the second question. Why are the odds of his getting back to Oak Grove so overwhelmingly small? Remember that this man is lost but has no desire to go to any particular place. From his starting position on the map, draw all the paths that he can potentially follow in 12 hours. The probability of his following any one of these paths is the same as for the others, but an overwhelmingly large number of the paths end up in Hattiesburg.

Since we gave Hattiesburg enough blocks to make it the largest city in the world, we might as well give its police department a patrol helicopter. A patrol officer has been following our traveler. The officer sees the traveler quite differently. He is not lost; he has a definite purpose. The more blocks there are for him to wander around, the more he seems to enjoy it. He arrived in Oak Grove, but he soon began moving around in Hattiesburg with so many blocks to wander around. The officer invents the term "tropy" for the number of city blocks the traveler may visit, or rather its logarithm. "The man in the brown Pontiac station wagon is in pursuit of tropy, which is why he traveled from Oak Grove to Hattiesburg. This man is after tropy and tropy alone." If there were another city in the area which offered him even more "tropy", he would move to that city in due time. His tropy therefore never decreases!

There are $10^{10} + 50 + 50$ blocks in the area, and our traveler's location at each block constitutes the microstate of his whereabouts. If we describe him as being in Petal, we will have specified his macrostate since he can be in Petal by being in any of its 50 blocks. He started his journey from one of the two smaller macrostates; after a while it is overwhelmingly probable that he will end up in the largest macrostate. The movement is one-way. He will not go back. There are too many ways to remain lost in Hattiesburg to prevent him from returning to Oak Grove. Of course, there is nothing special about being in Hattiesburg. If we were to arbitrarily designate 50 blocks of Hattiesburg including the block in which he is currently located and gave this area a new name, everything we have said about the chances of our traveler returning to Oak Grove would apply to the chances of his getting back to these 50 blocks as well.

If you can imagine the kind of world described by this example, you have entered the world of thermal physics. The brown Pontiac was meant to symbolize thermal systems like a gas or liquid confined in a container, or a piece of metal. What is it that causes these systems to posses thermal properties? *These systems consist of astronomically large number of particles.* Just to get an idea, take 1 cm × 1 cm × 1 cm of air. In that small and seemingly empty volume of space, there are as many as 10^{19} molecules. Because of this huge number of constituents, *the number of microstates in which they can exist turns out to be even more astronomical*; it is in the order of e^N where N is the number of particles. Shall we say that it is super-duper astronomical? It is this truly large number of microstates that gives rise to their thermal properties. Indeed the heart and soul of thermal physics is the number of microstates or equivalently what the police officer called 'tropy' which is a convenient abbreviation of the Boltzmann entropy defined by

$$S = k_B \ln \Gamma \qquad (1.1)$$

where k_B is called the Boltzmann constant and Γ is the number of microstates.

Just as the brown Pontiac wanders around from one city block to another, so does our thermal systems from one microstate to another. Since they change their states constantly rather than staying in one particular state, we may call Γ the multiplicity. What matters is not so much what they do in each microstate. As odd as it may sound, it is *how many different ways* they can do what they do.

All microstates may be divided into macrostates according to some convenient and physically meaningful way. For example, suppose that in macrostate A all molecules are in half of the container volume leaving the other half in vacuum, and in macrostate B molecules are everywhere leaving no part in vacuum. As we will find out later, B turns out to be much much larger than A. Put the molecules in the small macrostate A. In due time, they will move to the larger macrostate B and will never return back to A by themselves. Thus we may say that thermal systems are after entropy and entropy only.

The number of microstates Γ could also be called repertoire. We all strive for repertoire. Band directors and orchestra directors would all like to have a decent repertoire so that they can perform any of it at convenient times. Our football and basketball coaches would love to have several repertoires so that they can surprise their opponents. But these repertoires are in the order of 10 in most cases or in the order of 100 for musicians. The repertoire of thermal systems are in the astronomical order, and they do not keep them for show; they perform them all the time. The size of their repertoire only increases.

- P represents both probability and pressure. When it represents a certain probability, it comes with the word "probability" and as a function of one or two variables, like $P(U_s)$. Although pressure is dependent on many factors, it is plainly written as P.
- t represents time except when we talk about static critical phenomena in Chapter 8 where it represents the relative distance to the critical temperature, $t = |T - T_C|/T_C$.
- m represents both molecular mass and magnetic dipole moments, but nowhere do we talk about molecular systems and spin systems at the same time.
- $\beta = 1/k_B T$ appears in all Boltzmann factors like $\exp(-\beta E_\nu)$. In Chapter 8, it represents the order parameter exponent.
- S represents entropy everywhere except briefly in Sec. 8.2 where $S(q)$ represents a structure factor. Similarly, G represents the Gibbs free energy everywhere except briefly in Sec. 8.2 where $G(r)$ represents a correlation function. C represents the heat capacity everywhere except briefly in Sec. 4.8 where it represents the electric charge capacitance. There are a few more of this type.

Chapter 2

All You Need to Know to Read the Rest of the Book

Our purpose in the rest of the book is to elaborate on what was presented in the previous chapter and to transform it into something that can provide a physical foundation of thermal physics. It is my goal to do so in a painless and self-contained manner. To that end, we need some preparation. I hope that the last chapter did not mislead you to think that a few more jokes will do it. I will go through a number of items. In ordinary books, most of these would be at the end of the book as Appendices, but I will deviate from that custom. If you would go through the items now before we proceed, even if you do it casually, it will prove helpful when you encounter them in later chapters. I did not prepare them for one single readership. They contain some details, but these are for advanced readers and need not cause too much concern for beginners. At the minimum, the reader should acquire some sense as to what they are and how they work. As a simple guide, three stars ($\star\star\star$) mean that I would like you to go through it, two stars ($\star\star$) means that it is useful but not critical, and one star ($\star$) means that I believe that you already know it; if not take a moment to refresh yourself with it.

2.1 Several Taylor Series Expansions ($\star$)

We will quite often approximate $f(x_0 + x)$ with first few terms of its Taylor series expansion for small x. I will list three below:

$$\exp(x) = 1 + x + \frac{x^2}{2!} + \frac{x^3}{3!} + \cdots \tag{2.1}$$

$$\ln(1+x) = x - \frac{x^2}{2} + \frac{x^3}{3} - \frac{x^3}{4} + \cdots \tag{2.2}$$

$$f(x+\delta x, y+\delta y) = f(x,y)+f_x\delta x+f_y\delta y+\frac{1}{2!}(f_{xx}\delta x^2+f_{yy}\delta y^2+2f_{xy}\delta x\delta y)+\cdots$$

$$(2.3)$$

2.2 The Dirac δ Function ($\star\star$)

When we say that $f(x)$ is a function of x, we mean that given a value for its argument x it will give us precisely what the corresponding $f(x)$ is. The Dirac δ function $\delta(x-x_0)$ is not such a function. Instead it is defined by by what it should do under an integral sign. It should be zero everywhere except near $x = x_0$ where it makes such an infinitely high but infinitely narrow peak so that the area under the peak should come out to be unity:

$$\int_{-\infty}^{\infty} \delta(x-x_0)dx = 1.$$

$$(2.4)$$

If we use it with a well-behaving smooth function $f(x)$, since the peak is so narrow, it should satisfy

$$\int_A^B \delta(x-x_0)f(x)dx = \begin{cases} f(x_0) & \text{if } B < x_0 < A, \\ 0 & \text{otherwise}. \end{cases}$$

$$(2.5)$$

There are several analytical attempts to represent such characteristics. The most popular one is the following:

$$\delta(x-x_0) = \frac{1}{2\pi}\int_{-\infty}^{\infty} \exp\left[ik(x-x_0)\right]dk.$$

$$(2.6)$$

2.3 Gamma Function ($\star\star\star$)

The Gamma function is defined by

$$\Gamma(m) = 2\int_0^{\infty} \exp(-u^2)u^{2m-1}du.$$

$$(2.7)$$

By changing the integration variable it may also be rewritten as

$$\Gamma(m) = \int_0^{\infty} \exp(-t)t^{m-1}dt.$$

$$(2.8)$$

Notice that it is a well-defined function regardless whether m is an integer or not. We can see why this is a useful function once we perform the integration by parts:

$$\Gamma(m) = \left[-\exp(-t)t^{m-1}\right]_0^\infty + (m-1)\int_0^\infty \exp(-t)t^{m-2}dx$$
$$= (m-1)\Gamma(m-1). \tag{2.9}$$

If m is an integer, we may keep repeating this until we reach $\Gamma(1)$ to find

$$\Gamma(m) = (m-1)!\Gamma(1). \tag{2.10}$$

Since $\Gamma(1) = \int_0^\infty \exp(-t)dt = 1$, this means that for integer m

$$\Gamma(m) = (m-1)! \tag{2.11}$$

Thus if m is an integer, $\Gamma(m)$ reproduces the factorial function. If m is not an integer, $m!$ is not defined, but we may analytically extend it into the non-integer domain using $\Gamma(m+1)$. Thus $(1/2-1)! = \Gamma(1/2)$. We may practice the same and write $0! = \Gamma(1) = 1$; for the last equality, set $m = 1$ in Eq. (2.8) and carry out the integral.

We will use heavily the factorial function of very large m. The Γ function makes the task manageable. In particular, it is hard to approximate the bulky and awkward factorial function but the gamma function offers a convenient way to handle it.

In the meantime, let us find out how well $\Gamma(n+1/2)$ interpolates between $n! = \Gamma(n+1)$ and $(n-1)! = \Gamma(n)$ where n is an integer. Using the form of Eq. (2.12),

$$\Gamma(n + \frac{1}{2}) = 2\int_0^\infty \exp(-u^2)u^{2n}du \tag{2.12}$$

which we can calculate if we can handle the integral

$$I = \int_{-\infty}^\infty \exp(-ax^2)dx. \tag{2.13}$$

The trick is to multiply it with

$$I = \int_{-\infty}^\infty \exp(-ay^2)dy \tag{2.14}$$

to obtain

$$I^2 = \int_{-\infty}^\infty \int_{-\infty}^\infty \exp(-ax^2)\exp(-ay^2)dxdy \tag{2.15}$$

which is an integral covering the entire xy plane. Switch to the polar coordinates r and θ. Then $x^2 + y^2 = r^2$ and $dxdy = 2\pi rdr$. The integral is now evaluated to give

$$I^2 = \pi \int_0^\infty \exp(-ar^2)2\pi r = \frac{\pi}{a} \qquad (2.16)$$

and

$$I = \int_{-\infty}^\infty \exp(-at^2)dt = 2\int_0^\infty \exp(-at^2)dt = \sqrt{\pi}a^{-1/2}. \qquad (2.17)$$

The last line of this equation is going to be our workhorse equation because it will do so much for us. So let me rewrite the last line so that we can see it better.

$$2\int_0^\infty \exp(-at^2)dt = \sqrt{\pi}a^{-1/2}. \qquad (2.18)$$

Compare this with Eq. (2.12). The left-hand sides are different, but they can be made the same by merely differentiating Eq. (2.18) n times with respect to a, and then set $a = 1$; the differentiated right-hand side of Eq. (2.18) is the desired integral of Eq. (2.12). By trying first a few, we find the following pattern,

$$\Gamma(n + 1/2) = \frac{1.3.5.\cdots(2n-1)}{2^n}\sqrt{\pi}. \qquad (2.19)$$

Let me list some of the $\Gamma(n + 1/2)$ along with $\Gamma(n) = (n-1)!$

$$\begin{cases} \Gamma(1/2) = \sqrt{\pi} = 1.772 \\ \Gamma(1) = 1 \\ \Gamma(1 + 1/2) = \frac{1}{2}\sqrt{\pi} = 0.886 \\ \Gamma(2) = 1! = 1 \\ \Gamma(2 + 1/2) = \frac{1 \times 3}{2 \times 2}\sqrt{\pi} = 1.329 \\ \Gamma(3) = 2! = 2 \\ \Gamma(3 + 1/2) = \frac{1 \times 3 \times 5}{2 \times 2 \times 2}\sqrt{\pi} = 3.32 \\ \Gamma(4) = 3! = 6 \end{cases} \qquad (2.20)$$

It is clear that the Γ function does indeed smoothly extend $n!$ into the non-integer domain. In the limit of very large n,

$$\Gamma(n + 1/2) \approx \Gamma(n) \tag{2.21}$$

is a good approximation.

We will need Γ functions of very large n, and fortunately there is a very good and useful approximation. Let us start with

$$\Gamma(n + 1) = \int_0^\infty \exp(-x)x^n dx. \tag{2.22}$$

Note that the integrand is zero both at $x = 0$ and at $x = \infty$. Let us find where it is peaked.

$$\frac{d}{dx}[\exp(-x)x^n] = \exp(-x)x^{n-1}(n - x) = 0. \tag{2.23}$$

It is peaked at $x = n$. Taking the derivative once more, we find that the peak is quite sharp. So, if we approximate the integrand near the peak in an exponential form, it is most likely that the bulk of the integration will be captured. Thus let $x = n + \xi$ where $\xi \ll n$, and take

$$\ln[x^n \exp(-x)] = n\ln(n + \xi) - (n + \xi). \tag{2.24}$$

Since $\xi \ll n$,

$$\ln(n + \xi) = \ln[n(1 + \xi/n)]$$

$$= \ln n + \ln(1 + \xi/n)$$

$$\approx \ln n + \xi/n - \frac{1}{2}\frac{\xi^2}{n^2}, \tag{2.25}$$

where $\ln(1 + \xi/n)$ was Taylor expanded and then only terms up to the second order were kept.

Substituting this into Eq. (2.24),

$$\ln(x^n e^{-x}) = n\ln n + \xi - \frac{1}{2}\frac{\xi^2}{n} - n - \xi$$

$$= n\ln n - n - \frac{\xi^2}{2n}. \tag{2.26}$$

Exponentiating both sides,

$$x^n \exp(-x) \approx n^n \exp(-n)\exp(-\xi^2/2n) \tag{2.27}$$

and we have reduced the integrand into one exponential function as we intended. Moreover, because of the rapidly decreasing exponential function, relaxing the condition $\xi \ll n$ does little or no harm. This gives us the approximation

$$\Gamma(n+1) \approx n^n \exp(-n) \int_{-\infty}^{\infty} \exp(-\xi^2/2n)d\xi$$

$$= n^n \exp(-n)(2\pi n)^{1/2} \tag{2.28}$$

which is called Stirling's approximation. For $\ln \Gamma(n+1)$, we may ignore the last factor $(2\pi n)^{1/2}$ and write

$$\ln \Gamma(n+1) = \ln n! \approx n \ln n - n \tag{2.29}$$

which is also called Stirling's approximation. This equation will be kept very busy in later chapters. On one or two occasions, we will use Eq. (2.28) for a better accuracy.

2.4   The Gaussian Integrals ($\star \star \star$)

We will be dealing with quantities of the form

$$Z(a) = \int_0^\infty dx \, \exp(-ax^2) = \frac{1}{2} \left(\frac{\pi}{a} \right)^{1/2} \tag{2.30}$$

and their relationship to quantities of the form

$$I(n, a) = \int_0^\infty dx x^{2n} \exp(-ax^2). \tag{2.31}$$

I hope that you can easily see that

$$I(n, a) = \frac{\partial^n Z(a)}{\partial(-a)^n}. \tag{2.32}$$

We will practice this trick quite often. Please make sure that you can pull this trick out at anytime.

2.5   Hypersphere in d-Dimensional Space ($\star \star \star$)

Take a sphere in three dimensions. We know how to calculate its volume as a function of radius. We also know how to calculate the surface area of the sphere as a function of the radius. We normally use the term 'volume' only for objects in three dimensions. As you will see in the next chapter, however, we need to work in an imaginary space of very, very high dimensions where we need to speak of a sphere and calculate its volume and the surface area. Thus we wish to extend what we do in three dimensions to much larger dimensions. Well, before we attempt dimensions larger than three, we had better see if we can do the same in dimensions lower than three, namely, in two dimensions and in one dimension.

We shall be guided by a simple guideline. By volume, we will mean something which may be regarded as a measure of the totality of what is in the sphere. By surface area, we will mean something which may be regarded as a measure of the boundary that divides what belongs to the sphere and what does not.

In three dimensions, the volume is $4\pi r^3/3$ and the surface area is $4\pi r^2$. In two dimensions, we do not know how to look up or down; we only see east-west and north-south. The sphere is therefore reduced to a circle. Its volume is πr^2 and the surface area is $2\pi r$. In one dimension, we can only see east-west; the sphere is further reduced to a line extending from $x = -r$ to $x = +r$. Its volume is therefore $2r$ and the surface area is 2. Let us display all these so that we can find out if there is any pattern:

$$\begin{cases} \text{Vol} = 4\pi r^3/3, & \text{Surface} = 4\pi r^2, & d = 3 \\ \text{Vol} = \pi r^2, & \text{Surface} = 2\pi r, & d = 2 \\ \text{Vol} = 2r, & \text{Surface} = 2, & d = 1 \end{cases} \quad (2.33)$$

We see that volume is proportional to r^d and the surface area to r^{d-1} for $d = 3, 2, 1$. Also notice that the proportionality constant for the surface area in $d = 3$ is 4π which is the total (solid) angle subtended by the surface area. In $d = 2$, it is 2π which is again the total angle subtended by the surface area. We may claim this similarly for $d = 1$. In addition, if the volume is differentiated with respect to the radius, it gives the surface area.

Now our purpose is to extend these relationships to a multidimensional space. How do we calculate the volume and surface area of the 'hypersphere' as a function of radius r?

Consider a sphere in d-dimensions. Call its volume and surface area $V_d(r)$ and $A_d(r)$, respectively. Extending the pattern of relationship between the surface area and the radius in the lower dimensions, we write

$$A_d(r) = \Omega_d r^{d-1} \quad (2.34)$$

and, extending the relationship between the differential volume element and radius,

$$dV_d(r) = A_d(r)dr \qquad (2.35)$$

which gives

$$V_d(R) = \int_0^R A_d(r)dr = \Omega_d \int_0^R r^{d-1}dr = \Omega_d R^d/d. \qquad (2.36)$$

Now we need to determine Ω_d. One way is to extend the pattern that we noticed earlier to d-dimensions. I will leave this for you as an exercise and take a different route. We will find it more conveniently if we can think of something in d dimensions which can be calculated once involving Ω_d and once without involving it. Someone was clever enough to suggest

$$I_d = \int_0^\infty \exp(-r^2)dV_d(r). \qquad (2.37)$$

It is a weighted volume of the entire space which is of no interest by itself to us but serves us nicely because it may be calculated in the two ways mentioned above. First let us write it involving Ω_d by substituting Eq. (2.35) here to obtain

$$I_d = \Omega_d \int_0^\infty \exp(-r^2)r^{d-1}dr = \Omega_d \frac{\Gamma(d/2)}{2} \qquad (2.38)$$

where I hope you will recognize the Γ functions. Next, let us write I_d entirely in terms of the extended Cartesian coordinates $(x_1, x_2, x_3, \ldots, x_d)$, thus avoiding Ω_d. To do so, substitute

$$dV_d(r) = dx_1 dx_2 dx_3 \cdots dx_d \qquad (2.39)$$

and

$$r^2 = x_1^2 + x_2^2 + x_3^2 + \cdots + x_d^2 \qquad (2.40)$$

into Eq. (2.37) to obtain

$$Id = \int_{-\infty}^\infty dx_1 \int_{-\infty}^\infty dx_2 \cdots \int_{-\infty}^\infty dx_d \exp\left(-(x_1^2 + x_2^2 + \cdots + x_d^2)\right)$$

$$= \left(\int_{-\infty}^\infty \exp(-x^2)dx\right)^d = \left(\sqrt{\pi}\right)^d. \qquad (2.41)$$

Equating Eq. (2.38) and Eq. (2.41), we find

$$\Omega_d = \frac{2\pi^{d/2}}{\Gamma(d/2)}.$$
(2.42)

Upon substituting this into Eqs. (2.34) and (2.36), we obtain

$$A_d(R) = \frac{2\pi^{d/2} R^{d-1}}{\Gamma(d/2)}$$
(2.43)

and

$$V_d(R) = \frac{2\pi^{d/2} R^d}{\Gamma(d/2)}.$$
(2.44)

Notice that when $(d/2 - 1)$ is a positive integer we may write $\Gamma(d/2) = (d/2 - 1)!$ When $d/2 - 1$ is not an integer, we may still write it as a factorial function but what we really mean is $\Gamma(d/2)$ which does not require an integer for its argument. With that understood, we may finally write

$$A_d(R) = \frac{2\pi^{d/2} R^{d-1}}{(d/2 - 1)!}$$
(2.45)

and

$$V_d(R) = \frac{2\pi^{d/2} R^d}{(d/2 - 1)!}.$$
(2.46)

We will use these results in the next chapter.

Before we leave here, it is worthwhile to examine Ω_d. It is given by

$$\Omega_d = \frac{A_d(R)}{R^{d-1}} = \frac{2\pi^{d/2}}{(d/2 - 1)!}.$$
(2.47)

It is the generalized solid angle that subtends the surface area: for $d = 3$, $\Omega_3 = 4\pi$; for $d = 2$, $\Omega_2 = 2\pi$; for $d = 1$, $\Omega_1 = 2$.

2.6 Fourier Transformations (⋆⋆)

Let $g(\vec{r})$ be a function of position vector $\vec{r}$. Its Fourier transform $h(\vec{q})$ is defined by

$$h(\vec{q}) = \frac{1}{(2\pi)^{3/2}} \int d^3r \, g(\vec{r}) \exp(i\vec{r} \cdot \vec{q}),$$
(2.48)

where $\vec{q}$ is called a wave vector, and

$$\exp(ix) = \cos(x) + i\sin(x)\,, \tag{2.49}$$

and $i = \sqrt{-1}$.

The transform $h(\vec{q})$ tells us how much of a pattern there is which tends to repeat itself in the direction of $\vec{q}$ with the periodicity of $\lambda = 2\pi/q$. To see this, consider the following simple case in one dimension. Suppose that the function $g(x)$ is such that $g(x_n) = g(x_0)$ where $x_n = x_0 + n\lambda$ and $n = 1, 2, 3, \dots$. Now let $q = q_0 = 2\pi/\lambda$ and examine how $g(x_0), g(x_1), g(x_2)$, etc. contribute to $h(q)$. Since all $q_0 x_n$'s differ from each other only by an integer multiple of 2π which amounts to no difference for trigonometric functions, their contributions add up to make a large net contribution.

Given $h(\vec{q})$, we may find $g(\vec{r})$ by inverting the transformation,

$$g(\vec{r}) = \frac{1}{(2\pi)^{3/2}} \int d^3q\, h(\vec{q}) \exp(-i\vec{r}\cdot\vec{q})\,. \tag{2.50}$$

This equation says that the function $g(\vec{r})$ may be regarded as a superposition of many trigonometric functions. Each of the trigonometric functions maintains a different repeating pattern which contributes to $g(\vec{r})$ with a weight given by $h(\vec{q})$. Thus, if $h(\vec{q})$ is very large compared to the rest, it simply means that *there is a recurring spatial pattern of length scale of* $\lambda = 2\pi/q$. This is all we need to remember. I do not expect you to carry out any Fourier transformations while reading this book.

One can actually carry out the transformation to find out the prominent patterns using light scattering experiments where one lets light scatter from the sample, or neutron scattering experiments where one lets neutron beams scatter from the sample.

2.7 Functional Derivative ($\star\star$)

Suppose that A is a functional of $f(x)$, i.e. it is a function of function $f(x)$, like

$$A = \int_0^\infty f(x')^2 dx'\,. \tag{2.51}$$

Suppose that we change $f(x')$ by a slight amount to $f(x') + \delta f(x')$. Since the integration covers the entire range of x', from 0 to ∞, the change of function at each point could potentially contribute a different amount of change to A. In the limit of infinitesimal change $\delta f(x')$, what is the rate of change of A versus the change in the function at, say, $x' = x$, namely $\delta f(x)$? That is called the functional derivative, $\delta A/\delta f(x)$. Notice that this is not about the total change of A due to the changes in $f(x')$ everywhere. Rather it is the change at $x' = x$ only. So the sought-after change δA is given by

$$
\begin{aligned}
\delta A &= \int_0^\infty \left[\left(f(x') + \delta f(x') \right)^2 - f(x')^2 \right] \delta(x' - x) dx' \\
&= \int_0^\infty \left\{ 2f(x')\delta f(x') + [\delta f(x')]^2 \right\} \delta(x' - x) dx' \\
&= 2f(x)\delta f(x) + [\delta f(x)]^2 \, .
\end{aligned}
\tag{2.52}
$$

Hence, in the limit $\delta f(x) \to 0$,

$$
\frac{\delta A}{\delta f(x)} = 2f(x) \, .
\tag{2.53}
$$

2.8   The Method of Undetermined Lagrange Multiplier (⋆)

Suppose that F is a function of three independent variables x, y, and z. We wish to find out where, if any, F has a maximum, minimum or saddle point. Suppose that it has one at (x_0, y_0, z_0). Then

$$
\begin{aligned}
\delta F &= F(x_0 + \delta x, y_0 + \delta z, z_0 + \delta z) - F(x_0, y_0, z_0) \\
&= \frac{\partial F}{\partial x}\delta x + \frac{\partial F}{\partial y}\delta y + \frac{\partial F}{\partial z}\delta z = 0
\end{aligned}
\tag{2.54}
$$

where it is understood that the derivatives are taken at (x_0, y_0, z_0). Since $\delta x, \delta y$, and δz are independent and the above must be satisfied no matter what they may be, the three partial derivatives must each be zero. When the three equations are solved, we should find the extremum point.

Suppose that δx, δy, and δz are not independent because (x, y, z) are subject to a constraint, say,

$$
G(x, y, z) = 0 \, .
\tag{2.55}
$$

Then the above argument must be amended. The trick is to minimize

$Q = F + \lambda G$, where λ is an undetermined parameter. Thus we have

$$\delta Q = \delta F + \lambda \delta G$$

$$= \left(\frac{\partial F}{\partial x} + \lambda \frac{\partial G}{\partial x}\right) \delta x + \left(\frac{\partial F}{\partial y} + \lambda \frac{\partial G}{\partial y}\right) \delta y + \left(\frac{\partial F}{\partial z} + \lambda \frac{\partial G}{\partial z}\right) \delta z$$

$$= 0 \qquad (2.56)$$

and ensure that $\delta Q = 0$ by demanding

$$\left(\frac{\partial F}{\partial x} + \lambda \frac{\partial G}{\partial x}\right) = 0$$

$$\left(\frac{\partial F}{\partial y} + \lambda \frac{\partial G}{\partial y}\right) = 0$$

$$\left(\frac{\partial F}{\partial z} + \lambda \frac{\partial G}{\partial z}\right) = 0. \qquad (2.57)$$

We shall take advantage of the fact that we have a parameter which we can choose any way we wish, and we shall choose it so that the constraint is satisfied at the minimum point. So solve the three equations simultaneously and find the solution (x_0, y_0, z_0). The solution will have λ in it. Let λ be such that the constraint is satisfied at x_0, y_0 and z_0, namely, $G(x_0, y_0, z_0) = 0$. So it looked as if we were searching for the minimum point of $F + \lambda G$ rather than that of F, but since $\lambda G(x, y, z) = 0$ at the minimum, the minimum is that of F.

As an example, let $F(x, y, z) = x^2 + y^2 + z^2$ and $G(x, y, z) = x + y = 1$. The three equations to solve are

$$2x + \lambda = 0$$

$$2y + \lambda = 0$$

$$2z = 0$$

which give $(x_0 = -\lambda/2, y_0 = -\lambda/2, z_0 = 0)$. Now choose λ so that $x_0 + y_0 = 1$, which gives $\lambda = -1$. Substituting this back to x_0 and y_0, we find $(x_0 = 1/2, y_0 = 1/2, z_0 = 0)$. And it is a minimum.

2.9 Permutations, Combinatorial and Arrangements ($\star$)

Just like we had to ask how many ways there were for the brown Pontiac to be in the greater Hattiesburg area, we will repeatedly ask how many ways there are for a thermal system to possess, or to do, a certain thing. The multiplicity, to be called Γ, is at the heart of thermal physics. Most of the multiplicities that we will encounter are obtained through permutations and combinations. For the purpose of reference, let me write down two formulas that we will use:

(a) Suppose that we have N distinguishable objects. How many different ways are there to pick M out of them? The order in which they are selected is immaterial.

$$\Gamma(N, M) = \frac{N!}{(N - M)!M!} \tag{2.58}$$

(b) Suppose that we have N children and M identical pieces of candies. How many ways are there to distribute the M pieces of candies amongst the N children?

$$\Gamma(N, M) = \frac{(N - 1 + M)!}{(N - 1)!M!} \tag{2.59}$$

Proving (b) is not as trivial as (a). Call the children by their initials A,B,C, etc and label the candies 1,2,3, etc. One possible distribution is A(1,8),B(9,3,2,6),C,D(7),... which says that A got two candies (1 and 8), B got four candies (9,3,2 and 6), C did not get anything, D got one, etc. Now write this information without the commas and circular parenthesis like A18B9326CD7... Then we can tell how many ways there are:

$$\Gamma(N, M) = \frac{N(N - 1 + M)!}{N(N - 1)!M!} \tag{2.60}$$

which gives the desired result.

2.10   Distribution Function (⋆ ⋆ ⋆)

The foundation of thermodynamics is called Statistical Mechanics. The word 'statistical' often gives the wrong idea that it is just a statistics of data. We often perform such calculation of statistics with data obtained by laboratory experiments. But what we do in statistical mechanics is equivalent to generating our own data set over which we then take the mean value and the standard deviation, etc. When we take measurements of a certain physical quantity, the result is a time series which shows different outcomes at different times of measurement. The statistical mechanics provides us with various distribution functions which may be regarded as picking up each time one out of many possible outcomes and then adding it to the time series. It is like an abstract painter who is distributing dots of different colors to different parts of a canvas. In this case the distribution would be dictated by what the artist intends to create. In our case, it is dictated by the specific mathematical distribution function. That is what statistical mechanics provides. We will meet them soon. To make sure that we understand what they mean and what we can do with them, it would be a good idea to practice with some distribution functions that we encounter in our daily life.

Imagine a stick placed at a certain spot in a lake. The stick is equipped with instruments which allow us to read the instantaneous water height. The stick is marked $1, 2, 3, \ldots \infty$, starting from the bottom end and progressing each time we go up a distance of $\Delta y = 1$ cm. The stick is infinitely tall and can handle any tsunami, well supposedly anyway. Thus the height variable y can take on any one value at each measurement from the set $(Y_1, Y_2, Y_3, \ldots)$ where

$$Y_j = j \times \Delta y, \qquad j = 1, 2, 3, \cdots . \tag{2.61}$$

Record the height at the beginning of each minute over a long period of time. We will represent the time series as $(y_1, y_2, y_3, y_4, y_5, \ldots y_N)$ where the subscript merely refers to the time when the measurement was taken. Do not confuse the lower-case y's with the capital Y's; Y_3 refers to the third mark on the stick while y_3 is the actual reading taken at the third minute. To give an example, if the water level decreased for the first two minutes and then started rising, your data book will look like $(y1, y2, y3, \ldots) = (Y_{10}, Y_7, Y_6, Y_9, Y_{13}, \ldots)$.

Now let us do the statistics. The average is

$$<y> = \frac{1}{N} \sum_{i=1}^{N} y_i . \tag{2.62}$$

The fluctuation is defined as

$$\delta y^2 = \frac{1}{N} \sum_{i=1}^{N} (y_i - <y>)^2 . \tag{2.63}$$

By squaring the deviation from the average, all deviations are allowed to contribute to the fluctuation regardless of the direction of deviation.

Expand the square and obtain

$$\delta y^2 = <y^2> - <y>^2 , \tag{2.64}$$

where

$$<y^2> = \frac{1}{N} \sum_{i=1}^{N} y_i^2 \,. \tag{2.65}$$

What is often called standard deviation is $\sqrt{\delta y^2}$. The fluctuation of the water level may be regarded as a measure of the wind strength. Likewise, thermodynamic quantities fluctuate and their fluctuations are not insignificant ripples. They carry important signatures, as we will see later.

Now go through your data series and count how many times each mark appeared. Suppose that $Y_1, Y_2, Y_3, \ldots$ appeared, respectively, $N_1, N_2, N_3, \ldots$ times. Then

$$W_j = \frac{N_j}{\sum_{i=1}^{\infty} N_i} \tag{2.66}$$

gives the probability of finding the water level to be Y_j at any instant. The set $\{W_j, j = 1, 2, 3, \ldots, \infty\}$ may be called a discrete distribution function. It tells us how $Y_1, Y_2, Y_3, \ldots$ are picked each time to be **distributed** into the time series.

Now suppose that we have lost our original data series and all the statistical analysis we did with it. No problem! We can do the statistical analysis with the distribution function. In terms of $(W_1, W_2, W_3, \ldots W_\infty)$, the mean and fluctuation are given by

$$<y> = \sum_{i=1}^{\infty} W_i Y_i \tag{2.67}$$

and

$$\delta y^2 = \sum_{i=1}^{\infty} W_i Y_i^2 - \left(\sum_{i=1}^{\infty} W_i Y_i \right)^2 . \tag{2.68}$$

In fact, once we have the distribution function, we may say that we have everything we need. We can generate a time series which will be hard for anyone to distinguish from the actual measured data set; this will be discussed in the last chapter. The changing water heights are a result of a large number of very complex atmospheric effects plus some human actions. In addition, the geometry of the lake and the bottom topography would play very complex roles in shaping these effects. It is therefore utterly unthinkable for anyone to carry out a calculation to determine this distribution function without using a measured data set. Thermal systems are similar in many ways. Take a volume of space in air in front of you, a volume of 2 cm × 2 cm × 2 cm. The amount of energy in it or the number of air molecules in it changes with time. Now hear this. We can calculate the distribution of these changing physical quantities without using a measured data set. That is what thermal physics will do for you. Is that not awesome?

In order to study thermal physics, one should be very clear about the idea of a distribution function, what it means and what to do with it etc. So I urge you to do the next exercise which will do just that for you.

Exercise. *Suppose that the outcome of each measurement can only be one of three, i.e. Y_1, Y_2, or Y_3, and that $W_1 = 0.1, W_2 = 0.6$, and $W_3 = 0.3$. Generate a time series using a random number generator which generates randomly real numbers between 0 and 1. If you do not have one, you may use your local telephone directory. Open a page randomly and pick up one number randomly and read the last four digits of the telephone number but backward. If the chosen telephone number is 601-268-1976, the random number is 0.6791. Call it r.*

$$\text{If} \quad r < 0.1, \quad \text{pick } Y_1.$$
$$\text{If} \quad 0.1 < r < 0.1 + 0.6, \quad \text{pick } Y_2.$$
$$\text{If} \quad 0.1 + 0.6 < r < 0.1 + 0.6 + 0.3, \quad \text{pick } Y_3.$$

That is how Y_1, Y_2, and Y_3 are distributed in the time series that you are trying to construct. Repeat this many times to construct a time series which will look like $(Y_2, Y_1, Y_2, Y_2, Y_3, \cdots)$. Now calculate the distribution function using it. Is it the same as what we started with? You have just practiced a very simple Monte Carlo method.

Until now we treated a continuous variable approximately with a discrete model. Now let us treat them as a continuous variable. Suppose that the distribution function for a continuous variable s is $W(s)$. It means that if we make a measurement the probability that we will find any value between s and $s + ds$ is given by $W(s)ds$. Thus it follows that

$$<s> = \int_0^\infty sW(s)ds \tag{2.69}$$

and

$$\delta s^2 = \int_0^\infty s^2 W(s)ds - \left(\int_0^\infty W(s)sds \right)^2. \tag{2.70}$$

In the above, we assumed that $W(s)$ was normalized. If not, it will have to be normalized. For example, equation Eq. (2.69) should read

$$<s> = \frac{\int_0^\infty sW(s)ds}{\int_0^\infty W(s)ds} . \tag{2.71}$$

To generate a time series, construct

$$\mathcal{W}(s) = \int_0^s W(s')ds' \tag{2.72}$$

which stacks up the probabilities just like we did a moment ago. Call it the accumulated distribution function. If s is bound in the region $a < s < b$, $\mathcal{W}(s = a) = 0$ and $\mathcal{W}(s = b) = 1$; by the time s reaches b, the accumulated probability should reach 1. To generate a possible time series, generate a random number r and then find the value of s by solving

$$\mathcal{W}(s) = r . \tag{2.73}$$

Repeat this N times. This is just a continuum counterpart of what we did in the last exercise for a discrete case of $N = 3$.

Exercise. *Let y be distributed by $W(y) = \frac{1}{k}\cos ky$. Let $k = 0.1$ and let y be bound between 0 and π/k. Calculate $<y>$, δy^2, and the accumulated distribution function $\mathcal{W}(y)$. Using your calculator or computer, generate a possible data set for y, and then numerically compute the average and fluctuation again. The electrons emitted from some radioactive isotopes have an energy spread in a range with a spectrum resembling this distribution function.*

Exercise. *The most important distribution function in thermal physics, and for that matter in all other fields as well, is the Gaussian distribution function,*

$$W(y) = \frac{1}{\sqrt{2\pi b^2}}e^{-(y-y_0)^2/2b} . \tag{2.74}$$

Calculate the average $<y>$ and δy^2. You will then be able to tell what y_0 and b mean. If you can program on a computer, let $y_0 = 10.0$, $b = 2.5$. Compute the accumulated distribution function covering a range of y with increments of $\Delta y = 0.1$, and then generate a long 'time' series. Using thus generated sample data, compute the average $<y>$ and δy^2. Do they come out to be close to y_0 and b? This may appear to be a tedious and mundane exercise, but it will reward you well.

2.11   The Central Limit Theorem ($\star \star \star$)

We have already seen the central limit theorem in the above exercise, but it is so useful and important that we should study it in some detail.

Suppose that you collect samples, and each of them is a sum of many random variables. It does not matter to what extent they are random. They could be very random or just barely random, but they are random enough to make it impossible to predict the outcome of each sample. For example, suppose that a man can only move sideways along a line and he takes a step to the left or to the right randomly following a random number generator, or if you will allow me, that he is so drunk that he does not know whether he is taking his step to the right or to the left. He repeats this N times. The net displacement after his N random steps constitutes one sample. Suppose $N = 100$, meaning that for one sample we let him take 100 random steps. If he does this many, many times for a large number of samples, how would the net displacement be distributed? How would the net displacement fluctuate from sample to sample? The central limit theorem answers these two questions. You can guess that most of the times he will end up with little or no net displacement, but on some very rare occasions he may move 80 steps to the right, but a net displacement of 100 steps would be unlikely.

Consider the sum of similar random variables,

$$y = s_1 + s_2 + s_3 + s_4 + \cdots + s_N \,, \tag{2.75}$$

where s_i is random in magnitude, in sign, or in both, and N is large. According to the central limit theorem, the distribution function of y is given by

$$W(y) = \frac{1}{(2\pi\delta y^2)^{1/2}} e^{-(y-<y>)^2/2\delta y^2} \,. \tag{2.76}$$

Since $W(y)$ represents a probability, it is often written with the symbol $P(y)$ in the literature, and in fact we will do so whenever we wish to emphasize the probability nature. The fluctuation δy^2 turns out to be

$$\delta y^2 = \sum_{i=1}^{N} \delta s_i^2 \,, \tag{2.77}$$

where δs_i^2 is the fluctuation of his i-th step; just keep a record of his i-th step separately over many trials and calculate the fluctuation.

These results are remarkable for two reasons. First, the Gaussian function gives the probability distribution, no matter what kind of distribution function is distributing each of the variables s_i. Second, since y is a sum of N variables, its full range should go like N, and therefore it may appear that δy^2 should go like N^2. But it does not; according to Eq. (2.77), there are only N terms, not N^2 terms, and therefore it goes like N. Assuming that all the random variables are distributed by the same distribution function $w(s)$, then $\delta s_1^2 = \delta s_2^2 = \delta s_3^2 \cdots$, whence

$$\delta y^2 = N \delta s_1^2. \tag{2.78}$$

So Eqs. (2.76) and (2.78) are the answers to the two questions we posed.

The significance may be better appreciated if we consider the average rather than the sum of the random variables,

$$y = (s_1 + s_2 + s_3 + s_4 + \cdots + s_N)/N. \tag{2.79}$$

This is still a sum of N random variables, and the theorem still applies. All we need to do is to replace each s's with s/N. It is obvious that

$$<y> = <s_1> \tag{2.80}$$

and

$$\delta y^2 = N \frac{1}{N^2} \delta s_1^2 = \frac{1}{N} \delta s_1^2. \tag{2.81}$$

The average deviation $\delta y = (\delta y^2)^{1/2}$ goes like $1/\sqrt{N}$. It therefore vanishes in the limit of large N! The bell-shaped peak of the Gaussian function therefore becomes a sharp spike. The foundation of thermal physics is statistical in nature, but thanks to this theorem, thermal physics is a predictable science, as we will see in the following chapters.

Example. Throw a coin. If it lands on head, you win 1 point. If it lands on tail, you lose 1 point. The chance of landing head and tail are the same. Here s_1 represents your earnings on the first try, and s_2 that of your next try, and so on. They are not a continuous variable. They take on only two values: $s_i = 1$, or -1 and the chance of s_i taking on 1 point is $1/2$, the chance of it taking on -1 is also $1/2$. Now throw the coin N times. Let $M = s_1 + s_2 + \cdots + s_N$ be your total net earning. It follows that

$$<s_1> = 1 \times \frac{1}{2} + (-1) \times \frac{1}{2} = 0$$

$$\delta s_1^2 = 1^2 \times \frac{1}{2} + (-1)^2 \times \frac{1}{2} = 1. \tag{2.82}$$

Therefore $<M> = N<s_1> = 0$ and $\delta M^2 = N\delta s_1^2 = N$.
Your chance of winning M points is:

$$P(M) = \frac{1}{(2\pi N)^{1/2}} \exp(-M^2/2N). \tag{2.83}$$

What is your chance to win $M = 10$? Your chance after $N = 100$ trials is 0.024. Would your chance be better if you try $N = 1000$ times? No, your chance would then be 0.012. That is why the gaming industry thrives.

• Central Limit Theorem in greater detail

For beginners, what has been said should be adequate. Those who wish to know more about it may read the rest. We can easily prove Eq. (2.77). Let us suppose that all the variables, $s_1, s_2, s_3, \ldots$ are distributed by the same function $w(s)$. This is to simplify the steps and is not a necessary condition. What is necessary is that s_j's are statistically independent of each other. To put it mathematically, the probability of $s_1, s_2, s_3, \ldots, s_N$ taking on any set of values is given by

$$P(s_1, s_2, s_3, \ldots, s_N) = w(s_1)w(s_2)w(s_3)\ldots w(s_N) \tag{2.84}$$

where the distribution function $w(s)$ has been normalized:

$$\int w(s_i)ds_i = 1. \tag{2.85}$$

Let y represent the sum of the random variables. It then follows that

$$<y> = \int\int\cdots\int (s_1 + s_2 + \cdots + s_N)w(s_1)ds_1 w(s_2)ds_2 \cdots w(s_N)ds_N$$

$$= \int s_1 w(s_1)ds_1 + \int s_2 w(s_2)ds_2 + \cdots + \int s_N w(s_N)ds_N$$

$$= N<s> \tag{2.86}$$

where $<s> = <s_1> = <s_2> = \cdots = <s_N>$ and

$$\delta y^2 = \int\int\cdots\int ((s_1 - <s>) + (s_2 - <s>) + \cdots)^2$$

$$\times w(s_1)ds_1 w(s_2)ds_2 \cdots w(s_N)ds_N$$

$$= \int\int\cdots\int \sum_{i<j}(s_i - <s>)(s_j - <s>)w(s_i)ds_i w(s_j)ds_j$$

$$+ \int\int\cdots\int \sum_i (s_i - <s>)^2 w(s_i)ds_i$$

$$= N\delta s^2. \tag{2.87}$$

In the second line, all the terms for which $i \neq j$ vanish leaving only the self terms, and $\delta s^2 = \delta s_1^2 = \delta s_2^2 = \cdots = \delta s_N^2$. The statistical independence of different s_j's plays an important role here.

If y represents the average of N random variables, the above equation should read

$$\delta y^2 = N\frac{\delta s^2}{N^2} = \frac{\delta s^2}{N}, \tag{2.88}$$

which completes the proof.

The reason for the particular Gaussian form in Eq. (2.76) is not as straightforward as above. The following is only for advanced readers. All beginners should skip the rest and move on to the next topic.

The probability function $P(y)$ that we wish to find out must be such that

$$P(y)dy = \int' w(s_1)ds_1 \int' w(s_2)\cdots \int' w(s_N)ds_N \tag{2.89}$$

where the prime signs mean that the N integrations must be performed only in the region where $y < \sum_i s_i < y + dy$. This awkward condition may be put in with the Dirac δ function

$$\delta(y - \sum s_i) = \frac{1}{2\pi}\int_{-\infty}^{\infty} dk \exp\left[-ik\left(y - \sum s_i\right)\right]. \tag{2.90}$$

Inserting $\delta(y - \sum s_i)dy$ into Eq. (2.89), the prime signs may be removed and the result is:

$$P(y) = \frac{1}{2\pi}\int_{-\infty}^{\infty} \exp(-iky)Q(k)^N dk \tag{2.91}$$

where

$$Q(k) = \int w(s_1)e^{iks_1}ds_1. \tag{2.92}$$

In Eq. (2.91), due to the oscillatory behavior of $\exp(-iky)$, the integral comes predominantly from the small k region. Thus expand $\exp(-iks_1)$ and retain only terms up to the order of k^2. This gives

$$Q(k) \approx \int w(s_1)\left[1 + iks_1 - \frac{1}{2}k^2s_1^2\right]ds_1$$

$$= 1 + ik<s_1> - \frac{1}{2}k^2<s_1^2>. \tag{2.93}$$

This says that $Q(k)$ is not far from unity, which should also be true for $Q(k)^N$. Then we can further simplify the matter by taking the logarithm

$$\ln\left[Q(k)^N\right] \approx N\ln\left[1 + ik{<}s_1{>} - \frac{1}{2}k^2{<}s_1^2{>}\right]$$

$$\approx N\left[ik{<}s{>} - \frac{1}{2}k^2\delta s^2\right] \qquad (2.94)$$

where the logarithmic function is expanded and only terms up to the order up to k^2 are kept.

Thus we approximate $Q(k)^N$ with

$$Q(k)^N \approx \exp\left[N(ik{<}s{>} - k^2\delta s^2)\right] . \qquad (2.95)$$

Substitute this into Eq. (2.91). Then put together all the exponents. We have a k^2 term and a k term in an incomplete square form. Complete the square, which gives a Gaussian function to integrate. Carrying out the integration leads to Eq. (2.76).

2.12   Quantum Mechanics ($\star\star\star$)

• Single Particle

The initial seed of quantum mechanics was the idea of de Broglie who suggested that when a particle of mass m is in motion with momentum p, it should be described as a wave, a matter wave, and that its wavelength should be

$$\lambda = \frac{h}{p} \qquad (2.96)$$

where $h = 6.626 \times 10^{-34}$ Js the Plank constant.

Consider a particle of mass m confined in a cubic box of size $V = L \times L \times L$. The particle is free of any external field. The matter wave has to form a standing wave. The wavelength along the x, y, and z direction should then be[1]:

$$\lambda_x = \frac{2L}{n_x}, \qquad \lambda_y = \frac{2L}{n_y}, \qquad \lambda_z = \frac{2L}{n_z} \qquad (2.97)$$

---

[1]Check your introductory physics book. It will say that the wavelength of a standing wave on a string of length L tied at both ends is $\lambda = 2L/n$ where n is an integer.

where the three quantum numbers n_x, n_y, n_z can be any positive integer excluding zero:

$$n_x = 1, 2, 3, \cdots \qquad n_y = 1, 2, 3, 4, \ldots, \text{etc.} \qquad (2.98)$$

The corresponding x, y, and z component of the momentum are, respectively,

$$p_x = \frac{h n_x}{2L}, \qquad p_y = \frac{h n_y}{2L}, \qquad p_z = \frac{h n_z}{2L} \qquad (2.99)$$

which gives for the energy

$$\mathcal{E}_{n_x, n_y, n_z} = \frac{p^2}{2m} = \frac{\hbar^2}{2m} \left(\frac{\pi}{L} \right)^2 (n_x^2 + n_y^2 + n_z^2), \qquad (2.100)$$

where $\hbar = h/2\pi$. These are the only allowed values of energy for the particles and are called the energy eigenvalues.

We may obtain the same result in a more formal way by solving the energy eigenvalue equation which is also called the Schroedinger equation or simply the free particle wave equation,

$$-\frac{\hbar^2}{2m} \nabla^2 \psi(x, y, z) = E\psi(x, y, z). \qquad (2.101)$$

The solution is:

$$\psi_{n_x, n_y, n_z}(x, y, z) = A \sin(n_x \pi x/L) \sin(n_y \pi y/L) \sin(n_z \pi z/L) \qquad (2.102)$$

and the corresponding energy eigenvalue E is as given by Eq. (2.100). The single particle states are called orbitals following the practice of atomic physics; each orbital is characterized by three quantum numbers (n_x, n_y, n_z).

If the particle is in one of the eigenstates, say, $\psi_{n_x, n_y, n_z}(x, y, z)$, then we can say with certainty that the particle's energy is given by Eq. (2.100) and the wave function $\psi(x, y, z) = \psi_{n_x, n_y, n_z}(x, y, z)$ contains all the wave-nature of the matter wave and its physical properties.

If we make a measurement to find out where the particle is, it can be found anywhere as long as $\psi(x, y, z)$ is not zero. The probability of finding it in the differential volume element of $dxdydz$ at $\vec{r}$ is $|\psi(\vec{r})|^2 dxdydz$. Then it should be possible to represent the spread of the potentially possible region where the particle may be found as Δx, Δy, Δz. In other words, the volume of space in which there is a decent chance of finding the particle is written as $\Delta x \times \Delta y \times \Delta z$. Think of this as an effective volume of the cloud-like patch where $|\psi(\vec{r})|^2 dxdydz$ is not zero. But when we actually make the measurement, the outcome is always a sharp point located within the patch, and therefore, Δx, Δy, and Δz, are regarded as defining the uncertainty in position. For the same given wavefunction, if we now ask what is the particle's momentum, the uncertainty in momentum is similarly given by Δp_x, Δp_y and Δp_z. The two sets of uncertainties are related to each other by

$$\Delta x \Delta p_x = h, \qquad \Delta y \Delta p_y = h, \qquad \Delta z \Delta p_z = h \qquad (2.103)$$

which is called the Heisenberg uncertainty principle. To be precise, these are the relationships between the minimum uncertainties. Depending on how we define the uncertainty, this can be written in several different ways, but for our purpose it does not matter which one we choose; we have chosen this one because it is the most popular choice in books on thermal physics.

• Many Particles

Now suppose that we put N indistinguishable and non-interacting particles in the same box. The total energy would then be given by

$$E_{n_1, n_2, n_3, \cdots n_N} = \frac{\hbar^2}{2m} \left(\frac{\pi}{L} \right)^2 (n_1^2 + n_2^2 + n_3^2 + \cdots + n_N^2), \qquad (2.104)$$

where the subscript i refers to the particle label and $n_i^2 = n_{i,x}^2 + n_{i,y}^2 + n_{i,z}^2$ and all the quantum numbers $n_{i,x}$ etc. may take on any positive integer excluding zero. This is simply the sum of N single-particle energies. It consists of $3N$ integers. Here N is a very large number, but just as an exercise, let $N = 2$ and list the number of the 2-particle states corresponding to each value of energy. Next repeat the same with $N = 3$. Observe how many eigenstates there are corresponding to each value of energy. All the eigenstates corresponding to the same value of energy is said to be degenerate and their total number is called the degeneracy or the multiplicity. The number of degeneracy is truly astronomical if N is as large as 10^{23}! As it turns out, this huge number of degeneracy plays the central key role in thermal physics. All the principles and laws of thermal physics originate from this.

Alas, we have put labels on indistinguishable particles. Quantum mechanics demands that we cannot treat them as if they were carrying registration numbers; they cannot be distinguished from one another. We tagged the particles for convenience, and by doing so we will grossly over count the number of the energy eigenstates. It is important to understand the nature of this error so that we can find a way to correct it. If we tag the molecules and let each $\vec{n}_i = (n_{i,x}, n_{i,y}, n_{i,z})$ take on any allowed value and direction irrespective of the others, the N-particle states generated would include $A = (19, 1, 13, 12, \ldots)$ which says that the first particle is in the 19th single-particle orbital, the second in the first orbital, the third in the 13th orbital, and so on. Also included would be $B = (1, 19, 12, 13, \ldots)$ where the first four molecules have exchanged their orbitals while all the other molecules remain in the same orbitals. If the N molecules were distinguishable, A and B would be two distinct N-particle states. However, since these N molecules are indistinguishable, A and B do not constitute different N-particle states. The fact that the first four orbitals are occupied by "different" molecules is an artifact of the tagging action; the four molecules are identical molecules. We will discuss how the over counting error can be corrected in Chapter 3.

We will find a more sensible way to handle identical particles without tagging them in Chapter 9. The idea there is to focus on each orbital and count the occupation number and not to try to identify the occupants.

There is an alternative way of treating particles in a box. Instead of using the standing waves, we may use the plane waves. The plane waves extend everywhere even outside the box, but we normalize it so that what is inside the box may represent appropriately the particles in the box. The single particle energy now takes the form

$$\mathcal{E}_{n_x, n_y, n_z} = \frac{\hbar^2}{2m} \left(\frac{2\pi}{L} \right)^2 (n_x^2 + n_y^2 + n_z^2), \qquad (2.105)$$

where (n_x, n_y, n_z) can be any positive or negative integers including zero:

$$n_x = 0, \pm 1, \pm 2, \pm 3, \cdots \qquad n_y = 0, \pm 1, \pm 2, \pm 3, \pm 4, \ldots, \text{etc.} \qquad (2.106)$$

The corresponding energy eigenfunctions are

$$\psi_{n_x, n_y, n_z}(x, y, z) = A \exp(i2\pi n_x x/L) \exp(i2\pi n_y y/L) \exp(i2\pi n_z z/L) \qquad (2.107)$$

and the corresponding N-particle energy is given by

$$E_{n_{i,x}, n_{i,y}, n_{i,z}} = \frac{\hbar^2}{2m} \left(\frac{2\pi}{L} \right)^2 (n_1^2 + n_2^2 + n_3^2 + \cdots + n_N^2). \qquad (2.108)$$

From here the rest proceeds in the same way as with the standing wave method. The two methods yield the same results for all the issues of thermal physics that we will discuss. Some texts use the plane wave method while others use the standing wave method. We will stay with the standing wave method except when we discuss the Bose-Einstein condensation in Chapter 9. The reason is that the zero momentum state is allowed in the plane wave method, but not in the standing wave method. In the classical regime where we will use the standing wave method, it makes no difference whether the zero momentum state is allowed or not. But for certain atomic systems near absolute zero, the zero momentum state plays the most dominant role, whence there is the need to switch.

2.13   Thermal Physics on Computer

Once a student of mine asked me after a class near the beginning of the course. He asked "We are here just to learn the basic material of this subject. Why do you not teach us something simple and basic that we can do without making so many approximations? What Lonnie said is true. I guess that we will use the Stirling approximation in the next two chapters perhaps a dozen times. That is unfortunate, but there is no other way around. In that regard, thermal physics is different from other subject areas. How nicely does the story evolve in, for example, Mechanics! We start from the concept of inertia and then Newton's laws of motion. Then we turn it into work-energy relationship and then conservation laws, and then Lagrange equation and Hamilton equation, and so on. We have to make approximations later for some complex problems, but certainly not at the beginning. That is also the case for Electricity and Magnetism and all other areas. Unfortunately that is not so for the subject of Thermal Physics. The subject requires even at the basic level things that cannot be done simply and without making approximations. Then there is another matter very similarly unique for this subject. For all other subject matters, we have very helpful and even inspiring laboratory experiments which we do in the elementary physics laboratory course. What do we have there for thermal physics? Heat capacity measurements and perhaps expansion coefficient measurements. And even from this one or two laboratory experiments, what did you really learn? Were you inspired to learn that the heat capacity of copper was such and such? Unfortunately there is no simple experiment that we can do in the elementary physics laboratory course that can teach us the basic concepts. Was it not nice to map the electric field lines with a simple compass needle? The most basic concept in thermal physics is entropy, but there is not even one simple experiment which we can do to 'map' it.

Well, there are things we can do. Do it on computer! It would be safe to assume that you have one of your own or can use one at your university. What would it take you to use your computer for learning purpose? Learn C or FORTRAN or any other language. I would recommend that you start with FORTRAN. Some of your computer science friends and the computer wizards around you will tell you that FORTRAN is a useless language which is about to disappear soon. Do not listen to that. It is the easiest one to learn and as far as crunching numbers are concerned, it is great, and it will not go anywhere. You can learn it in a day. Once you have learned it, it will give you a distinctive advantage. For example, if you did the exercise on distribution functions, the notion of distribution function becomes a part of you and will offer you so much intuition. As was the case with the two exercise problems that I suggested, most of the problems, or the computer experiments, would require lots of algebra or simply counting, so much so that we have neither time nor patience. But computers have infinite patience and they run so fast that time is not an issue for our learning purpose. You can take advantage of all those fabulous free graphic tools. Plot the function you want to study. Change the parameters and see how the graph changes.

Chapter 3

Isolated Thermal Systems

3.1 Introduction

We will consider in this chapter thermal systems isolated from the rest of the universe. Once we learn how to handle isolated systems, we will be ready to move on to more familiar systems coupled to the atmosphere or other large reservoirs. The systems we shall study consist of extremely large numbers of particles, as many as 10^{23} or more. This is a very large number indeed. The total number of temporal ticks (say, once per second) since the Big Bang is estimated to be on the order of 10^{17}. Because these systems consist of such a large number of particles, they can exist in many different states, each of which may be regarded as a microstate. Or we may say that they can do what they do in in an overwhelmingly large number of ways, each of which may be regarded as a microstate. The number of possible microstates is astronomically large. The systems are not "stuck" in any one of the microstates. Instead, they move around from one to another constantly. Each system has something "thermal" in it which causes incessant movements. It could be the walls that confine the particles or it could be some truly minute impurities. We need not know the exact nature of the mechanism, but merely assume that such a mechanism exists for each thermal system. Since we know what the system does in each microstate, the only additional information we need in order to proceed is the number of times the system visits each microstate or the probability of visiting each microstate. That is where thermal physics takes off.

3.2   The Fundamental Postulate

The fundamental postulate says:

An isolated thermal system visits all of its microstates with equal frequency, i.e. without preferring any single one over the others. This is also called the *principle of equal a priori probabilities*. If there are as many microstates as Γ, the frequency of visiting microstate i is $f_i = 1/\Gamma$. Given a sufficient amount of time, the system will visit all of its microstates. By 'a sufficient amount of time' we mean no more than a fraction of a second, a time scale much larger than atomic time scales. Atomic time scales are on the order of 10^{-15} seconds. When we make measurements of thermodynamic quantities, the measurement may take only a small fraction of a second, but this is an eternity on the atomic time scales. This gives the atoms and molecules enough time to visit the bulk, if not all, of their microstates. The instruments measure the average of the microstates that the system visits during the measurement time.

Let us imagine what the instruments do when they take measurements of thermodynamic quantities. The instruments observe the system visiting different microstates and take an average over what the system shows in each microstate. If we imagine making a record of what the instruments observe even over a short measurement time span, it would still give us a very long time series of the microstates that the system visits in succession during the specified measurement time span. The distribution function that gives this time series is very simple:

$$W_j = f_j = \frac{1}{\Gamma} \quad \text{for all j}. \qquad (3.1)$$

What could be simpler than this?

All equilibrium properties may be written as an average over this time series. Suppose that we wish to measure A. Let A_i stand for what A would be when the system is in microstate i. The equilibrium average of A is then trivially given by

$$<A> = \frac{1}{\hat{N}} \sum_{j}^{\hat{N}} A_j \qquad (3.2)$$

where $\hat{N}$ is the total number of entries in the series. If we had the time series, we could calculate not only the equilibrium properties but also all non-equilibrium properties as well, but we do not. Notice, however, that as far as the equilibrium properties are concerned, the order in which the

A_j's are added in the above sum is immaterial. We also know that all the microstates show up in this series with equal frequency. These two facts make it possible for us to mimic what the instruments do without having to know the exact time series.

3.3   The Microcanonical Ensemble

For another way to visualize this, make as many replicas of the system as there are microstates. Each replica represents the same system but frozen in a different microstate. We have Γ such replicas, one for each microstate. For the average of quantity A, we can go through the ensemble of the replicas and find out what A is in each member and then take the average:

$$<A> = \frac{1}{\Gamma} \sum_i A_i , \qquad (3.3)$$

where the sum is over all the replicas, the microstates. This will be referred to as the method of microcanonical ensemble.

The microstates come in groups, and all the microstates in each group share the same characteristics to be called α in the rest of this section. It could be about the way the molecules are distributed in the volume or the way energy is distributed among the constituents, or any other physically meaningful gross characteristics. And these groups are called macrostates.

Now let us find out what the *equal a priori probability principle* says about the macrostates. Consider a gas confined in a volume. The huge pool of microstates includes one in which all the particles are gathered in one corner leaving the rest of the volume in vacuum. It also includes one in which the molecules are more or less scattered everywhere. The postulate gives the same probability to these two. How could this be? Yes, the postulate gives the same probability, but realize this. There are many, many microstates that possess the general characteristics of the former, and there are also many, many microstates that possess the same general characteristics of the latter. Since the postulate gives the same probability to each microstate in both groups, the question of likelihood or unlikelihood is determined by the size of the two groups. We will shortly start counting the number of microstates in various macrostates. Let me give you a preview. As unlikely as the first one seems, its size is still astronomically large. There are astronomically large number of ways molecules can huddle around near one corner leaving the rest void of molecules. But the size of the latter is, if you will allow me, even more super-duper astronomical. The size of the former is no match for the latter. If we allow the system a sufficient amount of time, the postulate says that the system will spend for all practical purposes all of the time in the latter spending little or no time in the former.

By the way, why do we bother about macrostates? There is no compelling reason to ask which microstate the system is in at any given instant. But there is a good reason to ask which macrostate because we would then know something about the gross characteristics of the system's behavior as opposed to the microscopic details which are washed away in equilibrium states.

In the above, macrostates were defined in terms of the way particles are distributed in a volume. Another way would be to define in terms of the way energy is distributed among the constituents. Consider the macrostate in which half of the molecules divide all the energy in the gas among themselves leaving no energy for the rest. Contrast this with one in which the total energy is divided more or less equally among all the molecules. The former sounds like an utterly unlikely scenario in a real world gas, and because of its size it is no match compared to the latter. So, if the system is allowed a sufficient amount of time, it will spend for all practical purposes the entire time moving around among the microstates of the latter.

Consider four macrostates A, B, C, and D. They are in the order of decreasing size; A is much larger than B which in turn is much larger than C and so on. These macrostates were defined with respect to a certain macroscopic physical characteristics to be called α. Suppose that the system is at the moment in C. May we predict that the system will remain forever in C or that it will move to another one. If it is the latter, to which? In order for the system to stay in C, the distribution function should keep on picking only the microstates in C, but since C is not as large as B, its luck may run out soon. The system is likely to jump to B soon, but not to D. It is also unlikely to make a big jump to A skipping B because it takes time for α to change to that of A; due to the initial history, A is effectively not accessible yet. Once it moves to B, it will then move to A soon. A is not just larger than the rest, it is overwhelmingly larger than the rest. Will the system then stay in A or will it return to C? It will stay in A to the extent that A is overwhelmingly larger than the rest. Once it lands on A, we say therefore that it has reached its equilibrium state.

We may summarize this important trend of movements in a very simple way. The size of the macrostate that the system is in will never decrease.

This is no trivial statement. Let me say it again but in more detail. **If we just watch the size of the macrostates that the system will visit as time progresses forward, the size will never decrease.**[2] The alert reader may find the statement too strong, and in a sense it is. Suppose that we define macrostates so that each covers an extremely narrow range of α. Then, since the neighboring macrostates will differ from each other only slightly, the movement would not always be in the direction of increasing size. Only when each macrostate covers a wide range will the movement be in the uni-direction of increasing size. By making the range wide, all the insignificant ripples are washed away leaving only the gross macroscopic behavior.

The term "equilibrium state" should be understood carefully. Contrary to what it means in mechanics, the system is not stuck in one microstate. It is moving around all the microstates in the largest macrostate and equilibrium means the totality of all the microstates it visits. When the system is not in the largest macrostate yet, we say that the system is in a non-equilibrium state.

3.4 Entropy and the Second Law of Thermodynamics

That being the case, the only thing that truly matters in determining how a thermal system will behave is Γ, the multiplicity (the number of microstates) of each macrostate. The natural logarithm of Γ of the macrostate that the system is in is called entropy:

$$S = k_B \ln \Gamma, \tag{3.4}$$

where $k_B = 1.38 \times 10^{-23}$ J/K is the Boltzmann constant. It contains a temperature unit which we have not yet discussed, but for now simply regard it as a constant.

The statement that *the size of macrostate never decreases* may now be translated into: **entropy can only increase and it never decreases spontaneously**. This is called the second law of thermodynamics or the *maximum entropy principle*; the first law will be introduced later.

---

[2]We will repeat this again and again throughout the book in various revised forms, but we can say something interesting right away. This thermal behavior distinguishes whether time is advancing forward or backward. That is not true in mechanics. Change time t to $-t$ in $F = ma$. The Newton's law remains invariant to the time reversal. Throw a ball with your right hand and catch it when it comes down with your hand. Make a video and run it backward. You will see that you are throwing the ball with your left hand and catching it when it comes down with your right hand. You cannot tell which is real.

Think of entropy as available options. Do we not try so hard in life to put ourselves in a position where we have more options? People migrate from one region to another in search for more job options. We tend to prefer big grocery stores because they offer more options for grocery. So do thermal systems. Sometimes we do not know how to exercise the blessing of given options, in which case the blessing becomes a burden and uncertainty. That is not a problem in the thermal world because they do not have to commit themselves to just one.[3] Instead, they incessantly change from one to another thus enjoying the options to the fullest extent. It is more like those who have their own wine cell loaded with an unlimited variety of wines — use this one today and that one tomorrow.

The idea of entropy became necessary because of the astronomical size of macrostates, which in turn is due to the fact that thermal systems consist of an astronomically large number of constituents. If all the matter around us consisted of a few or a few thousand particles, the notion of entropy would have never been born, and we would certainly not be speaking of the second law. Every issue that thermal physics addresses has its roots in the very fact that thermal systems are assemblies of very large number of particles. Let me give you an example. Make a closed box with any material for the walls. Inside the box, there are invisible electromagnetic waves bouncing between walls to form standing waves of various wavelengths. If we are interested in each of the standing waves or the manner in which they are reflected by the walls, etc., it is a problem of optics. If we we are interested in, however, the spectral energy distribution in various wavelength, the electromagnetic waves must be regarded as a collection of particles called photons. The number of these photons is astronomical, and therefore the issue becomes one of thermal physics.

Let me dramatize the role that entropy plays. Thermal systems are, so to speak, after entropy and entropy only. This is to say that if a thermal system lands on a small macrostate it will soon leave the macrostate because its 'options' (to remain in the same macrostate) are limited. Sooner or later it will land on the largest macrostate where the 'options' are limitless. This is the only principle that guides thermal systems. Entropy reaches its maximum possible value when the system finally reaches its dominant macrostate, and we say that the system has reached an equilibrium state. Displace the system from the equilibrium state with an external

---

[3]This has exceptions in amorphous materials like glasses which is stuck in one microstate.

interference. When the system finally reaches its dominant macrostate, we say that the system has reached an equilibrium state. Displace the system from the equilibrium state with an external interference. The displaced system will again find its way to its new dominant macrostate. As we will find out shortly, thermal systems have a built-in control system which does just that. This predictable direction of change of thermal systems is an obvious result of the postulate that all microstates are equally likely to be visited. Every branch of physics has its postulate, but none is simpler than this and, I would add, that none is more powerful than this one.

One way of displacing a thermal system from its equilibrium state is to impose a different set of constraints on the system. This demands that the system change its behavior. For example, consider a gas confined to a cylinder by a mobile piston. Fix the piston position, let the gas reach an equilibrium state, and then displace the gas from the equilibrium by moving the piston so as to allow more space for the molecules. The typical microstates in the largest macrostate before the system was displaced are no longer typical. For the displaced system, there are new typical microstates consistent with the changed constraint. Once the system reaches its new largest macrostate, it never spontaneously returns to the initial macrostate as it is no longer the largest macrostate. If the system did, it would be quite spectacular as the molecules would then appear as if they wanted to stay away from a certain volume of space even though there is no dividing wall. So, the question of how entropy changes with volume is an important issue, and we shall address it shortly.

Several features about entropy should be noted. To that end, let me remind you of a basic probability theory. The probability of two independent events A and B both occurring is equal to the product of each event occurring regardless of the other, i.e. $P(A \text{ and } B) = P(A) \times P(B)$. This is because if there are as many as Γ_A ways A can occur and Γ_B ways B can occur, then there are as many as $\Gamma_A \times \Gamma_B$ ways A and B can both occur. Contrast this to $P(A \text{ or } B) = P(A) + P(B)$.

The multiplicity that defines entropy may be thought of as the number of different ways the system can do what it does. For example, suppose that a system consisting of non-interacting molecules are confined to a box and do two unrelated things α and β at the same time; let α represent the action of arranging themselves in various parts of the box and β the action of distributing energy among themselves. The total number of ways of doing both α and β is given by $\Gamma = \Gamma_\alpha \Gamma_\beta$, where Γ_α represents the number of ways of doing α and Γ_β the number of ways of doing β. Upon

taking the logarithm, we get

$$S = S_\alpha + S_\beta. \tag{3.5}$$

The additivity also holds if the system consists of two uncoupled parts A and B. The multiplicity for the total is given by $\Gamma = \Gamma_A \Gamma_B$, and therefore we again have $S = S_A + S_B$. Because of the additivity, we may regard entropy as something that particles possess like energy.

Carrying out an exact calculation for Γ, can be quite difficult, but we can get away with large errors. This is because Γ is very large, but what is meaningful is not Γ itself but its logarithm. Suppose that $\Gamma = 10^{10^{23}}$, but we overestimated it 100-fold, 1000-fold or even 10^{23}-fold. What is the consequent error in entropy? Since $\ln(10^{23}\Gamma) = 23 \times \ln 10 + \ln \Gamma$, the error is only $23 \times \ln 10$, which is negligible compared to $\ln \Gamma = 10^{23} \ln 10$. Take it as saying "You do not have to be a math wizard. Just do your best and everything will be fine."

3.5 Entropy of Ideal Gas: Mobile Particles

The idea of entropy was recognized before Boltzmann, but its physical meaning was far from clear. Boltzmann clarified its meaning, but much of his thermal physics was based on an atomistic view of matter. This is remarkable because Boltzmann worked before the time of Rutherford and Schrödinger. Boltzmann's atomistic point of view ran contrary to the prevailing tendency in his time to think of everything, except falling apples, as a kind of hydrodynamic phenomenon. He was clearly ahead of his time, and for that reason, his idea met with vicious attacks from some of his contemporaries. If he had lived just three more years, he would have seen his ideas vindicated. As it turns out, Boltzmann left behind him a powerful new science. Equation (3.4) is attributed to him as his equation. Let us pay tribute to him by using it to calculate the entropy of an ideal gas. Shown in the picture below is his equation inscribed on his tombstone in Vienna; here W means the same as our Γ.

Fig. 3.1 Boltzmann's equation.

Ideal gas molecules do not interact with one another. Real molecules do, but in the gas phase, because they are so far away from one another, it makes no difference if we ignore the molecular interactions. They are "ideal" because we can calculate lots of things that we would not be able to if they were interacting with one another. By starting our journey first with an ideal gas, we are deviating from the general custom. A more popular way is to start with magnetic systems which are much easier to handle than ideal gases and then to apply what is learned from magnetism to gases at some convenient point. I preferred our way because gases and liquids are far more ubiquitous around us. For that reason, we devoted some additional effort in Chapter 2. I hope that you will feel well awarded after this chapter.

To calculate entropy, we first have to declare what microscopic information each microstate is going to tell us. In the first method, we will declare how the particles are distributed in the given volume and how the particles share the given amount of energy, all within the limitation of Heisenberg uncertainty principle. In the second method, we will follow as quantum mechanics dictates. According to quantum mechanics, thermal systems can only exist in certain discrete number of states or the energy eigenstates. These eigenstates are, however, vastly degenerate, and our Γ will simply be the number of degeneracy corresponding to the energy given to the system before they were thermally isolated. Quite delightfully, these two methods yield the same answer.

3.5.1 With Position- and Momentum-Definite Microstates

- Single Particle

We begin by learning how to construct and count microstates for just one molecule in a box of volume V. To specify the position of the molecule, divide the volume of the box into three-dimensional cubic blocks, each with volume $(\Delta x)^3$. The molecule can be in any one of the blocks, and therefore the total number of the positional microstates is equal to the number of the blocks,

$$\Gamma_{\vec{r}} = \frac{V}{(\Delta x)^3}. \tag{3.6}$$

Next, turn to the momentum of the molecule. If the molecule is given energy U, the momentum must satisfy

$$U = (p_x^2 + p_y^2 + p_z^2)/2m = p^2/2m \quad \text{or} \quad p = \sqrt{2mU} \tag{3.7}$$

where m is the mass of the molecule. It tells us that the momentum vector must lie on the surface of a sphere of radius p. We cannot construct microstates using the surface because we would then be specifying certain components of momentum precisely without acknowledging any uncertainty. This would be inconsistent with what we did in the spatial cubic blocks above. So we multiply the surface by Δp_x to make a shell. We then make three-dimensional momentum blocks in the shell. The number of momentum blocks in the shell is:

$$\Gamma_{\vec{p}} = \frac{4\pi p^2 \Delta p_x}{(\Delta p_x)^3} = \frac{4\pi \left(\sqrt{2mU}\right)^2 \Delta p_x}{(\Delta p_x)^3}. \tag{3.8}$$

We can now specify the microstates by indicating in which spatial block the particle is, thus indicating the whereabouts of the particle, and in which momentum block the particle is in, thus indicating how fast it is moving. We treat these two as being independent within the limitation of the Heisenberg uncertainty principle, and therefore the total number of microstates is simply the product of the total number of spatial blocks and that of the momentum blocks, i.e.

$$\Gamma = \Gamma_{\vec{r}}\Gamma_{\vec{p}} = \frac{V}{(\Delta x)^3} \frac{4\pi \left(\sqrt{2mU}\right)^2 \Delta P_x}{(\Delta P_x)^3}$$

$$= V 4\pi \left(\frac{\sqrt{2mU}}{h}\right)^2 \frac{1}{\Delta x} \tag{3.9}$$

where we have used the uncertainty principle

$$\Delta x \Delta p_x = h \tag{3.10}$$

which is the minimum uncertainty that quantum mechanics allows. The more precise minimum uncertainty should have $\hbar/2$ on the right-hand side but, as we will see shortly, the factors 2π and $1/2$ are negligible when we take natural logarithm for entropy.

We cannot proceed any further without choosing Δx. What should we choose for Δx? If we are asked to give the precise location of an ant, the uncertainty of our answer would be in units of 'mm'. If we have to give the specific position of an elephant, our uncertainty would be greater than 1 meter. For waves, the wavelength is the extent of the spread, and for the matter wave of our thermal particles, the de Broglie wavelength λ_D is the extent of the spread and provides the lower bound of Δx. So we choose

$$\Delta x = \lambda_D \,, \qquad (3.11)$$

where

$$\lambda_D = \frac{h}{p} = \frac{h}{\sqrt{2mU}} \,. \qquad (3.12)$$

We may now proceed. Notice that in the last line in Eq. (3.9) what is in the parenthesis is the inverse of λ_D, whence

$$\Gamma = V4\pi \left(\frac{\sqrt{2mU}}{h}\right)^3$$

$$= 4\pi \left(\frac{V}{\lambda_D^3}\right) \,. \qquad (3.13)$$

The last line is interesting. If we have to picture the particle as a matter wave, it would look like a puffy looking patch of cloud of size λ_D^3. How many can we pack them into volume V? The answer is in the parenthesis. Apart from the front factor 4π, that is the number of microstates of one molecule. It is interesting that the city blocks and momentum blocks merged into spacial blocks but of size λ_D^3.

Let me go back to Eq. (3.6) and the first line of Eq. (3.9) to answer the question: How many single particle microstates are there for a particle confined in a box of volume V with momentum anywhere between p and $p + dp$? Regarding dp as Δp_x, the answer is:

$$\Gamma(p) = \frac{V4\pi p^2 dp}{h^3} \qquad (3.14)$$

where $V4\pi p^2 dp$ is the volume of the phase space[4] containing the microstates in question. So the number of microstate is equal to the phase space volume divided by h^3. That should definitely be remembered. So let me write it in words:

$$\text{Number of Single Particle Microstates} = \frac{\text{Phase Space Volume}}{h^3} \,. \qquad (3.15)$$

---

[4]It means the combined space of the spacial space spanned by $\vec{r}$ and the momentum space spanned by $\vec{p}$ so that one point in that 6-dimensional space can tell us precisely where the particle is and how fast it is moving.

As we will see shortly, how to treat identical molecules or atoms as such without tagging them is an important issue. One way is to break the rule and tag them as if they were distinguishable and then correct the resultant overcounting error. Another way is to regard multi-particle microstates as various sums of single particle microstates occupied by various number of occupants, in which case there is no need to ask the identity of the occupants. We will use Eq. (3.14) when we follow the later way in a later chapter. Of course, a single particle cannot make a thermal system, but it has prepared us to tackle N particles.

• N-Particles

Now we are ready to handle N ideal gas molecules confined in the same box. Draw cubic spatial blocks just as we did above, and place the molecules in them one by one. Because the particles do not interact with one another, it is also possible to put them all in one block. The total number of ways to accommodate N guests in the spatial blocks is

$$\Gamma_{\vec{r}} = \left[\frac{V}{(\Delta x)^3} \right]^N. \tag{3.16}$$

Turning now to the momentum part, we may proceed as we did for one molecule, but the radius of the shell p is now given by

$$U = (p_1^2 + p_2^2 + p_3^2 + \cdots + p_N^2)/2m = p^2/2m \quad \text{or} \quad p = \sqrt{2mU}. \tag{3.17}$$

We need to draw a shell in the $3N$-dimensional momentum space, make blocks, and then count the number of blocks in it. This calls for the surface area of the shell. We learned in Chapter 2 that the surface area of a sphere of radius r in $3N$ dimensions is not $4\pi r^2$. It is given by

$$A_{3N}(r) = \frac{2\pi^{3N/2} r^{3N-1}}{(3N/2 - 1)!}. \tag{3.18}$$

which leads us to

$$\Gamma_{\vec{p}} = \frac{1}{N!} \frac{2\pi^{3N/2} \left(\sqrt{2mU} \right)^{3N-1} \Delta p_x}{\left(\frac{3N}{2} - 1 \right)! (\Delta p_x)^{3N}} \tag{3.19}$$

where the division by $N!$ is to correct the over counting error committed by tagging the particles in Eq. (3.17) as 1,2,3, etc. We also learned in Chapter 2 that quantum mechanics demands that identical particles be treated as such to the letter. We tagged them for convenience as if they were carrying some kind of identification numbers, which results in an over counting error. Let us see whether the error has been corrected. When the N momenta are all different, we over count the number of blocks systematically by a factor of $N!$. When the N momenta are the same, tagging the molecules does no harm. For the rest of the cases, the severity of the over counting is in between these two extremes. However, if we assume that $\boxed{U \text{ is very large}}$, then the first group constitutes the overwhelming majority. Thus the error may be corrected by dividing the count by $N!$. Let us keep track of what we assume as we progress. All assuming statements will be boxed or underlined.

Putting the two counts together, we have

$$\Gamma = \Gamma_{\vec{r}}\Gamma_{\vec{p}} = \left[\frac{V}{(\Delta x)^3}\right]^N \frac{2\pi^{3N/2}(\sqrt{2mU})^{3N-1}\Delta p_x}{N!(3N/2-1)!\,(\Delta p_x)^{3N}}$$

$$= \frac{V^N}{N!\,(3N/2-1)!}2\pi^{3N/2}\left(\frac{\sqrt{2mU}}{h}\right)^{3N-1}\frac{1}{\Delta x}. \qquad (3.20)$$

All those ΔX and Δp have been taken care of by the Heisenberg uncertainty principle. But unfortunately one ΔX is left. So, we have to choose Δx. This is problematic because all the particles are not moving around with the same amount of energy. The first attempt would be to use the average energy U/N, but the de Broglie wavelength is not a linear function of energy, and therefore there is no guarantee that the average de Broglie wavelength is equal to just one of the average energies. Nevertheless, it is reasonably safe to assume that $\lambda_D(N) \sim \frac{1}{\sqrt{N}}$. We can then do better without any additional assumption. We will choose for Δx

$$\Lambda_{th} \equiv \frac{h}{\sqrt{2mU/aN}} \qquad (3.21)$$

where a is a constant unknown to us at the moment, but all we need to know is that it is a constant independent of U and N.

Substituting $\Delta x = \Lambda_{th}$ into the last equation for Γ, we have

$$\Gamma = \frac{V^N}{N!(3N/2-1)!}2\pi^{3N/2}\left(\frac{\sqrt{2mU}}{h}\right)^{3N}\frac{1}{\sqrt{aN}}$$

$$= \frac{2V^N}{N!\,(3N/2-1)!}\left(\frac{2\pi mU}{h^2}\right)^{3N/2}\frac{1}{\sqrt{aN}}. \qquad (3.22)$$

To calculate the corresponding entropy, calculate the logarithm of Γ. Throw away -1 from $(3N/2 - 1)$ and approximate the logarithm of the two factorials using Sterling's approximation $\ln N! = N \ln N - N$. Retain only terms of order N like $N \ln N$ and $3N/2 \ln(3N/2)$, and throw away all the lower order terms like $(1/2) \ln a$ and $(1/2) \ln N$.

$$S = k_B N \left\{ \ln \left[\frac{V}{N} \left(\frac{4\pi mU}{3Nh^2} \right)^{3/2} \right] + \frac{5}{2} \right\}. \tag{3.23}$$

This is called the Sackur-Tetrode equation. This is $k_B \ln \Gamma$, but it is in the order of N. Then imagine how large Γ must be. It is truly astronomical. That is how many the microstates that the system has for visiting. That is how many options that they have to move around, and they are jumping around from one to another as if they treasure every one of them.

We cautiously divided U with Na in Eq. (3.21). Well, as you have noticed, that N did not make the cut, let alone a. That is what I meant when I said earlier that when calculating Γ we can get away with an N-fold error. If we did not divide U by N, it would have been regarded as a $\sqrt{N}$-fold error. But we have seen that such an error does not make even a dent.

3.5.2 *With Energy-Definite Microstates*

Now let us calculate the entropy a second way using the energy-definite microstates fully based on quantum mechanics. We shall use the standing wave method discussed in Chapter 2.

• Single Particle

Let us first practice the calculation with a single particle. According to quantum mechanics, a particle of mass m confined in a cubic box of volume $L \times L \times L$ can only have one of the following discrete values of energy given by

$$\mathcal{E}_{n_x, n_y, n_z} = \frac{\hbar^2}{2m} \frac{\pi^2}{L^2} (n_x^2 + n_y^2 + n_z^2), \tag{3.24}$$

where the quantum numbers take on any positive integers:

$$n_x = 1, 2, 3, \ldots \quad n_y = 1, 2, 3, 4, \ldots, \quad n_z = 1, 2, 3, \ldots. \tag{3.25}$$

Draw a three-dimensional quantum number space spanned by n_x, n_y and n_z axes. The quantum states constitute a simple cubic lattice, there being a microstate corresponding to each lattice site. The lattice spacing is the same along the three axes, $\Delta n_x = \Delta n_y = \Delta n_z = 1$. Our task is to count the number of the lattice sites that lie within the shell corresponding to a given value of energy U. The radius of the shell should satisfy

$$n^2 = n_x^2 + n_y^2 + n_z^2 = \frac{2mUL^2}{\pi^2 \hbar^2}. \qquad (3.26)$$

The alert reader finds that, since the left-hand side can only be selected integers, we may not be able to give an arbitrary amount of energy to the system before it is thermally isolated. Strictly speaking, that is true, but in reality it is virtually impossible to isolate it so that there is no broadening in energy. Not only are there always sources which broaden the system energy, the energy itself is subject to uncertainty. So, it is entirely permissible to regard the surface of the shell as being subject to some uncertainty.

The count is then given by

$$\Gamma = \frac{1}{2^3} \frac{4\pi n^2 \Delta n_x}{(\Delta n_x)^3}$$

$$= \frac{1}{2^3} \frac{8\pi mUL^2 \Delta n_x}{\pi^2 \hbar^2 (\Delta n_x)^3}$$

$$= \frac{1}{2^3} \frac{8\pi mUL^2}{\pi^2 \hbar^2}, \qquad (3.27)$$

where we took advantage of the symmetry and let the shell extend to all quadrants and then corrected the over counting by dividing the count by 2^3; the quantum numbers are limited to positive inters only and thus the shell only to the first quadrant. This is definitely different from what we obtained earlier in a position and momentum definite way.

A record keeping is in order. Since the thickness Δn_x is equal to unity and was not infinitesimal, what we called shell in the above is actually quite thick. The above is only an approximation because the volume of the real shell is given by

$$\frac{4\pi(n+1)^3}{3} - \frac{4\pi n^3}{3} = \frac{4\pi}{3}(3n^2 + 3n + 1). \qquad (3.28)$$

Since only $3n^2$ was taken ignoring $3n + 1$, we implicitly assumed that the quantum number radius n is very large which is equivalent to the earlier assumption that the energy is very large.

• N Particles

Now we are ready to handle N identical molecules. The microstate energies are given by

$$E_{n_1,n_2,\ldots,n_N} = \frac{\hbar^2}{2m} \left(\frac{\pi}{L}\right)^2 (n_1^2 + n_2^2 + n_3^2 + \cdots + n_N^2), \tag{3.29}$$

where $n_i^2 = n_{i,x}^2 + n_{i,y}^2 + n_{i,z}^2$. We have three quantum numbers for each particle and they can be any positive integer. Draw a $3N$-dimensional quantum number space. The N-particle microstates constitute $3N$-dimensional simple cubic lattice and our task is to count the lattice sites that lie in the shell of radius n given by the radius should now be determined by

$$n^2 = n_1^2 + n_2^2 + \cdots + n_N^2 = \frac{2mUL^2}{\pi^2\hbar^2}. \tag{3.30}$$

$$\Gamma = \frac{1}{N!} \frac{2\pi^{3N/2} \left(\sqrt{2mUL^2/\pi^2\hbar^2}\right)^{3N-1} \Delta n_x}{2^{3N} (3N/2 - 1)! (\Delta n_x)^{3N}}$$

$$= \frac{\sqrt{\pi}V^N}{N! (3N/2)!} \left(\frac{2\pi mU}{h^2}\right)^{3N/2} \tag{3.31}$$

where $N!$ is again to correct the over counting error by tagging the identical particles as if they carried identification numbers $1, 2, 3$, etc. The factor 2^{3N} is there to correct the over counting error committed by letting the shell extend to all quadrants. For the last line, we threw away -1 from $3N$ and $3N/2$; N is so large that taking -1 away from it does not even warrant the $\approx$ sign. Comparing the result with what we obtained earlier in the semi-classical way, the two differ only by the factors of 2 and $\sqrt{\pi}$, which are negligible when we take logarithm. The two results for the entropy are therefore in agreement. It was ignoring -1 in the above that has brought the agreement. In the semi-classical way, we specified both position and momentum at the same time. Even though this was done within the limit of the uncertainty principle, there was no guarantee that such treatments may be justified. It has been! This is great because it is so convenient to be able to speak of the spatial part and the kinetic part separately. I will leave it as an exercise for you to repeat the above treatment using the plane wave method mentioned in Chapter 2.

Two assumptions have been made for convenience. Check the two underlined statements. They make the same assumption that energy U has to be very large. Since this and the next chapters are devoted to classical ideal gases, there is a need to clarify how large the energy has to be to ensure that the system is in the classical regime. We will do so soon.

3.6 The Fundamental Thermodynamic Equations

With Eq. (3.23) in hand, we now know at least for gases how entropy changes with energy, volume and the number of particles. In fact this is all we need to understand thermal behavior of gases. We will substantiate this statement in the remainder of this chapter, but let us first see how we may best describe the way entropy changes with energy U, volume V and the number of particles N. To this end, let us assume that entropy is given by

$$S = S(U, V, N). \tag{3.32}$$

As Eq. (3.23) shows, entropy is a monotonically increasing function of energy U. This is because the total energy U is a sum of individual particle energies and different microstates differ from each other by the way the total energy is distributed among the constituents. The more energy there is to divide up, the more the ways to divide them up. Since it is a monotonic function, it may be inverted to give

$$U = U(S, V, N). \tag{3.33}$$

These equations are referred to as the fundamental thermodynamic equations. For ideal gases, the Sackur-Tetrode equation, i.e. Eq. (3.23), is the most fundamental thermodynamic equation. It carries the grand title because all thermodynamic properties of gas may be calculated from it, as we will see below.

3.7 Temperature, Pressure and Chemical Potential

Let us now see how entropy is affected when the variables change. To that end, take the differential of Eq. (3.33):

$$dU = \left(\frac{\partial U}{\partial S}\right)_{V,N} dS + \left(\frac{\partial U}{\partial V}\right)_{S,N} dV + \left(\frac{\partial U}{\partial N}\right)_{S,V} dN. \tag{3.34}$$

This says: If we had given the system before we isolated it just a little more energy (dU), and a few more particles (dN) and allowed just a little more volume (dV), it would have this much more entropy (dS). Given

dU, dV, and dN, the resultant dS is determined by this equation. Playing big roles in making this determination are the three partial derivatives. For example, if $\partial U/\partial S$ is large, it means that changing U has little or no effect on entropy. Since knowing these partial derivatives makes things so convenient, we treat them as variables. They then need to be represented by symbols with big titles. If we had the privilege to choose the titles, we surely would come up with something grand, perhaps using some exotic living language, but we do not have that privilege. Their titles come in very familiar words, but please do not just as yet jump to the conclusion that you know everything there is to be learned about them. The three partial derivatives are abbreviated as follows with titles:

$$\left(\frac{\partial U}{\partial S}\right)_{V,N} \equiv T, \qquad \text{Temperature;} \tag{3.35}$$

$$\left(\frac{\partial U}{\partial V}\right)_{S,N} \equiv -P, \qquad \text{Pressure;} \tag{3.36}$$

$$\left(\frac{\partial U}{\partial N}\right)_{S,V} \equiv \mu, \qquad \text{Chemical Potential.} \tag{3.37}$$

I urge you to pretend that you have never heard of these names before. Until you are convinced that they are what they are supposed to be, do not call them by their title. Call T 'tee', μ 'miu', and P 'pea'. With these abbreviations, Eq. (3.34) can be written as follows

$$dU = TdS - PdV + \mu\,dN, \tag{3.38}$$

or

$$dS = \frac{1}{T}dU + \frac{P}{T}dV - \frac{\mu}{T}dN, \tag{3.39}$$

where the coefficients may be identified with

$$\frac{1}{T} = \left(\frac{\partial S}{\partial U}\right)_{V,N} \tag{3.40}$$

$$\frac{P}{T} = \left(\frac{\partial S}{\partial V}\right)_{U,N} \tag{3.41}$$

$$-\frac{\mu}{T} = \left(\frac{\partial S}{\partial N}\right)_{U,V}. \tag{3.42}$$

The three equations above define three most important concepts, temperature, pressure and chemical potential. Focusing on the temperature, let us see what the definition says. For ideal gases, entropy depends on energy as shown in Fig. 3.2. The curve itself is determined by the physical nature of the system. If we allow a certain value of energy before we isolate the system (from the rest of the universe), the slope of the curve at that value of energy defines the temperature; the larger the slope, the lower the temperature. Since the slope is large at low energy, a low energy means a low temperature and a higher energy a higher temperature.

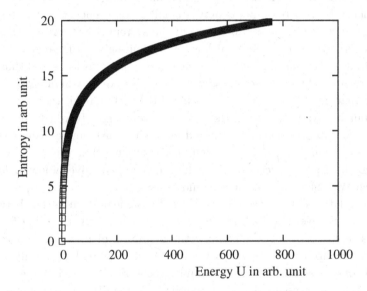

Fig. 3.2   The figure depicts the relationship between entropy and energy. The range of energy has been chosen so as to magnify the large slope at low energy and the decreasing slope with increasing energy.

To add some color to these important concepts, think of them in terms of investment. Suppose that you are going to buy a certain stock. Would you then not like to know the amount of return that you will earn for each dollar of investment? Let us call it the yield. In the investment world, it is a good positive yield that makes investors happy. In the thermal world, it is a good gain in entropy that makes systems happy. There are three ways to invest: give the system a little more energy, a little more volume, or add one more particle. The coefficients appearing in Eqs. (3.39) through (3.42) give the corresponding yields, i.e. the gains in entropy per each unit of investment: $1/T$ is the yield for investing in energy, P/T for investing in volume, $-\mu/T$ for investing in the particle number N. So, when T is very low, investing in energy will award the system with a great yield, but not so when the temperature is high. Similarly, at a given temperature, when the pressure is high, investing in volume brings a great yield, but not so when the pressure is low. In a similar manner, at a given temperature, if μ is negative, adding a particle brings a handsome yield, but not so if μ is positive.

Focus just on the energy investment, and see how different the yield is depending on the energy. Is it not stunning how big the yield is when the energy is low. If we allow only a small amount energy to the system before we isolate it, the temperature of the system is low. Adding a little more energy to such a system at low temperature gives a huge yield. Contrast that with what happens when the system energy is large and thus the temperature is high. The yield is very poor. So, we may summarize it by saying that investing in energy pays big if the system is at a low temperature, but it pays very little if the temperature is high. If we translate this into the language of business, it would mean that the company whose stock you bought will give you a very high yield assuming that you are a typical hard-working poor college student while the company gives a low yield for the rich Wall Street folks. That would be cool.

The relationship between S and V has the same pattern as that between S and U. So, we may look at the same figure but with the label for the U axis replaced by V. To examine the definition of pressure, consider a system of a fixed number of particles and fixed energy. If we allow the system only a small amount of volume, the slope $\partial S/\partial V$ is large, which means that the pressure is high. Give the system a little more volume, and the yield is huge. By contrast if we give the same system a large amount of volume, the slope is low and therefore the pressure is low. Give to such a system a little more volume, and the investment does not pay much; the yield is very poor. So, prepare a system at a low temperature but at a high pressure and give it a little more energy or volume. You will make the system very happy as the yield soars through the ceiling.

The relationship between S and N is not so simple. We shall learn how we may prepare a system at a low or high chemical potential. For the moment, let us examine the yield $-\mu/T$ and find a proper way to understand μ. We are here talking about whether or not the system will increase its entropy when we add one more particle. Suppose that the yield is positive, i.e., $\mu < 0$. It would then mean that the particles are hospitable to the added one, but the added particle is identical to the rest and the positive yield simply means that they are all hospitable to each other. What then does a negative yield, i.e., $\mu > 0$, mean? It should be taken as a measure of the hostility among themselves in the system. If μ/T is positive and large, investing in such a system with one additional particle would be a highly ill-conceived idea; the total entropy will go down, meaning that their happiness level goes down. It is like a hostile place where people do not like each other and therefore the place is simply unlivable. A big warning may be in order. We are here talking about ideal gas particles which do not interact with each other. The hostility or hospitality are a pure thermal effect and have nothing to do with whether or not the particles are pulling each other or repelling each other. If μ is still positive but smaller, the yield is better. It is like people migrating from a highly unlivable place to a less unlivable place. For the chemical potential, the issue is which is less of an evil. As we will shortly find out, the minus sign in the yield $-\mu/T$ is to legitimize the word 'potential' in its title; the chemical potential is actually like a potential.

3.8 The Approach to Equilibrium

Entropy provides the direction of all thermal processes, but we cannot take advantage of it unless we know the fundamental thermodynamic equations of all the substances involved in the process. There is a short-cut. Instead of the fundamental equation itself, if we know its partial derivatives, we will then know all the pertinent yields which can serve us even more conveniently. To understand this important point, let me dramatize the second law one more time. Thermal systems are driven in search for more and more entropy, and that is all they want to do; they just want to increase their entropy as much as possible. Those partial derivatives tell the system how to achieve that goal.

Imagine a cylinder which is divided into two chambers by a partitioning piston, as depicted in Fig. 3.3. Assume that the cylinder wall and the piston are both made out of insulating, non-expanding and non-permeable material. The two chambers contain two different gases. The piston is firmly fixed at a position and the two gases are at their respective equilibrium states. Call the gas on the left A and the gas on the right B. Let the partial derivatives be T_A, P_A and μ_A for chamber A and T_B, P_B and μ_B for B. Assume that $T_A > T_B$, $P_A > P_B$, and $\mu_A > \mu_B$. Let us now suppose that we have a way of changing the characteristics of the piston one by one

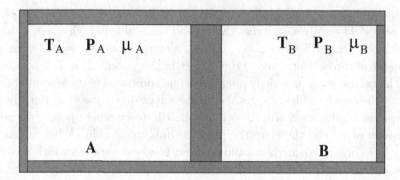

Fig. 3.3   A gas is confined in two parts of a cylinder, A and B, divided by a piston-like middle wall.

to disrupt the equilibrium. Then we will see what roles these coefficients play while the system is moving toward its new equilibrium state.

3.8.1   *Temperature*

The position of the partition remains fixed, but now imagine it turning into a thermally conducting wall which allows an exchange of energy between the two sides. What happens then? Assume $T_A > T_B$. The combined system of A and B still constitutes a thermally isolated system, but each of them is no longer isolated and no longer in equilibrium. Since entropy is a monotonic function of energy, both A and B will want to have more energy to increase their own entropy. However, since the combined system is isolated, one part can increase its energy only at the expense of the other part. It is then crucial that the side receiving energy increases its entropy more than what the donor loses. Which side must give and which side must receive? From Eqs. (3.40) and (3.23), we have

$$\left(\frac{\partial S}{\partial U}\right)_{V,N} = \frac{1}{T} = \frac{3Nk_B}{2U} \tag{3.43}$$

which says that if T is very low, even a small amount of additional energy can bring a huge increase in entropy. The side for which T is lower should therefore receive energy, namely, B. **If a system at a low T receives energy, the resultant increase in entropy is huge. If such a system loses energy, the resultant loss in entropy can be devastating. By contrast, if a system at a high T receives that same amount of energy, the resultant increase in entropy is very small, and if it** loses that amount of energy, the resultant loss in entropy is hardly noticeable.

As energy flows from A to B, Eq. (3.43) also tells us what happens to each part. The second line says: T increases with U; a small amount of energy means a small value of T, a low temperature, and a large amount of U means a large value of T, a high temperature. Therefore the receiving side will increase its temperature while the donor side lowers its temperature. The energy flow will stop when the two temperatures match; the combined system has now reached thermal equilibrium. Are you now convinced that T as defined by Eq. (3.40) is indeed what temperature is supposed to be?

If you prefer equations to words, assume that A's energy will increase by ΔU and B's will decrease by the same amount. Then

$$\Delta S = \frac{1}{T_A}\Delta U + \frac{1}{T_B}(-\Delta U) = \left(\frac{1}{T_A} - \frac{1}{T_B}\right)(\Delta U). \qquad (3.44)$$

If the net entropy change is to be positive as demanded by the second law, it follows that ΔU has to be negative, meaning that energy will flow from A to B.

Think of $1/T$ as the molecules' appetite for energy. When two different systems, or two different parts of the same system, have different degrees of appetite for energy, the party with more appetite takes energy from the party with less appetite. This is not a runaway process, however. As the energy is exchanged, the receiving party loses its appetite while the donor party gains it, which ultimately leads to an equilibrium state where both parties have the same amount of appetite. This built-in safety system brings displaced thermal systems back to equilibrium and is called the Le Châtelier's theorem. Molecules know when to receive energy and when to give it, all for the good of the total system to which they belong. Moreover, they know when to increase their appetite and when to decrease their appetite. We may say that thermal physics is a science of giving and receiving.

3.8.2 *Pressure*

If we leave the system as it is, nothing will happen further and the two parts will remain in their respective equilibrium states. Both chambers are at the same temperature, $T_A = T_B$. Now suppose that the partition becomes mobile so that the volume of the two sides, A and B, can be exchanged. What then happens? Assume $P_A > P_B$. Entropy is a monotonically increasing function of volume, and therefore each side will want to increase its volume to increase its own entropy. But since the total volume of the cylinder is fixed, one can increase its volume only at the expense of the other. In order to ensure that the total entropy will increase, the receiving side must increase its entropy more than the donor loses. Which side must give and which side must receive?

From Eqs. (3.41) and (3.23), it follows that

$$\left(\frac{\partial S}{\partial V}\right)_{U,N} = \frac{P}{T} = \frac{k_B N}{V}.$$ (3.45)

Equation (3.45) gives us the answer. The side for which $\partial S/\partial V$ is larger should receive, i.e. A which has the higher pressure. In order to increase their combined total entropy, B should shrink so that A can expand. **If a system at a high pressure is given more volume to expand into, the resultant increase in entropy is huge. If such a system loses volume, the resultant loss in entropy can be devastating. By contrast, if a system at a low pressure receives that same amount of volume, the resultant increase in entropy is very small, and if it loses that amount of volume, the resultant loss in entropy is hardly noticeable.** I hope that the reader is convinced that P as defined by Eq. (3.41) is indeed what we have known all along as pressure.

Equation (3.45) also tells us that the exchange of volume results in a change in pressure. It says that the pressure P decreases with increasing volume. As the partition moves to increase A's volume, A's pressure decreases while B's increases. The volume exchange, or the volume "flow", will continue until the two pressures match; the system has now reached mechanical equilibrium. Think of P as the molecules' appetite for volume.

3.8.3   Chemical Potential

Both thermal and mechanical equilibrium have been reached, hence $T_A = T_B$ and $P_A = P_B$. To continue the story for N, imagine that the partition now becomes a permeable wall through which molecules on one side can migrate to the other. Since we assumed that chamber A is at a higher chemical potential than B,

$$\mu_A > \mu_B, \qquad \text{or} \qquad -\frac{\mu_A}{T} < -\frac{\mu_B}{T}.$$ (3.46)

The combined system of A and B is still an isolated system but now is in a non-equilibrium state. How will it reach an equilibrium state? Since molecules are neither annihilated nor created in the cylinder, if one side is to increase its population, it can do so only at the expense of the other. To ensure that the total entropy increases as a result of the migration, which side should give and which side should receive? The first equation says that chamber A is more hostile than chamber B, and the second equation says that the entropy yield is higher for chamber B than for chamber A. So, particles will be migrating from A to B. Chamber B gains far more entropy by the migration than chamber A loses.

To see how the yield changes with N, let us calculate the chemical potential. From Eqs. (3.42) and (3.23), it follows that

$$\left(\frac{\partial S}{\partial N}\right)_{U,V} = -\frac{\mu}{T} = k_B \ln \left\{ \frac{V}{N} \left(\frac{4\pi mU}{3h^2 N}\right)^{3/2} \right\} . \qquad (3.47)$$

The first line says that, if $-\mu/T$ is high, adding even just a few more particles can bring a big increase in entropy while the second line says that $-\mu/T$ decreases when N increases. So, if $-\mu/T$ is large, the system entices particles to move in, but the yield decreases with increasing N. This is a remarkable self-control system. My city Hattiesburg does that very well. For the past two decades there has been a steady increase in population, and for a good reason. We may again think of $-\mu/T$ as the molecular analogue of the 'livability' or μ/T as the molecular analogue of 'unlivability' which tends to encourage its population to leave. As the population increases, the unlivability gradually increases very much like in the molecular world.

Let me dramatize it again. High chemical potential means high un-livability. **If a system at a high chemical potential decreases its population, its entropy goes up by a huge amount. Should such a system increase its population, it will suffer a devastating loss in entropy. By contrast, if a system at a low chemical potential receives that same number of molecules, the resultant loss in entropy is hardly noticeable, and if it loses its population by that number of molecules, the resultant gain in entropy is very small.** So, the final answer is: since unlivability is higher at A than at B, particles will migrate from A to B. According to the second line of Eq. (3.47), the unlivability of A will then decrease with decreasing N while that of B will increase. When the two match, the migration stops, and we say that the system has reached its diffusive equilibrium.

To summarize, if we know T, P and μ of a thermal system away from equilibrium, we can decide the direction of the non-equilibrium process without having to ask anything about the composition or its fundamental thermodynamic equation. Of course, these quantities were not introduced initially for this purpose. We are able to enjoy the logical inter-connections thanks to many luminaries like Boltzmann, Gibbs and Landau.

The human world works in a similar way. Hattiesburg was at the top of many lists of the most livable small cities in the U.S. about 15 years ago. Since then it has experienced a steady increase in population. It is still a very livable small city, but not as much as it used to be. Needless to say, the brown Pontiac contributed to the decreased livability.

3.8.4 *Multiple Traffic Control Systems*

We have changed the nature of the partition between the two chambers stage by stage in order to highlight the individual roles of T, P, and μ, respectively. What would happen if we were to convert the insulating, fixed, and impermeable partition all at once into a conducting, mobile and permeable wall? The three kinds of flow will take place simultaneously. The energy traffic will be controlled by $1/T$, the volume traffic by P/T, and the particle diffusion traffic by $-\mu/T$. The energy traffic is coupled to the concurrent volume and particle traffic, and therefore all three traffics are coupled to each other.

What we called A and B up to this point can be viewed as two adjacent parts of a large thermal system out of equilibrium. What causes a system to be out of its equilibrium state? The answer: U, V, and N are not properly distributed among the different parts of the system. To deal with this situation, we have to regard T, P and μ as representing local properties. There may be too much energy in one part; if so, energy must flow from that part to the rest. Molecules in one part may be unduly compressed; if so, that part needs to expand at the expense of the rest. There may be too many molecules in one part; if so, they must diffuse away from that part. The traffic should be directed to increase the total entropy. Energy should flow along the direction in which T decreases most rapidly, namely, $-\nabla T$. Volume should flow along the direction in which P/T increases most rapidly, namely, $\nabla(P/T)$. Particles should migrate along the direction in which μ/T decreases most rapidly, namely, $-\nabla(\mu/T)$.

Draw a map showing the equi-surfaces of T, P/T, and $-\mu/T$. We have three topographic maps, one for the energy traffic, one for the volume traffic, and one for the particle traffic. The three equi-surfaces are constantly changing to ensure that the traffic will result in increasing the entropy and guiding the system to its new equilibrium state. In other words, the molecules have built-in safety control systems: automated thermostats for temperature control, automated valves for pressure control, and automated border control systems for migration control. This is again the Le Châtelier's theorem.

What would it be like if molecules were in a state in which they did not have this built-in control system? The temperature would decrease with increasing amount of energy, the pressure would increase with increasing volume, and the chemical potential would decrease with increasing population. Their appetite for energy would increase with an increasing amount of energy, and their appetites for V and N would increase likewise. These molecules would then gobble up everything in the universe. Worse yet, should the safety system be restored after a while, their return to a normal state would be dangerously explosive. To give a human analogue, think of a man whose appetite increases with an increasing amount of food in his stomach. Molecules will never lose the safety system. We can force the molecules into a non-equilibrium state. In Chapter 14 we will see what the molecules do under such a condition.

3.8.5 Some Reflections

We actually derived two rather important results earlier. The reader should be familiar with them, and so let us stop here and reacquaint with them in a more reflective way.

Going back to Eq. (3.43), notice

$$U = 3N \times \frac{k_B T}{2}. \tag{3.48}$$

Each particle has 3 degrees of freedom, meaning that its momentum has three components. The total number of degrees of freedom of N particles is then $3N$, and this equation shows, rather remarkably, that each degree freedom contributes to the total energy the same amount, $k_B T/2$. This will be discussed again later under the title of the equipartition theorem.

Knowing how energy changes with temperature, we may easily calculate the specific heat,

$$C_V = \left(\frac{\partial U}{\partial T}\right)_V = 3N \times \frac{k_B}{2}, \tag{3.49}$$

where V is held fixed to keep L the dimension of the box and thus the microstate energies fixed. The result says that the specific heat is a constant independent of temperature, each degree of freedom contributing $k_B/2$. Our ideal gas molecules was assumed to have only translational degrees of freedom. As it turns out, measuring specific heat is a good way to probe the inner workings of thermal systems. For the ideal gas in its classical regime, the result simply says that the molecules are moving around in three different directions. The only thing pertinent for specific heat is the degree of freedom. How simple!

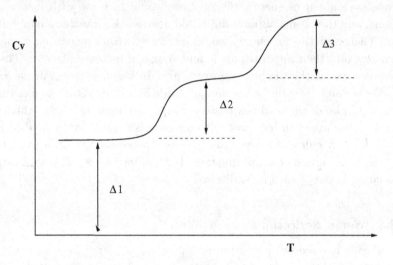

Fig. 3.4   The general pattern of the specific heat of diatomic gas molecules. See the text for the jumps, Δ_1, Δ_2 and Δ_3.

This point is best illustrated by diatomic molecules such as Hydrogen, Oxygen and Nitrogen molecules. For these molecules, there are two additional modes of motion which bring in more degrees of freedom: the two atoms can rotate like a rigid rod with two degrees of freedom, and they can also vibrate along the direction of their axis, stretching the bond between the two atoms and shrinking it periodically. The vibrational mode brings in two additional degrees of freedom, the kinetic degree of freedom and the potential degree of freedom. These motions, however, require more energy than the translational motions, and therefore they remain frozen until the temperature is sufficiently high; the rotational mode is first activated and then the vibrational mode at an even higher temperature. The specific heat at constant volume therefore changes with temperature as depicted in Fig. 3.4. When the rotational mode is activated, the right-hand side of Eq. (3.49) becomes $(3+2)Nk_B/2 = 5Nk_B/2$, and when the vibrational mode is also activated, the right-hand side becomes $7Nk_B/2$. Measurements do confirm this pattern of change. Thus in Fig. 3.4 the three jumps are related by $\Delta_2 = \Delta_3 = \Delta_1 2/3$.

At sufficiently high temperatures particles try to maximize their entropy or their multiplicity by mobilizing all possible modes of motions. As the temperature approaches zero, they become instead more and more picky, and at exactly zero temperature the specific heat becomes zero! Much of what we now know about the low temperature physics has been gathered by measuring specific heat.

Going back to Eq. (3.48) again, there are those who wish to take it as the definition of temperature. If you take the view that temperature may be defined as the average kinetic energy divided by $(3/2)Nk_B$, how would you handle the thermdoynamics of magnets and other non-mobile thermal systems? It is true that even in the latter systems the constituents are vibrating and therefore there is a kinetic degree of freedom, but their magnetic properties are due to the orientational potential energy of each magnetic dipole moment against the internal and external magnetic fields. How would you handle this? You will be forced to fudge with the fundamental concept of temperature. There is only one definition for temperature which can be applied regardless of whether the constituents are mobile as in our ideal gas or non-mobile as in magnets, and each should stand on its own.

Also going back to Eq. (3.45), there is another item which should definitely be familiar:

$$PV = Nk_BT\,, \tag{3.50}$$

which shows how the two partial derivatives or thermal coefficients, P and T, are related. It is called the ideal gas equation of state. This is how we physicists write it, but chemists and engineers write it as

$$PV = nRT\,, \tag{3.51}$$

where n is the number of moles and R is called the gas constant given by $R = N_A \times k_B$; N_A is the Avogadro number.

We first learn this in the introductory physics course using a mechanical argument, which may have misled you to believe that thermal physics is just a special case of mechanics. The consequence of this type of confusion can be devastating.[5] One then tries to search for a mechanical interpretation of entropy, which ends in a frustration. The usual outcry is that

---

[5]No one suffered more than this author early in his youth as explained in the preface.

entropy is 'mysterious', by which one really means that he could not find a satisfactory mechanical interpretation for it. Yes, there was a time when it was mysterious, but that was more than a century ago before Boltzmann came along. Entropy and temperature are not mechanical quantities. They are thermodynamic quantites reflecting the uniqueness of thermal systems not shared by mechanical systems.

The most important result we have worked out so far is the Sackur-Tetrode equation for entropy which we would like to display here,

$$S = k_B N \left\{ \ln \left[\frac{V}{N} \left(\frac{4\pi mU}{3Nh^2} \right)^{3/2} \right] + \frac{5}{2} \right\}, \qquad (3.52)$$

which shows how entropy changes with energy U, volume V and the particle number N. The three thermodynamic parameters T, P/T and μ/T were defined as the rate of change of entropy with respect to changes in U, V and N, respectively. Nature always wants more and more entropy. These coefficients indicate how that tendency will manifest itself when U, V and N change, which is why they are so convenient to work with.

In terms of temperature, the entropy may be written as

$$S = k_B N \left\{ \ln \left[\frac{V}{N} \left(\frac{2\pi mk_B T}{h^2} \right)^{3/2} \right] + \frac{5}{2} \right\}. \qquad (3.53)$$

The chemical potential will play an important role when we discuss how matters in different phases come to equilibrium in Chapter 12. Let me rewrite it here also in terms of the temperature:

$$\mu = k_B T \ln \left[\frac{N}{V} \left(\frac{2\pi mk_B T}{h^2} \right)^{3/2} \right]. \qquad (3.54)$$

We had to introduce the uncertainties Δx and Δp_x. By using the Heisenberg uncertainty principle, all of them were paired up and entered the calculation as one factor of Planck constant h for each pair, and we did not have to specify the uncertainties. One Δx was, however, left unpaired and we attempted to extend for it the single particle de Broglie wavelength λ_D to Λ_{th}, the appropriate wavelength for each of those N particles sharing a total energy U. $\Lambda_{th} \sim 1/\sqrt{N}$, that small additional N-dependence was proven to be negligible, and we left the issue without determining the constant a. However, since the Planck constant remains in the calculation, and since the only reason for h is the uncertainties, h may show up in the final result in a form of uncertainty. Indeed that is the case! Look at Eqs. (3.52) and (3.53) and see where h is. Examine the argument of the logarithmic function containing the Planck constant:

$$\frac{V}{N}\left(\frac{4\pi mU}{3Nh^2}\right)^{3/2} = \frac{V/N}{\left(h/\sqrt{4\pi mU/3N}\right)^3} .$$ (3.55)

We may identify

$$\Lambda_{th} = \frac{h}{\sqrt{4\pi mU/3N}} = \frac{h}{\sqrt{2\pi mk_BT}}$$ (3.56)

and the constant that we left uncertain is $4\pi/3$. History has shown that whenever a physical phenomenon is pursued via the right track there emerges the characteristic length scale. This is important because one then knows the length scale in which the phenomenon is taking place but also by identifying that length scale one knows the most dominating player in the phenomenon. If you took the first semester quantum mechanics, look at the wave functions of a Hydrogen atom. You see a_0 all over, which is the Bohr radius. We have just identified ours and that is the thermal de Broglie wavelength Λ_{th}.

How could the uncertainty play the dominating role? We may imagine the matter wave associated with each particle as a puffy looking object like a patch of cloud. Then Λ_{th}^3 may be thought of as the size of the puffy object. You see, our ideal gas molecules are literally point particles with no size, which is why when we calculated the number of ways of putting them in the city blocks we allowed the possibility of all of them occupying the same block. The matter wave associated with them, however, does have a definite size. How ironic that what we introduced as an uncertainty ends up providing something definite! Thus Λ_{th} is their effective size, and it is no wonder then that it plays the dominating role. More on this shortly.

Inspecting the entropy and the chemical potential above, it is apparent that

$$S = k_BN\left\{\ln\left(\frac{V/N}{\Lambda_{th}^3}\right) + \frac{5}{2}\right\} ,$$ (3.57)

$$\mu/T = -k_BN\ln\left(\frac{V/N}{\Lambda_{th}^3}\right) .$$ (3.58)

So, what determines the entropy is the ratio

$$\frac{(V/N)^{1/3}}{\Lambda_{th}} \equiv \frac{l}{\Lambda_{th}}$$ (3.59)

where l is the linear size of the volume per particle and measures the typical inter-particle distance. Let us then see how the multiplicity Γ depends on Λ_{th}. Since $S/k_B = \ln \Gamma$, by exponentiating the entropy eequation, Eq. (3.57), we find

$$\Gamma = \left(\frac{l}{\Lambda_{th}} \right)^{3N} e^{5N/2} . \tag{3.60}$$

It is then of interest to remember if the same relationship holds between the multiplicity count and the (single particle) de Broglie wavelength λ_D. Let me bring Eq. (3.13) here. The multiplicity count for a single particle is given by

$$\Gamma_1 = 4\pi \left(\frac{L}{\lambda_D} \right)^3 . \tag{3.61}$$

Although the relationship is not exactly the same, there is a strong consistent message. If l/Λ_{th} or L/λ_D is large, i.e. if the ratio of the linear size of the free space and the wavelength is large, then there is a large multiplicity and thus a large entropy. Repeating it again, assuming a very small Λ_{th} means that the multiplicity is large, namely, that the number of microstates is very large.

The ratio appearing in the above equations essentially asks: how many of those puffy patches may be packed in the volume per particle. With this in mind, recall the assumptions we made to get here. The two underlined statements said that we assume a large energy, i.e. a high temperature. A high temperature means a very small thermal de Broglie wavelength. We have just found out that the question of how small the de Broglie wavelength is should be judged in comparison with $l = (V/N)^{1/3}$. In fact raising the density has the same effect as raising the temperature. So, we may now make the two boxed statements we made earlier more precise with

$$\boxed{\frac{l}{\Lambda_{th}} \gg 1} . \tag{3.62}$$

which says that the inter-particle distance should be much greater than the thermal de Broglie wavelength. In other words, the puffy looking patches should be far away from each other so that they do not touch each other.

If the puffy patches touch and overlap each other, on the other hand, i.e. if

$$\frac{l}{\Lambda_{th}} \le 1 , \tag{3.63}$$

then, the classical ideal gas becomes an ideal quantum gas which will be the topic of Chapter 8.

Next turn to the chemical potential. In the classical regime, it is clear that $\mu < 0$ and

$$\exp(\mu/k_B T) \ll 1. \tag{3.64}$$

In the opposite quantum regime, on the other hand, $\mu > 0$, and the classical ideal gas becomes an ideal quantum gas which will be the topic of Chapter 8.

Let us also remember that assuming Eq. (3.62) is equivalent to assuming that the number of microstates is very, very large .

3.9 Macrostates

3.9.1 *The Largest Macrostate*

We emphasized how important the size is for macrostates several times in previous sections. Now that we have Γ as a function of N, V and U, we are ready to construct macrostates for our ideal gas system and see if the size distribution is as we promised. Let me bring the equation here:

$$\Gamma(U, V, N) = \frac{\sqrt{\pi} V^N}{N!(3N/2)!} \left(\frac{2\pi m U}{h^2} \right)^{3N/2}. \tag{3.65}$$

Macrostates may be constructed in a number of ways. One way would be as follows. Divide the particles into two groups. N_A molecules in part A and the remainder $N_B = N - N_A$ are in part B. Macrostates may then be defined according to the way the two parts share the total available volume and energy. Consider a macrostate in which part A holds energy U_A and volume V_A while part B holds $U - U_A$ and $V - V_A$. From Eq. (3.65), the multiplicity of that macrostate goes like this:

$$\Gamma_{AB}(U, V, U_A, V_A) \sim V_A^{N_A}(V - V_A)^{N_B} U_A^{3N_A/2}(U - U_A)^{3N_B/2}. \tag{3.66}$$

Is there a particular value of U_A for which the value of Γ is maximum? Differentiating Eq. (3.66) with respect to U_A, we find that the answer is yes, and that the maximum occurs when

$$\frac{U_A}{N_A} = \frac{U_B}{N_B}. \tag{3.67}$$

Is there an optimum value for V_A which maximizes Γ? The answer is again yes; the maximum occurs when

$$\frac{V_A}{N_A} = \frac{V_B}{N_B}. \tag{3.68}$$

In the largest macrostate, energy and volume are divided between the two parts in proportion to their respective numbers N_A and N_B. We may repeat this argument with more partitions for further refinement. We conclude that in the largest macrostate, energy and volume are more or less uniformly distributed in all parts of the system.

3.9.2 *How Dominant?*

Next, let us find out how dominant the largest macrostate is. This question can easily be answered if we assume that $N_A = N_B = N/2$. We do not lose any generality with this assumption since each of the halves can be divided into halves again and again. In the largest macrostate, $U_A = U/2 \equiv \hat{U}$ and $V_A = V/2 \equiv \hat{V}$. Now consider a macrostate in which U_A deviates from $\hat{U}$ by x, namely,

$$U_A = \hat{U} + x, \qquad U_B = \hat{U} - x. \tag{3.69}$$

The size of the corresponding macrostate is as such:

$$\begin{aligned}
\Gamma_{AB}(x) &= \Gamma_0(V, N)(\hat{U} + x)^{3N/4}(\hat{U} - x)^{3N/4} \\
&= \Gamma_0(V, N)(\hat{U}^2 - x^2)^{3N/4} \\
&= \Gamma_0(V, N)\hat{U}^{3N/2}(1 - x^2/\hat{U}^2)^{3N/4}
\end{aligned} \tag{3.70}$$

where all the factors independent of U are lumped into Γ_0. We shall use the symbol Γ_0 in this sense below without any further explanation. Because N is very large, Γ_{AB} decreases quite rapidly with increasing x. When a function changes this rapidly, it is best to slow it down by taking the logarithm and then approximate it with its first order Taylor series term,

$$\begin{aligned}
\ln \Gamma_{AB} &= \Gamma_0' + \frac{3N}{4} \ln \left(1 - \frac{x^2}{\hat{U}^2}\right) \\
&\approx \Gamma_0' + \frac{3N}{4} \left(-\frac{x^2}{\hat{U}^2}\right).
\end{aligned} \tag{3.71}$$

Exponentiate the result to obtain

$$\Gamma_{AB}(x) = \Gamma_0'' \exp\left(\frac{-x^2}{2\delta x^2}\right), \qquad (3.72)$$

where

$$\delta x^2 = \hat{U}^2/(3N/2). \qquad (3.73)$$

To see how x is distributed, let us convert this into a distribution function by normalizing it. To do so, integrate the above over x,

$$\int_{-\infty}^{\infty} \Gamma_{AB}dx = \Gamma_0''(2\pi\delta x^2)^{1/2}. \qquad (3.74)$$

By dividing Eq. (3.72) with this, we obtain the distribution function of x:

$$W(x) = \left(\frac{1}{2\pi\delta x^2}\right)^{1/2} \exp\left(\frac{-x^2}{2\delta x^2}\right), \qquad (3.75)$$

which we recognize from the central limit theorem in Chapter 2. Notice how narrow the width of the Gaussian peak is. From Eq. (3.73), since N is an astronomically large number, the distribution is hardly bell-shaped; it is for all practical purpose a spike. That is how the largest macrostate, namely, that corresponding to $x = 0$ dominates.

How close then do we have to go to $x = 0$ for a tangible chance of finding a competing macrostate? Let $N = 10^{23}$ and try $x = 1.14 \times 10^{-11}\hat{U}$. Compare its size with with that of the largest macrostate corresponding to $x = 0$. Its multiplicity is given by

$$\Gamma_{AB}(x = 1.14 \times 10^{-11}\hat{U}) = \Gamma_{AB}(x = 0)e^{-10}, \qquad (3.76)$$

which is smaller than the largest one by a factor of e^{10}. Therefore the chance of the system visiting it is e^{-10}, or one out of 22026. If we check the system once every second for 6 hours, we have a chance of witnessing the deviation. That is not bad, but the deviation is only one part out of 10^{11} which is not much of a deviation. Of course, if we go further away from $x = 0$ than this one, we will have a tangible deviation, but its size is truly negligible. On the other hand, if we go even closer to $x = 0$ than this one, there will be macrostates which are large enough to compete with the largest one, but they are hardly different from the largest macrostate, and therefore it does no harm if we simply ignore them.

Exercise. *Repeat the above for volume.*

When a gas is given a certain volume, it takes up all the given space, no matter how large the volume may be. Why? Why do the molecules not reserve some part of the space for guests or for some special purpose like we do? Let us define the macrostates according to y, the amount of space left vacant out of the given volume V. Since the size of the macrostate corresponding to y goes like $(V - y)^N$, $y = 0$ clearly constitutes the largest macrostate. Let us consider the macrostate B in which molecules occupy only 99.9 percent of the given volume leaving 0.1 percent empty, i.e. $y = 0.001V$. How small is this macrostate compared to the largest one? Eq. (3.31) says that the multiplicity ratio is $(0.999V/V)^N = 0.999^N$. With $N = 10^{23}$, the ratio is 4.3×10^{-19}. Do we have a chance of observing the system in B? That is, is there any chance of seeing the molecules spontaneously staying away from 0.1 percent of the available volume? To have any reasonable chance of seeing this, we would have to devote at least the length of time since the Big Bang checking the system once every second.

To summarize, at the thermodynamic limit of large N, it is safe to take the largest macrostate and ignore the rest to study equilibrium properties. This is a virtue; if it were not true, thermal physics would not be a predictable science.

3.10   First Law of Thermodynamics

Let us write Eq. (3.38) once more

$$dU = TdS - PdV + \mu\,dN\,. \tag{3.77}$$

It is called the First Law of Thermodynamics. It is simply a statement of the energy conservation law for a system. As we did before, imagine that we go back to the isolated system and add just a little more energy (dU), allow a little more volume (dV) and add a few more particles (dN), then the entropy will change by dS given by this equation. But because entropy is a monotonically increasing function of energy, it is permissible to change the order in the above and say that, if we change entropy by dS, volume by dV and the number of molecules by dN, then energy will change by dU. I hear the alert reader asking 'How do we change entropy?' Hold the question and read on. Examine how each term contributes to a change in energy. The system can receive or give away energy in three different ways: (a) by changing its entropy for the TdS term, (b) by changing its volume for the PdV term, or (c) by changing its population number for the $\mu \, dN$ term. It is important to understand the difference between (a) and (b).

When energy increases via (a), we say that heat has entered the system. The energy increases without any change in volume and therefore without any change in the energy units. There is now more energy in the same units to be divided up among the same number of molecules, which means an increase in entropy. The increased entropy is kinetic entropy. If we put a hot brick next to another physical system, the vibrational kinetic energy of the brick molecules is transferred to the system for TdS. If we send a light beam into a cup of water so that the water molecules absorb the light energy, the light energy is transferred to the water for TdS.

When energy changes via (b), we say that work has been done to the system. It is important to note that this occurs **without changing entropy** or N. How can energy change without changing entropy? Consider a gas held in a cylinder by a piston. Imagine that we push the piston very slowly. We are performing work to the system. What happens to the work? The change in volume causes a change in all the $\mathcal{E}$'s. When energy changes in this way, entropy does not change. To see this, imagine that we have 1111 dollars in a bag to be divided up among 4 people. In the bag are one 1000-dollar bill, one 100-dollar bill, one 10-dollar bill, and one 1 dollar bill. Suppose that you are going to just grab whatever you can in the bag and hand it over to one of the four, and repeat the action until there is nothing left in the bag. How many different ways are there to divide up the money? Hold your answer for a moment and let me ask you one more question. What if the bag has one 100-dollar bill, one 10-dollar bill, one 1 dollar bill, and one dime? It is clear that the answer is the same for both questions. Equations (3.23) and (3.29) confirm this intuition. Notice that V and U enter as $VU^{3/2}$ into Eq. (3.23) to determine S. Since Eq. (3.29) says that U scales with V like $V^{-2/3}$, $VU^{3/2}$ does not change when V changes.

If we compare our wealth with energy, we may also say that our wealth can change in two similar ways. At this very moment I have in my pocket one ten dollar bill, one 1-dollar bill, two quarters, two dimes, and four pennies. I can become richer in two ways. (a) I may win a prize to bring a few more quarters and dimes into my pocket, the "heat" way. (b) No additional money goes into my pocket, but thanks to the hard *work* of the nation's economy, I may find one day that each of my four pennies can buy what costs today one dollar, and likewise each of my two dimes can buy what costs today twenty dollars, and so on, the "work" way. I rely on the "work" way, namely (b), which is why I have two large jars full of pennies. A better way is to buy some stocks and let the company do the *work*, but we know that this is a risky proposition because the company may do the work in the wrong direction, decreasing their $\mathcal{E}$ instead of increasing it. I have just read a physicist's autobiography. Apparently he became brilliant with stocks and rare art pieces because his stockbroker once proposed that they exchange roles! It is good to know that some of us can do better than just collecting pennies.

Returning back to the first law, several comments and remarks are in order. Here and in the rest of this book, the symbols such as dU, dS, or ΔU, ΔS, etc. represent changes brought about by reversible processes, and by a reversible process it means that the system is in an equilibrium state at every step of the way. When a system is disturbed and thrown off its equilibrium state, it takes time for the system to reach its new equilibrium state; the necessary time is called the relaxation time. So if a change takes place faster than the relaxation time allows, it is an irreversible process. Unless stated otherwise, we will always be dealing with reversible processes and dU and ΔU symbols will be used without the qualifiers 'equilibrium' and 'reversible'. The more general statement of the first law should read

$$dU = \bar{d}Q + \bar{d}W \tag{3.78}$$

where $\bar{d}Q$ is just a small amount of energy entered as a heat, not a differential of Q. Given U for a system for which we do not know the past history, there is no way to find out how much entered as a heat and how much as work. It is like being asked how much the water in a pond entered directly through raindrops and how much through the water streams that feed it.

Also, depending on how heat is given, the change in entropy can be different. If it is given reversibly

$$dS = \frac{\bar{d}Q}{T} \tag{3.79}$$

while if it is given irreversibly,

$$dS > \frac{\bar{d}Q}{T}. \tag{3.80}$$

The latter may appear counterintuitive. Entropy can increase without requiring heat, for example, by chemical reaction. When a gas expands into vacuum, it does not require heat or work but entropy does increase, the spatial entropy. Thus Clausius noted that for any natural change from state 1 to state 2,

$$S_2 - S_1 > 0. \tag{3.81}$$

If, on the other hand, the same change takes places reversibly or more appropriately if it can happen reversibly,

$$S_2 - S_1 = 0 \tag{3.82}$$

which is because the change can happen just as well from state 2 to 1. Let me reiterate that throughout the book the symbols like dU and ΔU represent changes, unless specified otherwise, brought about by a slow reversible process which ensures that the system remains in equilibrium at every stage of the process. I am a firm believer that one should understand equilibrium thermal physics before attempting the much more complex non-equilibrium irreversible processes, and thus, except in Chapter 14, the main focus of this book is on equilibrium thermal physics.

Finally, returning back to Eq. (3.77), look at the work term $-PdV$. Suppose that we are doing this work by pushing a piston of cross sectional area of A with force F. Then, as we knew all along, we see that $P = F/A$, namely, P is the force per area. This should be taken as a welcome outcome in spite of its thermodynamic definition, and it should not be taken as meaning that pressure is a mechanical quantity. In the same vein, notice Eq. (3.48) where energy is measured with temperature. There are situations where for necessity we take advantage of this outcome. That is what one does when one performs Molecular Dynamics computer simulations. But it would be badly misleading to take it as meaning that temperature is a mechanical quantity. They both should be taken as thermodynamic quantities.

3.11 Extensive and Intensive Variables

All thermodynamic quantities may be divided into two groups, extensive and intensive. Examples of extensive quantities are U, S, V, and N. They are called extensive because if we add two systems, the extensive variables of the two systems add for a combined total. Energy is additive as are entropy, volume, and particle numbers. We can speak of a total for these quantities. By contrast, the temperatures of two added systems do not add, nor do P and μ. Temperature, pressure, and chemical potential are called intensive variables. The product of an extensive variable and an intensive variable is extensive; PV, TS, and μN are extensive quantities.

Bearing this in mind, rewrite the first law of thermodynamics for a system,

$$dU = TdS - PdV + \mu\,dN. \tag{3.83}$$

Now increase the size of the system λ times. The corresponding first law should read

$$d(U\lambda) = Td(S\lambda) - Pd(V\lambda) + \mu\,d(N\lambda). \tag{3.84}$$

Carrying out the differentials one step further, we have

$$\lambda dU + U\,d\lambda = \lambda(TdS - P\,dV + \mu\,dN) + (TS - PV + \mu N)d\lambda. \tag{3.85}$$

Since λ is arbitrary, the coefficients of $d\lambda$ must match to give the following useful relationship

$$U = TS - PV + \mu N = N(TS/N - PV/N + \mu), \tag{3.86}$$

where the second line shows once again that U is an extensive variable. Since the three intensive variables T, P and μ are respectively derivatives of U with respect to its independent variables S, V and N, it should be possible in principle to eliminate all of the intensive variables from the second line. When this is done, we may predict that the result should be in the form of

$$U = N f_1(S/N, V/N). \tag{3.87}$$

For this reason we say that U is a homogeneous function of the first order. To put it another way, it means that the function $U(S, V, N)$ should be such that

$$U(\lambda S, \lambda V, \lambda N) = \lambda U(S, V, N). \tag{3.88}$$

Similarly we may now safely say that the entropy should be in the form of

$$S(U, V, N) = N f_2(U/N, V/N) \tag{3.89}$$

and in fact Eq. (3.52) confirms this prediction. We will introduce more extensive variables later and will return to this type of discussions for them.

We can obtain one more useful relationship. Consider Eq. (3.86) for a system. Suppose that the system goes through a process which changes everything, T, S, P, V, N and μ. Take the differential of Eq. (3.86) allowing all the variables to change on the right-hand side. No matter what that process is, however, the first law must hold. Therefore dU can only be what Eq. (3.83) says; the rest must go. In other words,

$$SdT - VdP + Nd\mu = 0,\qquad(3.90)$$

which is called the Gibbs–Duhem relationship. It says that changes can occur in temperature, pressure, and chemical potential, but they are subject to this constraint. I do not like this derivation because it is so brutally mathematical. When we introduce Gibbs free energy, we will be able to rederive it from its extensivity.

3.12   Entropy of Mixing

Until now, we have considered systems consisting of single species. Nature abounds with systems consisting of many different species, and in fact it is very difficult to purify them to obtain a system of single species. Pure air is already a mixture of nitrogen and oxygen molecules and a trace of others, but the real air we breathe contains a host of other undesirable substances. We all know how hard it is to purify water. Then there are man-made mixtures of elements, or alloys which are designed so as to manipulate their mechanical, electrical or magnetic properties to our advantage. Desirable or undesirable, for all these systems entropy of mixing is a vital aspect which we need to understand.

Let us first consider an alloy made up of A atoms and B atoms. Their total number is N, of which $(N - t)$ are A atoms and t are B atoms. How many different ways are there to seat these atoms on N distinct lattice sites? The answer is

$$\Gamma(N, t) = \frac{N!}{(N - t)!t!}.\qquad(3.91)$$

The entropy of mixing is given by

$$\begin{aligned}
S_m(N, t)/k_B &= (\ln(N!) - \ln(N - t)! - \ln t!)\\
&\approx N \ln N - N - (N - t)\ln(N - t) + (N - t) - t \ln t + t\\
&= -(N - t)\ln(1 - x) - t \ln x\qquad(3.92)
\end{aligned}$$

where x is the fraction of the B atoms, $x = t/N$. For the second line we used the Stirling approximation and for the last line we broke up $N \ln N$ into $(N-t) \ln N + t \ln N$. We may regard the first term as the contribution of A atoms and the second term that of B atoms.

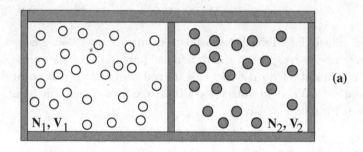

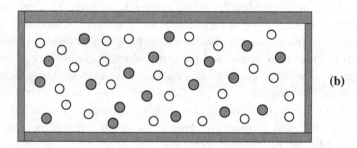

Fig. 3.5 (a) Two different gases are separated unmixed by a middle wall. (b) The middle wall has been removed to allow them to mix.

Although we obtained the above result for atoms in solid phase, we never had to specify the characteristics of the lattice sites, and therefore it may be applicable to mixtures in liquid phase and gas phase as well. Let us try to see if it is. Consider mixed gases. Two unmixed ideal gases 1 and 2 are separated by a mobile middle wall as depicted in (a) of Fig. 3.5. Let the volume of each compartment be V_1 and V_2 and their numbers N_1 and N_2, respectively. The temperature and pressure are the same in both compartments. Since the pressure is the same, it may appear that the presence of the middle wall is superfluous, or is it? So we remove the middle wall as shown in (b). What happens? The particles want to mix to increase their entropy. The entropy of mixing is defined as the total entropy of the final mixed gases minus that of the total of the two initial unmixed gases. In the absence of the dividing wall, we know what happens. Both species of gas molecules spread into the total volume $V = V_1 + V_2$ while the temperature and pressure remain the same.

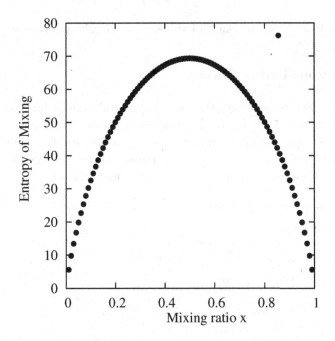

Fig. 3.6   The entropy of mixing as a function of the mixing ratio x.

Remember that the two gases consist of ideal gas molecules. They do not interact with each other and therefore any two or more particles can occupy the same spot. This makes the calculation easy. For the total multiplicity $\Gamma = \Gamma_1 \times \Gamma_2$, the calculation for Γ_1 would proceed just like before with $N = N_1$ and $V = V_1 + V_2$ to give a Sackur-Tetrode equation, and similarly for Γ_2. We have

$$
\begin{aligned}
S_m/k_B &= (S_{mixed} - S_{unmixed})/k_B \\
&= \sum_i N_i \ln(V/N_i) - \sum_i N_i \ln(V_i/N_i) \\
&= \sum_i N_i \ln(V/V_i) \\
&= -\sum_i N_i \ln x_i \,,
\end{aligned}
\tag{3.93}
$$

where

$$x_i = \frac{N_i}{N_1 + N_2} = \frac{PV_i/k_BT}{PV/k_BT} = \frac{V_i}{V}. \tag{3.94}$$

So we arrive at the same result for the entropy of mixing for gases.

3.13   Entropy-Driven Forces

Think of entropy as something thermal systems always thirst for. No matter how much they may have, they always want more. Their never-ending thirst for entropy can be strong enough to manifest itself as a force. In other words, we must exert force to suppress what molecules do out of their desire for more entropy. Let me give you four examples.

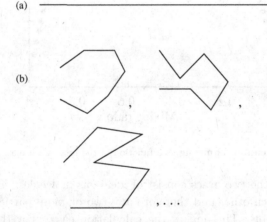

Fig. 3.7   A rubber band consisting of 6 elementary segments: (a) completely stretched, (b) partially curled.

3.13.1   *Rubber Bands*

Rubber bands consist of long flexible chain molecules. A good, simple model for each chain is to regard it as a large number of elementary, non-interacting segments. Take one long chain and stretch it to its maximum possible length. This can be done in only one way: all the elementary segments have to be aligned parallel to each other so that they do not bend, as shown in Fig. 3.7(a). Now bring the two ends closer to each other to a certain distance. As Fig. 3.7(b) shows, this can be done in many different ways. Thus it is sensible to define the macrostates according to the end-to-end length. Clearly the macrostates become larger with decreasing end-to-end length. Therefore the entropy-hungry chains will want to curl for greater entropy. When we stretch a rubber band, we feel the counter-force of the rubber band that opposes our action. It is a force whose origin lies in nature's attempt to maximize entropy. We will know how to model this thermodynamic force by the end of Sec. 3.15.

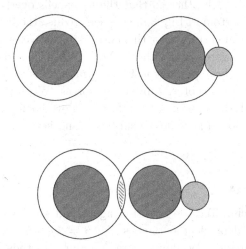

Fig. 3.8 Large molecules (shaded dark) and small molecules (shaded gray). Around each large molecule is a shell into which no small molecule can put its center. When two large molecules come close to each other as shown in the bottom figure, the two shells overlap. The cross-hatched area is the saved space which results in reducing the loss of entropy.

3.13.2 *Colloidal Suspensions*

Colloidal particles are very large hard-sphere molecules. If these large particles are suspended in a solution of small molecules, the size difference creates another entropy-driven force. Suppose that we add one such large molecule to a solution of small molecules. Since the volume of the space occupied by the large molecule is no longer available for the small molecules, the large molecule is depriving the small molecules of some of their positional entropy. The large molecule or the small molecules can do nothing about it. Now add one more large molecule to the solution. The small molecules lose even more positional entropy. This time, however, they can do something together to minimize the loss of entropy. Since the small molecules also have a certain size, the volume of the space that each of the large molecules takes away is the volume of the large molecule plus a shell-shaped space covering the large sphere. The thickness of the shell is equal to the radius of the small molecules, and therefore no small molecule can put its center within the shell shown in the top of Fig. 3.8. If all the small molecules and the two large molecules conspire to bring the two large molecules close to each other so that the two shells overlap even slightly as shown in the bottom of Fig. 3.8, they can reduce the wasted space and the resultant loss in entropy. For this entropy incentive, will the two large molecules come towards each other? Yes! When colloidal particles are put in a solution, they spontaneously aggregate. This provides one of the most dramatic examples of entropy-driven forces. When the colloidal particles are charged, this entropy-driven force can overcome the electric force, and when one colloidal particle is held below another, it can overcome the gravitational force.

3.13.3   *Osmosis*

Look at the Sackur-Tetrode equation again and see how it depends on V. The volume-dependent part goes like $S = k_B N \ln(V/N)$. Molecules will therefore spread to occupy as large a volume as they are allowed, to maximize their entropy, no matter what the shape of the volume may be. Let me tell you something which is very similar to this tendency. It is a great pleasure to own a home, but it is no pleasure to do the maintenance work. Those who have tried know rather painfully what ants and water do. If there is any crack just about anywhere, water and ants will get in and they will spread for no good reason as far as they can, no matter how tortuous the path may be. Such suffering is not limited to home owners. Those who make a vacuum chamber and maintain it suffer the same trouble. If there is any way for air to get in, the air will to ruin the chamber.

The phenomenon of osmosis originates from the same tendency, but unlike what has been complained about above it is a fascinating phenomenon. Biologists discovered long ago that if you fill up an animal bladder with a weak solution, say, wine, and put it in a vat of water, then water will gradually sip through the bladder. The resultant increase in the internal pressure can be large enough to burst the bladder. Plant roots have a similar membrane which bring in water. To demonstrate this, prepare a U-shaped glass vessel with two open-ended columns. Insert the middle part

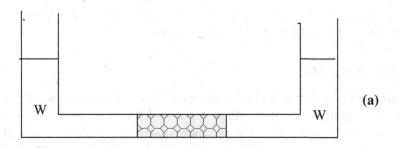

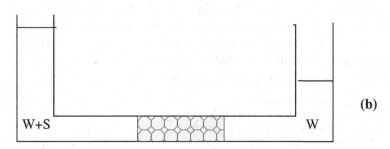

Fig. 3.9 The general pattern of the specific heat of diatomic gas molecules. See the text for the jumps, Δ_1, Δ_2 and Δ_3.

that connects the two tubes with a semipermeable membrane which allows water to pass through but not any solute molecules such as sugar. To familiarize the reader with the terminology of biologists and chemists, when we put sugar into a cup of water, water is called solvent and sugar solute. So fill up the columns with water (W). The water level is the same at both ends. Now pour sweetened sugar water (S) into the left column. See what happens. As Fig. 3.9 shows, more and more water now flows from the right to left to make the water levels uneven. There is apparently entropic force which pushes the water molecules to the left side. The added pressure is called the osmotic pressure, which can be measured by the height difference between the two columns. The solvent molecules want to spread so as to occupy the maximum possible allowed volume, and the solute molecules will likewise spread so as to mix with the solvent molecules. The migration of water into the tube has the additional incentive of getting the entropy of mixing. That is why the water molecules seep through the membrane as if they were after the solute molecules! We will return to this issue later when we are prepared to calculate the osmotic pressure.

3.13.4 *Ideal Gas*

Figure 3.10 shows a gas held in a cylinder with a mobile piston on its right end. If we wish to keep the volume fixed, we know that we have to push the piston to the left to keep the gas from expanding, but never to the right to keep it from shrinking. The gas always pushes the piston to the right for more volume. They want more volume for more positional entropy. Afterall, Eq. (3.23) says "if you want more entropy, expand your volume". These molecules are banging on the wall with their own body, the only tool they have, for more entropy, just like the Germans who banged on the Berlin Wall with a sledge hammer for more freedom. We will find in Sec. 6.3 that this force is not only a real force, it has a potential energy associated with it. The work that we do when we push the piston to the left to reduce the volume goes into this potential.

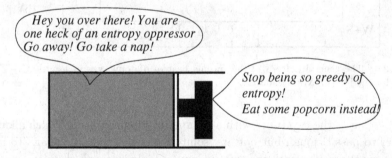

Fig. 3.10   The gas molecules inside the cylinder always want to push the piston out to increase their entropy, but the piston does not let them do that. The gas molecules do not like that.

Both rubber bands and the ideal gas exhibit elasticity for the same reason, but their manifestations are different; rubber bands do not want to be stretched, while the ideal gas does not want to be squeezed. The origin of the elasticity of springs is entirely mechanical.

Finally it should be mentioned that all the above examples except the rubber bands can be argued for by using kinetic arguments. So the colloidal force can be argued for by saying that when two large molecules come close toward each other, they are more frequently banged by small molecules which tend to push the two toward each other than those in between them which try to bang them to push them further apart. For osmosis, one may argue that since there are more water molecules banging from the right side of the membrane than there are on the left side, there is a better chance for water molecules to pass through the membrane from the right side. For the last ideal gas example, I tried to raise my voice because I know many readers learned about the gas pressure using a kinetic theory in the introductory general physics class. It is true that the gas pushes the piston out because the molecules bang on the piston wall outward and they do not know how to attach a rope to the wall and try to pull the wall in. But we cannot carry out thermal physics entirely on the basis of kinetic theory. Such arguments alone without the corresponding thermodynamic argument can potentially mislead the readers to think that thermal physics is just an extension of mechanics. It is wise to take the concept of pressure as a thermodynamic variable rather than a mechanical variable. As we will do later, we can calculate the osmotic pressure from our laws of thermal physics. I have seen many authors arguing the kinetic argument for osmosis, but I have never seen anyone who calculates the osmotic pressure using the same kinetic arguments.

Having said all these, I should also emphasize that there are issues for which a kinetic theory is an appropriate tool. Yes, there are issues for which we have to consider the individual particle's motion, how they interact with each other, and how they help transport particles or energy from one region to another in a non-equilibrium states. If one simulates the motion of particles in equilibrium, which is possible if one knows how the particles interact with each other, one can actually calculate various transport coefficients such as heat conductivity and viscosity, etc.

3.14 Non-Mobile Thermal Systems

Until now, we have studied thermal physics of mobile particles, namely, the ideal gas molecules which freely move around. Now we turn to thermal systems consisting of non-mobile objects. Two good examples are magnets and rubber bands. Magnets are solids and there is nothing flying around in it like the ideal gas molecules. Rubber bands belong to another group of systems in which the constituent objects have a unique geometrical shape which plays the key role. Another good example of this category is the liquid crystal, but they are not as simple as rubber bands. In spite of these differences, they all follow the same basic principles that guided us through the long sequence of sections devoted for ideal gas molecules. Actually, it is much easier to study these systems. We have another opportunity to practice what we have already learned but in different languages, which I hope will prove refreshing.

3.14.1 *Magnets*

Let us take the magnet stuck on your refrigerator door. It is a metallic solid. The atoms that make up the magnet are sitting on the lattice sites of the solid and act like tiny little magnets to be called magnetic dipole moments. The familiar magnetic properties are a result of a collective cooperative orientation of these moments. So, the issue is how the constituents are oriented, not where they are or how fast they are moving. When they conspire to orient themselves along the same direction, they can get stuck on the refrigerator door by themselves. Then they are said to be in a ferromagnetic phase. If the temperature is high enough, however, they do not act collectively and lose their magnetic properties. They are then in a paramagnetic phase. This state of matter is analogous to the gas phase of mobile particles. Their thermal behavior is going to be our subject of this section.

These dipole moments can orient themselves in various directions, which is their only degree of freedom. Their state may be described with a number of models. We shall adopt the Ising model. Draw any lattice, and put (Ising) spins on the lattice sites. The spins can orient themselves only in two ways. To specify the orientation, we associate with each spin a spin variable s, which can only be $+1$ or -1. If $s = +1$, it is said to be "up", and if $s = -1$, it is said to be "down".

Let us assume for now that the spins do not interact with each other; they are ideal spins. The spins in a real magnet do interact with each other in various ways and to various extent, but the ideal spin model is an attempt to capture the physics of paramagnets, and it does well. Just like ideal gases the spins are, however, coupled to a random thermal source which causes them to flip up and down, and therefore the system wanders around incessantly among the sea of its microstates. According to the fundamental postulate or the equal *a priori* principle, the spins visits all of the microstates with equal frequency.

The first item of our interest is the number of their microstates. Since the spins only flip up and down, the appropriate description of microstates would be to specify the spin configuration with $(s_1, s_2, s_3, \ldots, s_N)$; because the spins are located only at localized sites, we may tag them without committing any overcounting error! The net spin

$$s = (s_1 + s_2 + s_3 + \cdots + s_N) \tag{3.95}$$

measures the extent to which spins are aligned. Let us find out how many microstates there are corresponding to a given value of s.

If the net spin is to be s, the number of up spins and down spins must be

$$N_{up} = \frac{N+s}{2}, \quad \text{and} \quad N_{down} = \frac{N-s}{2}. \tag{3.96}$$

How many ways are there to divide N spins into N_{down} of down spins and N_{up} of up spins so that the total net spin comes out to be s? The answer is

$$\Gamma(N, s) = \frac{N!}{\left(\frac{N+s}{2}\right)! \left(\frac{N-s}{2}\right)!}. \tag{3.97}$$

Here s can be as large as N, but proceed assuming that it is much smaller than N. Because the number of microstates corresponding large s is so overwhelmingly small that treating them approximately does no harm. The three factorials are approximated using Stirling's approximation, $\ln N! = N \ln N - N$, which gives

$$\ln \Gamma(N, s) = N \ln N - \left(\frac{N+s}{2}\right) \ln \left(\frac{N+s}{2}\right) - \left(\frac{N-s}{2}\right) \ln \left(\frac{N-s}{2}\right). \tag{3.98}$$

Now break up $N \ln N$ into $(N_{up} + N_{down}) \ln N$ and then combine them with the remainder to obtain,

$$\ln \Gamma(N, s) = -\left(\frac{N+s}{2}\right) \ln \left(\frac{(N+s)/2}{N}\right)$$
$$-\left(\frac{N-s}{2}\right) \ln \left(\frac{(N-s)/2}{N}\right). \tag{3.99}$$

The two logarithmic functions are then in the form of $\ln(1+x)$. Since x is small, they may be approximated with $\ln(1+x) = x - x^2/2$. After some straightforward algebra, we find

$$\Gamma(N, s) = 2^N e^{-s^2/2N}. \tag{3.100}$$

This is correct as far as the s dependence goes, but it is incorrect in its N dependence in the front factor. Correct the problem by normalizing it so that the sum over all possible values of s may come out to be the correct number 2^N. Or, if you would like, you may wish to use the more accurate version of Stirling's approximation $\ln N! = \frac{1}{2}\ln(2\pi) + \left(N + \frac{1}{2}\right)\ln N - N$ on Eq. (3.97) to begin with. Either way, the final result is

$$\Gamma(N, s) = \frac{2^N}{\sqrt{2\pi N}} e^{-s^2/2N} = \Gamma(N, 0) e^{-s^2/2N}. \qquad (3.101)$$

The corresponding distribution function is

$$W(s) = \Gamma(N, s)/2^N = \frac{1}{\sqrt{2\pi N}} e^{-s^2/2N}. \qquad (3.102)$$

This is as expected from the central limit theorem. Indeed $s = s_1 + s_2 + s_3 + \cdots + s_N$ is a sum of N random spins and N is very large. Paramagnets provide a perfect case where the conditions required for the central limit theorem are met to the letter. This is in fact identical to the case of a drunken man whose step is sometimes forward (up spin) and sometimes backward (down spin). The fluctuation $\delta s^2 = N$ is in order of N, not of N^2, as the theorem says in Chapter 2. The relative deviation per spin is $\sqrt{\delta s^2}/N = 1/\sqrt{N}$, which vanishes in the large N limit. This is the second time we have recovered the predictions of the theorem, but this one is more helpful than the previous one we had for ideal gases because here we may precisely specify the micro and macro states; you may recall the less than precise specifications like 'molecules huddle around a corner'. It is clear that the maximum count is at $s = 0$. To see how the count decreases with increasing s, it is convenient to rewrite the multiplicity in terms of the fractional magnetization s/N. Let us take the case of $s/N = 10^{-10}$, i.e. the ratio is one out of 10^{10} of the possible maximum. What does Eq. (3.101) say? Assuming $N = 10^{20}$, $\Gamma(N, s) \sim \exp(-s^2/2N) = \exp(-10^{10})$. The macrostate in question is hardly different from $s = 0$, and yet its size does not come anywhere near that of the largest one.

Now let us turn on an external magnetic field. Then the spins will acquire energy. Always assume that $\vec{H}$ is directed in the z-direction. Then the spin coordinate s_i may be regarded as representing the z-component of the magnetic dipole moment that it represents. When $s_i = 1$, its energy is $-mH$. When $s_i = -1$, its energy is mH where m is the magnetic dipole moment that the spins represent. It pays to be in the same direction as the field and costs to be in the opposite direction. We should study here how this type of energy affects their unlimited thirst for entropy. The single spin microstate energy is therefore given by

$$\mathcal{E}(s_i) = -s_i m H \tag{3.103}$$

and the N spin energy by

$$E(s_1, s_2, s_3, \ldots, s_N) = -msH . \tag{3.104}$$

Thus, if the system is given energy U before it is isolated, its net spin s is also fixed at

$$U = -smH . \tag{3.105}$$

Visualize the spins as flipping up and down like kids in a big swimming pool, but the up spins and down spins vary in coordination so that $N_{up} - N_{down}$ remain fixed at s. Their thermal properties are determined by the number of ways they can do so, i.e. by the entropy given by

$$\begin{aligned}
S(N, s)/k_B &= \ln \Gamma(N, s) \\
&= N \ln N - \left(\frac{N+s}{2}\right) \ln \left(\frac{N+s}{2}\right) \\
&\quad - \left(\frac{N-s}{2}\right) \ln \left(\frac{N-s}{2}\right) .
\end{aligned} \tag{3.106}$$

Since $U = -msH$ and $M = ms$, this equation gives entropy as a function of energy or as a function of magnetization.

Differentiating it with respect to the net spin s, we have

$$\frac{1}{k_B} \frac{\partial S}{\partial s} = \frac{1}{2} \ln \left(\frac{N-s}{N+s}\right) . \tag{3.107}$$

The temperature is given by

$$\frac{1}{k_B} \frac{\partial S}{\partial U} = -\frac{1}{2mH} \ln \left(\frac{N-s}{N+s}\right) = \frac{1}{k_B T} \tag{3.108}$$

or

$$\frac{1}{T} = -\frac{k_B}{2mH} \ln \left(\frac{N-s}{N+s}\right) . \tag{3.109}$$

Inverting this, which requires only a short and straightforward algebra, we find that

$$s = N \tanh(mH/k_B T) \tag{3.110}$$

and therefore

$$U = -msH = -mHN\tanh(mH/k_BT) \tag{3.111}$$

and

$$M = ms = Nm\tanh(mH/k_BT). \tag{3.112}$$

At room temperature, $x = mH/k_BT$ is quite small, and for small x, $\tanh(x) \approx x$. Thus, to a good approximation, the magnetization changes with H and T like

$$M \approx Nm^2\frac{H}{k_BT} \tag{3.113}$$

which gives the susceptibility

$$\chi = \frac{\partial M}{\partial H} \sim \frac{1}{T}. \tag{3.114}$$

The inverse temperature dependence is called Curie's Law. The susceptibility is one of several important response functions. When magnetic field is applied to the system, the system responds with magnetization, and the susceptibility indicates how well the spins align themselves along the direction of the field. Curie's law says that the response diminishes with increasing temperature inversely.

Another important response function is the specific heat, which we find

$$C_H = \left(\frac{\partial U}{\partial T}\right)_H = NmH\,\text{sech}^2(mH/k_BT)\frac{mH}{k_BT^2}. \tag{3.115}$$

In the high temperature limit, since $\text{sech}(x) = 2/(e^x + e^{-x})$ and $x = mH/k_BT \to 0$, $\text{sech}(x) \to 1$, and therefore the specific heat goes like

$$C_H \sim \frac{1}{T^2}. \tag{3.116}$$

In the low temperature limit, on the other hand, $x \to \infty$, and $\text{sech}(x) \to e^{-x}$ for which the diverging $x = mH/k_BT$ is no match and the specific heat vanishes. How different from the ideal gases! All these differences originated from the different single spin and N spin microstate energies. For ideal gas molecules, the single particle microstate energies constitute for all practical purpose a continuous spectrum from zero to infinity. By contrast the single particle spin microstate energy is finite and thus the energy spectrum does not start from zero, which makes it hard for the spins to get excited when T

is near zero. This is why the specific heat vanishes in the low temperature limit. As we will find out later, quantum systems behave similarly in the low temperature limit.

Although paramagnets are so different from gases, entropy was calculated from the same definition, as was the temperature. One big difference is that volume of the magnets plays no role; it is the net spin or the magnetization M that matters, and how much volume of space the spins occupy is totally irrelevant. Since volume plays no role, we do not speak of pressure. The spins are safely there without any need of confining walls. Instead of pressure, we have the magnetic field. Just like pressure dictates on the volume of gases, so does the magnetic field on magnetization. When work is performed to a gas by applying a pressure, all single particle and N particle microstate energies change. Equations (3.103) and (3.104) show that the same is true when a magnetic field is applied to spins. With this in mind, since $\partial S/\partial V = P/T$ for ideal gases, let us see if the suspected analogy holds. Since

$$\frac{\partial S}{\partial M} = \frac{1}{m}\frac{\partial S}{\partial s}, \tag{3.117}$$

from Eqs. (3.107) and (3.109), we can read the answer right away,

$$\left(\frac{\partial S}{\partial M}\right) = -\frac{H}{T} \tag{3.118}$$

which shows that M is indeed to magnets what V is to gases, and that H is to magnets what P is to gases. The minus sign says that, while the entropy increases with volume in gases, the entropy decreases with increasing M in magnets.

Exercise. When an external field is turned on, the total spin is fixed at s. Now divide the spins into two groups A and B. The multiplicity then may be written as

$$\Gamma_{AB}(s, N) = \sum_{s_A} \Gamma_A(s_A, N_A)\Gamma_B(s - s_A, N_B), \tag{3.119}$$

where $N_B = N - N_A$ and each term defines a different macrostate. Given s, N, and N_A, discuss how s_A is distributed and how the macrostates are sized in comparison to the largest one.

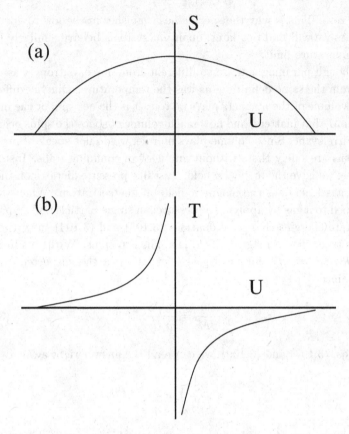

Fig. 3.11   Figure (a) depicts the relationship between entropy and energy while that in (b) depicts the relationship between temperature and energy.

3.14.2   *Negative Temperature*

Return back to Eq. (3.106) and examine it with s replaced by $s = -U/mH$ to find out how entropy changes with energy. U ranges between $-NmH$, which occurs when all the spins are parallel to the field, and NmH, which occurs when all the spins are anti-parallel to the field. As the energy increases from the lowest possible value $-NmH$, entropy increases reaching its maximum at $U = 0$. As U further increases into its positive region, entropy decreases and reaches zero at $U = NmH$, as depicted in Fig. 3.11.

Now return back to Eq. (3.109) and examine it with s again replaced by $s = -U/mH$ and find out how the temperature changes with U. The temperature vs. energy relation is depicted in Fig. 3.11. It is notable that the temperature is negative for positive U. The negative temperature simply means that entropy decreases with increasing energy or that the system can actually increase its entropy by losing energy. As the energy increases from the lowest possible value $-N\epsilon$ to $U = 0$, the temperature increases from $+0$, reaching $+\infty$ when $U = 0$. Once U enters its positive range, the temperature starts from $-\infty$ and then increases, reaching $T = -0$ when U reaches mNH. There are therefore four limit temperatures. In order of increasing U, they are: $+0$, $+\infty$, $-\infty$, -0.

Are the four limit temperatures listed in order of increasing "hotness" as well? If B is hotter than A, it means that if A and B come into thermal contact, B will be the donor of energy. Assume that A is a normal thermal system like an ideal gas while B is a spin system at a negative temperature. If A and B are brought into thermal contact, which one should be the donor of energy? The answer is clear: B should be the donor. This is true, no matter how high A's temperature is as long as it is positive. When B gives energy to A, it is not just A that benefits in terms of entropy; B also increases its entropy by doing so. What a happy situation! Thus the spin system at a negative temperature is always the donor which means that it is actually "hotter" than even the normal system at $T = +\infty$! Therefore, the answer to the question is "yes", and the spin system at $T = -0$ is at the hottest temperature. There is no significant physical difference between $T = +\infty$ and $T = -\infty$ for the spin system; the spin system only needs to lose an infinitesimal amount of energy to "cool" its temperature from $T = -\infty$ to $T = +\infty$.

What causes the temperature to be negative? Why does the ideal gas not show a negative temperature? It is the single spin energy which is limited only to two finite values $-mH$ and mH, one positive and one negative. The N spin energy is therefore limited to the finite range from $-NmH$ to NmH. In order for the spins to have the maximum energy, all spins have to be down, allowing the lowest possible multiplicity, i.e. unity, and therefore the entropy is zero. When the spins have zero energy, on the other hand, since about half of them are up and the rest down, there are many ways to divide the N spins into two groups each of size $N/2$, allowing the largest multiplicity. So, as the energy increases from zero to the maximum limit, entropy decreases, reaching zero at the maximum energy. This is an impossible scenario for ideal gas particles. Their single particle energy is always positive and has no upper limit. The N particle energy is therefore a sum of positive numbers covering from zero to infinity, hence the multiplicity increases with increasing energy.

Common laboratory spin systems can be driven into the paramagnetic phase with a negative-temperature, but they are cooled down to a positive temperature because they are coupled to a normal system at a positive temperature. The spins are sitting on the crystal lattice sites, but the lattice also has a vibrational mode of energy for which the N-particle energies have no limit and therefore the temperature is always positive. To achieve a robust paramagnetic spin system at a negative temperature, it is necessary that the system is not coupled, or coupled but very weakly, to a normal system at a positive temperature.

In 1951, Purcell and Pond found such a system in an assembly of nuclear spins. Its coupling to the lattice vibrational mode is weak, which allows a short time span in which the nuclear spin system is effectively isolated from the vibrational mode. In the intervening decades, their technique is so advanced that it is now used even for a practical application in radio astronomy. It works like this. Shower the system with an electromagnetic wave. The spins are in the lower energy region. As we will learn thoroughly in the next chapter, the energy in the electromagnetic wave may be thought to be possessed by a large number of particle-like quanta called photons. If the frequency is chosen right for the electromagnetic wave, the photons cause the spins do the followings. The spins absorb these photons to jump to the higher positive energy region. In the language of laser, this is a population inversion. Once the spins are in the higher energy region, the incoming photons are no longer absorbed. Instead, they stimulate the excited spins to emit identical photons, which brings the spins back to the lower negative energy regime, and the process repeats. In this way, a weak radio signal of astronomical interest may be amplified. Who would have thought that such a noble purpose can be served by a paramagnet which cannot stick to anything, let alone vertically on a refrigerator door?

Exercise. *Suppose that we have N parking spots each of which can be occupied by a student's car or by a professor's car. Assuming that N is very large, what is the multiplicity of accommodating n students and $N - n$ professors? How does the entropy associated with these arrangements change with n?*

Exercise. *Suppose that we are raising funds to promote thermal physics. We have N potential donors each of whom may donate 0, 1, 2, 3, ... or as much as ∞ dollars. What is the multiplicity of raising n dollars from this group? How does the entropy change with n? This is a difficult counting* *problem, but we will need to know the answer for a real physics problem in Sec. 4.10. For a trick, examine carefully*

$$\left(\frac{1}{1-t} \right)^N = \left(\sum_{s=0}^{\infty} t^s \right)^N \equiv \sum_n g(N,n) t^n. \qquad (3.120)$$

The multiplicity that we seek is given by $g(N,n)$. To obtain $g(N,n)$, differentiate $\sum_n g(N,n) t^n = [1/(1-t)]^N$ n times and then set $t = 0$, which gives

$$g(N,n) = \frac{(N+n-1)!}{n!(N-1)!}. \qquad (3.121)$$

3.15 Rubber Bands

The most prominent property of rubber bands is their elasticity. The tension increases with increasing length, but what is truly remarkable is that the tension also increases with temperature! If we warm up a rubber band, it does not expand like metallic rods; it shrinks! How does this happen?

The rubber bands consist of long chain molecules called polymers. Each chain in turn consists of monomers which are linked together to make a long flexible chain. Let us visit this remarkable thermal system with a very simple model. We will treat the chain as small rigid rods linked together in series so that each rod can bend freely. Their orientational degree of freedom is again the focus. Depending on how each unit rod is oriented at joints with respect to its preceding neighbor, the length of the chain L from one end to the other can be vastly different. The key question is then how many different ways (called confirmations) there are for the rods to orient themselves to keep the distance L at a given value. For a realistic model, each unit rod should be allowed to take any orientation with respect to its preceding neighbor, but we can simplify the mathematics while retaining the essential element of physics with the following one-dimensional model. In this simplified model, each unit rod can only point to the positive z direction or to the negative z direction. Call the total number of the up rods N_{up} and that of the down rods N_{down}. Since the rods are tied together in a particular order in series, and since none of them breaks its links and fly away, the rods may be treated as being distinguishable. The length of the chain is then given by

$$L = l(N_{up} - N_{down}) \qquad (3.122)$$

where l is the length of each unit rod. How many different ways are there for the chain to curl so that its length from one end to the other is L? That number is the same as the multiplicity that we computed for spins. So we know the answer:

$$\Gamma(N, L) = \frac{2^N}{\sqrt{2\pi N}} \exp\left(-(L/l)^2/2N\right) = \Gamma(N, 0)e^{-L^2/2l^2N} \qquad (3.123)$$

which, to be precise, should be divided by 2 because for any configuration ending up on the positive z axis, there is the same one ending on the negative z axis. But we will omit that correction because it has no consequence. For entropy, we have

$$S_{rods}(N, L) = k_B \ln \Gamma(N, 0) - k_B \frac{L^2}{2l^2N} \qquad (3.124)$$

where the S_{rod} is the entropy due to the orientational degrees of freedom of the rods. Another source of entropy is the vibrational degrees of freedom of the molecules (or atoms) that make up the rigid rods which will be represented by $S_{molecule}$. The total entropy is

$$S(T, L) = S_{rod}(L) + S_{molecules}(T) \qquad (3.125)$$

where it is crucial to notice that (a) S_{rod} is dependent only on L, and that (b) $S_{molecules}$ is dependent only on T. The statement (a) is clear from Eq. (3.124). For the statement (b), as we learned earlier for ideal gases, the entropy due to the translational or vibrational degrees of freedom increases with increasing internal energy, but the internal energy U may be measured with T, whence the statement. When we throw a baseball, the atoms and molecules in it do not know that the ball is moving. The molecules in the rods do not know how the rod is oriented. They only know the temperature. Similarly, how hard the molecules are giggling and wiggling in the rod has absolutely no effect on the orientational degrees of freedom. Prepare two cannon balls, one cold and one hot. Shoot them from the same cannon. Their trajectories are identical.

Now we are prepared to address the question we raised earlier. Suppose that we stretch the chain by ΔL. The work done to it is $F\Delta L$, where F is the tension. It should be noted that, unlike when we squeeze a gas, the direction of the force and that of the elongation are the same. So, the first law takes the form

$$\Delta U = T\Delta S + F\Delta L. \qquad (3.126)$$

Here the two Δ symbols in ΔU and ΔS are meant to represent the consequences of stretching the chain on U and S, respectively. Let us then first examine ΔU. The internal energy resides with the atoms and molecules in the rigid rods and has nothing to do with how the chain has been curled. Thus $\Delta U = 0$. Turn now to ΔS. Since $S_{molecules}$ has nothing to do with how the rods are oriented, it has no role to play here, and the change in entropy is entirely due to S_{rod}, i.e. $\Delta S = \Delta S_{rod}$. Putting all these into the equation, we have

$$\Delta U(L) = T \left(\frac{\partial S_{rods}(L)}{\partial L} \right)_T \Delta L + F \Delta L = 0 \qquad (3.127)$$

or

$$F = -T \left(\frac{\partial S_{rods}(L)}{\partial L} \right)_T . \qquad (3.128)$$

Carrying out the differentiation using Eq. (3.124), we arrive at

$$F = k_B T \frac{L}{l^2 N} . \qquad (3.129)$$

This says that (a) the tension increases with increasing length L, explaining the elastic property, and that (b) the tension also increases with temperature, answering the question we have been after.

The heat term represents the amount of heat that has to be taken away from the chain in order to keep the temperature T from changing. The work performed to stretch the chain is converted into the internal energy of the atoms that make up the rods, which then raises the temperature of the rods. If the rods are truly isolated thermally from the rest of the universe as assumed in this chapter, the energy will stay there with an elevated temperature. Asking, as we did above, to stretch the chain but without changing the temperature is an impossible proposition. We implicitly assumed a heat transferring mechanism as a part of the isolated system like air. In the presence of air the excess heat will be transferred to the air and the temperature will fall back to the initial temperature. This can easily be tested. Get a big rubber band that you can safely stretch without breaking it. Feel the temperature by touching it on your lips. While the rubber band is still on your lips, stretch it quick. You can feel the rising temperature. What we did in the above with the first law describes a slow stretching process, a process so slow that the chain stays in equilibrium at temperature T throughout. In other words, the stretching process was assumed to be an isothermal reversible process. We will discuss the same problem again in a later chapter.

A frequent question asks why the rod entropy S_{rods} is in the heat term when it does not depend on the internal energy. Whether it should be there or not should be determined by whether or not the multiplicity of the states in which the system possesses energy U involves the multiplicity Γ_{rod}. The total multiplicity is given by $\Gamma = \Gamma_{molecules} \times \Gamma_{rods}$, whence the presence of S_{rods} in the heat term.

If you still feel uneasy, I will repeat with an ideal gas exactly what has been done above to the ideal rods. We have an ideal gas confined in a piston of volume V and at temperature T. We are going to change the volume by ΔV. For these mobile particles, the first law says

$$\Delta U = T\Delta S - P\Delta V \tag{3.130}$$

where ΔS is the change in entropy due to the change in volume and ΔU is likewise the change, if any, in the internal energy due to the changed volume. The entropy consists of the spatial part which depends on V and the kinetic part which only depends on T and not on V. Similarly, the internal energy U depends only on T and not on V. If you are not sure about it, consider an ideal gas leaking into vacuum. The molecules are increasing their volume to change their spatial part of entropy but without changing their kinetic energy (the real interacting gases do change their kinetic energy, but ideal gases do not). Therefore $\Delta U = 0$ and $\Delta S = \Delta S_{spatial} = \Delta k_B N \ln V$, and the first law gives

$$P = T \left(\frac{\partial S_{spatial}}{\partial V} \right)_T \tag{3.131}$$

which, upon carrying out the differentiation, gives

$$P = T \frac{k_B N}{V}. \tag{3.132}$$

It gives the pressure necessary to keep the molecules in volume V and at temperature T.

So, we may call Eq. (3.129) the equation of state of ideal rods. Similarly, we may regard Eq. (3.113) as the equation of state of ideal spins. What did we do to arrive at these key equations? In spite of all the differences between these systems, the same 'equal *a priori* probability principle' guided us. Entropy and then temperature were defined with the same equation. We tested the guiding principle with a seemingly-silly question: why do air molecules in the room do not form a pocket of vacuum here and there? This was like Newton asking: why do apples fall? The same guiding principle carried us further to obtain many key results which answer many not-so-silly questions. We are not done yet. There are much more left to be learned.

Exercise. *Suppose that we have N fire flies. All of them have synchronized clocks and at the beginning of each tick, each fly decides to emit light or to stay dark during the ensuing time interval of one second. Treat all the flies as being distinguishable. In one second time interval taken at any given time, what is the most likely number of lighting flies? What is the chance of finding half that number?*

3.16 Schottky Defects

At the absolute zero of temperature, the atoms in solids occupy orderly arranged lattice sites. As the temperature is raised, they vibrate about their equilibrium sites, and some of them are completely displaced from their sites leaving vacancies. There are several different types of such point defects, one of which is called the Schottky defect. As the defects alter the physical properties of the solid, it is a matter of concern how the defects depend on the temperature. Just like the above two examples, the essential physics of point defects may be captured with a simple model and the same permutation formula.

Consider N lattice sites occupied by N atoms. How many ways are there to create n vacancies? The answer is

$$\Gamma = \frac{N!}{(N-n)!n!} \tag{3.133}$$

and

$$\ln \Gamma = N \ln N - (N-n)\ln(N-n) - n \ln n. \tag{3.134}$$

Because the atoms are bound to the sites, it costs energy to remove them. Let ϵ be the energy necessary to create each vacancy. When there are n vacancies, the system energy is then given by

$$\mathcal{E} = \epsilon n. \tag{3.135}$$

Notice that unlike in the cases of magnets and polymer in the earlier two examples, the rest $(N-n)$ atoms do not affect the vacancy microstate energy equation. Suppose that we had to expend total energy U to create n vacancies before we isolated the system from the rest of the universe. The vacancy entropy is

$$S = k_B \left[N \ln N - (N - U/\epsilon) \ln(N - U/\epsilon) - (U/\epsilon) \ln(U/\epsilon) \right] . \quad (3.136)$$

Their temperature is given by

$$\frac{1}{T} = k_B \left\{ \frac{1}{\epsilon} \ln \left(\frac{N\epsilon - U}{U} \right) \right\} . \quad (3.137)$$

Inverting this for U, we find

$$U = \frac{N\epsilon}{e^{\epsilon/k_B T} + 1}$$
$$n = \frac{N}{e^{\epsilon/k_B T} + 1} . \quad (3.138)$$

At $T = 0$, $n = 0$ and there is no defect. Slightly above $T = 0$, a good approximation is $n/N = e^{-\epsilon/k_B T}$. See how sensitive the defect is to the temperature! Typically, at room temperature $\epsilon/k_B T \approx 40$ and $n/N \approx 10^{-17}$. Needless to say, this is a very useful equation for material scientists. We will return to this example later again.

It has been a long chapter, but it has covered just about everything basic. If my repeating remarks 'isolated from the rest of the universe' caused a discomfort to you, Chapter 4 is devoted to thermal systems in contact with a large energy reservoir like the atmosphere. Chapter 4 will be also long, but it together with this chapter contains the core of thermal physics. So read on.

Chapter 4

Systems in Contact with a Thermal Reservoir

4.1 Introduction

So far we have studied isolated thermal systems. We are now ready to move on to study thermal systems exposed to the atmosphere or other forms of thermal reservoirs with which the system can exchange energy. Since the combined system, namely, the system of interest plus the reservoir, is isolated from the rest of the Universe, the fundamental postulate still guides the thermal behavior of the combined system. But we only wish to study the system of interest without getting involved with the microscopic details of the reservoir. Could this be too much to ask?

To what extent should the reservoir influence the thermal behavior of the system? When the combined system (the system plus the reservoir) is in equilibrium, the total energy should be distributed more or less uniformly throughout both the reservoir and the system, as we argued in Chap. 3. Then, since the reservoir is so much larger than the system, it should always hold the bulk of the total energy and therefore maintain a fixed temperature. The energy in the system is no longer fixed because energy may flow out of the reservoir into the system or into the reservoir from the system. The fundamental thermodynamic equation still holds but only for the combined system. The same is also true for the fundamental postulate and the maximum entropy principle. Thus the big question is: what do the system and the reservoir do to maximize the total entropy of the combined system? Could it be that the system may still remain lost among its own microstates? If not, is there a certain recipe that the system may follow to do its part to help maximize the total entropy?

The fundamental postulate and the fundamental thermodynamic equation served us well for isolated thermal systems. But even for isolated systems, everything was not as rosy as we may have painted. The fundamental thermodynamic equation states that the energy of an isolated system is a function of S, V, and N. Among the three variables, entropy is the most difficult to control. We have a fairly good idea about what one has to do to control N and V. But how do you control S? We will have to add more energy or reduce it, but how would you know that you did so just right amount as you intended? It would be no problem if we know the exact relationship between energy and entropy as we do for ideal gases, but we do not for most of the substances we deal with in laboratory. It is therefore desirable to work with another function, if there is one, for which entropy is not one of the independent variables. As far as isolated thermal systems are concerned, there is no other function to work with. No matter how inconvenient the situation may be, we cannot get rid of entropy because it is the heart of thermal physics; without it there is no thermal physics. For systems in thermal contact with a reservoir, on the other hand, it turns out that we have legitimate grounds for asking for another function to work with. This does not mean that we can completely get rid of entropy. We will be able to work with a different "energy-like" function for which entropy is not one of the independent variables. Entropy can be calculated from the energy-like function if we wish, and of course we do. We have a lot to find out in this chapter. Let us begin.

4.2   Helmholtz Free Energy

A simple idea attributed to Legendre provides the function that we are searching for. From a function $f(x, y)$, we can switch to $g(u, y)$ where the unwanted variable x is replaced by $u = (\partial f / \partial x)_y$. The procedure is simple:

$$g(u, y) = f(x, y) - ux. \qquad (4.1)$$

Now differentiate g to obtain

$$dg = v \, dy - x \, du, \qquad (4.2)$$

which shows that g is indeed a function of u and y. Here $v = (\partial f / \partial y)_x$. So we do have the option of removing entropy, but then we have no choice but to introduce $(\partial U / \partial S)_{V,N} = T$ as a new variable. This is what we wanted because T is determined by the reservoir and we do not wish to know any thing about the reservoir except that it maintains a constant temperature.

The new energy-like function that we wish to have is

$$F = U - TS, \tag{4.3}$$

which is called the Helmholtz free energy. So far this is all mathematics. We shall examine what the entropy principle says about it.

4.3 The Minimum Free Energy Principle

Let us write the total entropy of the combined system as follows

$$S_{R+s}(U_R, U_s) = S_R(U_R - U_s) + S_s(U_s), \tag{4.4}$$

where the subscript symbol s represents the system and R the reservoir. A total amount of energy U_R is to be shared between the reservoir and the system, and Eq. (4.4) represents what the total entropy of the combined systems would be if the reservoir allowed the system to have energy U_s. Let us now find out what the maximum entropy principle says about the idea of the reservoir sharing energy with the system. Assume that both the reservoir and the system are normal in the sense that entropy is a monotonically increasing function of energy. The two terms on the right side of Eq. (4.4) are then in competition: maximizing S_s requires a large U_s while maximizing S_R requires a large $U_R - U_s$ or a small U_s. We appear to have here a delicate optimization problem.

Since $U_R \gg U_s$, we may safely approximate the reservoir term in Eq. (4.4) with the first two terms of the Taylor series to obtain

$$S_{R+s}(U_R, U_s) = S_R(U_R) + \left(\frac{\partial S_R}{\partial U_R}\right)(-U_s) + S_s(U_s)$$

$$= S_R(U_R) - [U_s/T - S_s], \tag{4.5}$$

where we recognize $U_s/T - S_s = F/T$. Notice that the two extensive variables appearing in F/T are both those of the system. This means that we can enforce the maximum entropy principle by simply minimizing the Helmholtz free energy of the system without having to know anything about the reservoir except that it maintains a fixed temperature. In this way, the maximum entropy principle of the combined system is transformed into a minimum free energy principle of the system.

Here is a rather serious brother-to-brother conversation. A big brother to be called R and a little brother to be called S meet after a long separation.

The little brother wants to live by his big brother and to share the fortune that the big brother inherited from their parents while he disappeared from the family.

S: Hey my big brother! I want to live with you and share your fortune. May I?

R: You certainly may but our family lives by one very important principle, and so long as you do your part to follow that principle, I will be happy to have you back.

S: What is that principle?

R: You should not claim too much nor too little. You have to claim just the right amount to maximize our family happiness, and when I say family, I include you. You have to convince me that you are capable to exercise that kind of judgment.

S: Hey, you have to treat me with more respect. You cannot treat me like you used to. I learned some cool physics called Thermal Physics, according to which all I have to do is to minimize the free fortune. The free fortune appropriately balances two factors, the pain that I will inflict on you by taking some of your fortune away and the happiness that it will bring to me.

R: That makes a good sense. I am moved. You've got my permission, and we have a deal.

S: Not until you will promise that you will do your part.

R: Uhhhh, what's daaat, Mister Einstein?

S: Just don't do any thing stupid to drain away the huge family fortune. Just keep it where it is. Do I have your word?

R: Yap!

The two brothers lived happily for 127 years.

Let us examine Eq. (4.5) and clarify the two terms in F/T. The first term U_s/T comes from the reservoir and represents the entropy that the reservoir loses by allowing the system to have energy U_s. It is the price in entropy that the reservoir pays for sharing the energy with the system. The second term S_s, on the other hand, represents the entropy contribution that the system is able to offer in return thanks to the energy U_s it is given. Thus F/T is the net cost, which is why F should be minimized. Figure 4.1 depicts how the maximum entropy principle is transformed into the minimum free energy principle.

In Eq. (4.4), we speak of a specific value of reservoir energy $U_R - U_s$ and a specific value of system energy U_s as if the system and the

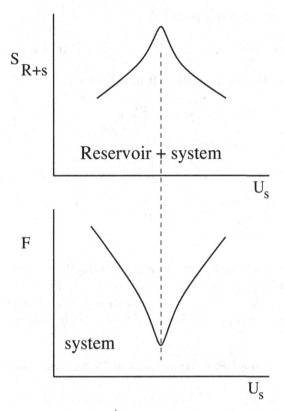

Fig. 4.1   The top figure shows the total entropy of the combined system vs. the system energy U_s while the bottom figure shows the free energy of the system vs. U_s. The two figures are equivalent.

reservoir were two disconnected systems. This is based on what we learned in the previous chapter. There are many macrostates of the combined system corresponding to many different distributions of the total energy U_R between the reservoir and the system. Among them we only need to worry about the largest macrostate. The system energy U_s appearing in Eq. (4.4) may be regarded as the system energy in the largest macrostate. This does not mean that we will ignore the fluctuations in the system energy; U_s is actually the average energy of the system.

The free energy of the system, F, is an energy-like function for which entropy is not one of its independent variables. Entropy is therefore a quantity to be calculated from F along with pressure and chemical potential.

To this end, we take a differential for F:

$$dF = dU - T\,dS - S\,dT = -S\,dT - P\,dV + \mu\,dN\,, \qquad (4.6)$$

where the first law has been used. It follows that

$$S = -\left(\frac{\partial F}{\partial T}\right)_{V,N} \qquad (4.7)$$

$$P = -\left(\frac{\partial F}{\partial V}\right)_{T,N} \qquad (4.8)$$

$$\mu = \left(\frac{\partial F}{\partial N}\right)_{T,V}. \qquad (4.9)$$

Exercise. *When a system in thermal contact with a reservoir is pushed out of equilibrium, the three partial derivatives S, P, and μ play crucial roles in guiding the system towards its new equilibrium state. Carry out the argument which may proceed as we did for thermally isolated systems.*

4.4   The Modified Postulate and the Boltzmann Factor

In the previous section, we transformed the maximum entropy principle that applies to the combined system, i.e. the system of our interest plus the reservoir, into a minimum free energy principle that applies only to the system. We now wish to repeat this at a deeper level. The fundamental postulate states that the combined system visits every one of its microstates with equal frequency. We wish to translate this fundamental postulate which applies to the combined system into a statement which is only true of the system. Thus, we ask: how often does the system visit each of its microstates irrespective of what the reservoir does?

Since the reservoir and the system are two different systems, we may characterize the microstates of the combined system as (i, j), where i refers to the microstate of the system of interest and j the microstate of the reservoir. Let the microstate energies corresponding to i and j be E_i and $\tilde{E}_j$, respectively. Since the combined system is isolated, the total energy must remain fixed at U_R, that is,

$$E_i + \tilde{E}_j = U_R. \qquad (4.10)$$

For a given microstate i for the system, all the microstates of the combined system satisfying this condition are visited with equal frequency. Therefore the frequency of the system visiting i is given by the number of the combined microstates (i, j) that satisfy Eq. (4.10), namely, the size of the macrostate of the reservoir corresponding to $U_R - E_i$, or the number of ways the reservoir can accommodate what the system wants to do,

$$f(E_i) \sim \Gamma_R(U_R - E_i). \tag{4.11}$$

Let us examine $\Gamma_R(U_R - E_i)$ in detail. It decreases quite rapidly with increasing E_i, and therefore it is wise to convert it into a more slowly changing function by taking the logarithm,

$$\Gamma_R(U_R - E_i) = \exp[\ln \Gamma_R(U_R - E_i)]$$

$$= \exp\left[\frac{S_R(U_R - E_i)}{k_B}\right]. \tag{4.12}$$

Since $U_R \gg E_i$, the reservoir entropy $S_R(U_R - E_i)$ may be approximated with the first two terms of its Taylor series

$$S_R(U_R - E_i) = S_R(U_R) + \left(\frac{\partial S_R}{\partial U_R}\right)(-E_i)$$

$$= S_R(U_R) - \frac{E_i}{T}. \tag{4.13}$$

Substituting Eq. (4.13) into Eq. (4.12), we find

$$\Gamma_R(U_R - E_i) \sim \exp\left[-\frac{E_i}{k_B T}\right], \tag{4.14}$$

where the first term of Eq. (4.13) has been dropped because it is a constant independent of E_i which we do not have to display in a proportionality statement. The frequency of the system visiting its microstate i is therefore

$$f(E_i) \sim \exp\left[-\frac{E_i}{k_B T}\right], \tag{4.15}$$

where $\exp(-E_i/k_B T)$ is called the Boltzmann factor. This result again shows that we do not have to know anything about the reservoir except that it maintains a constant temperature T.

Finally, converting the proportionality relationship into an equality relationship, we may write

$$f(E_i) = \exp \frac{[-E_i/k_B T]}{Z}, \qquad (4.16)$$

where the normalization factor

$$Z = \sum_i \exp \left[-\frac{E_i}{k_B T} \right] \qquad (4.17)$$

is called the partition function. The fundamental postulate for the combined system has been transformed into the following statement which involves only the system. **The system visits each microstate with a frequency proportional to the Boltzmann factor.** For the purpose of reference, this will be called the modified postulate. Apparently this is what the system actually does, but from the macroscopic point of view of thermodynamics, it appears as if the system is trying to minimize its free energy. Or conversely, we may state that this is what the system has to do in order to minimize its free energy. The visiting frequency is substantial for all states up to $k_B T$, but drops off quite rapidly for $E_i > k_B T$. Since we will often have to write the Boltzmann factor, let us write it as $\exp(-\beta E_i)$ where $\beta = 1/k_B T$.

We may take the average of the energies of the visited microstates according to the frequency of visits as follows:

$$U = \bar{E} = \sum_i E_i \exp[-\beta E_i]/Z$$

$$= \frac{1}{Z} \frac{\partial Z}{\partial(-\beta)}$$

$$= k_B T^2 \frac{\partial}{\partial T} (\ln Z). \qquad (4.18)$$

This average energy $\bar{E}$ may be identified with what we called U_s in the previous section. From now on U means $\bar{E}$.

The partition function Z was introduced to serve the relatively minor role of normalization, but as we have already seen above it becomes the master as every thing is related to it. Now that we have already related energy to Z, let us turn to entropy. But since entropy may be determined from F, let us turn to F starting from

$$U = F + TS$$

$$= F - T\frac{\partial F}{\partial T}$$

$$= T^2\left(\frac{F}{T^2} - \frac{1}{T}\frac{\partial F}{\partial T}\right)$$

$$= T^2\frac{\partial}{\partial T}\left(\frac{-F}{T}\right). \tag{4.19}$$

Comparing the last line with Eq. (4.18), the free energy F may be identified with $-k_B T \ln Z$. In other words,

$$Z = \exp[-\beta F]. \tag{4.20}$$

Thus if we know Z, we have in hand everything we want to know about the thermal behavior of a system. Equation (4.20) may be regarded as connecting the microscopic world which we specify with microstates and the macroscopic world which we describe with F.

Entropy was introduced originally for isolated systems as the natural logarithm of the total number of accessible microstates. The fundamental postulate states that an isolated system visits its microstates with equal frequency. Since systems in contact with a thermal reservoir do not visit all of their microstates with equal frequency, it is very curious in what shape entropy will then show up here. Because we have just identified F in terms of Z, we have to turn to Eq. (4.7) which gives

$$S = -\left(\frac{\partial F}{\partial T}\right)_{V,N}$$

$$= k_B \sum_i \left[\ln Z + \frac{E_i}{k_B T}\right]\frac{\exp(-E_i/k_B T)}{Z}$$

$$= -k_B \sum_i f(E_i) \ln f(E_i). \tag{4.21}$$

As a consistency check, substitute Eq. (4.16) for f along with $Z = \exp(-\beta F)$ into Eq. (4.21). The result is:

$$S = (-F + U)/T. \tag{4.22}$$

Thus the relationship between the entropy given by Eq. (4.21) and the free energy we identified as $-k_B T \ln Z$ is the same as that between the entropy

S_s we associated in the previous section with U_s and the free energy we introduced there as the net entropy cost. With U identified as U_s, this means that the entropy given by Eq. (4.22) is the same as $k_B \ln \Gamma_s(U_s)$. This is remarkable because Eq. (4.21) refers to all microstates of different energies while the latter, $k_B \ln \Gamma_s(U_s)$, only refers to those of energy U_s. It again shows that although the system visits all macrostates, we can still meaningfully single out one value of energy. With entropy given in the form of Eq. (4.21), the current formalism for systems in contact with a thermal reservoir maintains this remarkable continuity and consistency with the previous formalism for isolated systems. Moreover, Eq. (4.21) reflects the modified fundamental postulate according to which the system does not visit all the microstates with an equal frequency. What more can you ask?

In the meantime, I would like to make sure that you are convinced that Eq. (4.21) does give the only correct description of entropy for systems in contact with a thermal reservoir. We will shortly derive Eq. (4.21) from the very definition of entropy that we adopted at the beginning, but it will be a good idea to do some thought exercise. I will have you consider a plausible and seemingly logical idea of entropy and then argue why it must be rejected. I shall argue as follows. It is true that these systems do not visit their microstates with equal frequency. However, we can still proceed in the same spirit as we did for isolated systems; we can still count the microstates but weigh each count according to the frequency with which the system visits it. The weighted sum is none other than the partition function Z. Since Z is to the present systems what Γ is to isolated systems, and since $S = k_B \ln \Gamma$ for isolated systems, it seems logical to say that entropy should be given by $k_B \ln Z$ for the present systems. Entropy thus defined would, however, be inconsistent with F because the logarithm of Z does not give entropy; instead it gives the free energy F. The idea sounds logical but has to be rejected.

Rather, we should say that the role played by $\ln \Gamma$ for isolated systems is played by $\ln Z$ for the present system. Indeed, $\ln \Gamma$ gives the entropy whose "maximum" principle guides isolated systems, while $\ln Z$ gives the free energy whose "minimum" principle guides systems in thermal contact with a reservoir. The change from "maximum" to "minimum" is not significant; had we defined the free energy as the net gain rather than the net cost, we would be speaking of a maximum free energy principle instead of the minimum free energy principle. Bearing that in mind, look at Eq. (4.21). It is a weighted sum, but the logarithm is taken for each weighted count of the microstates before they are summed. This should sound just as "logical" as the false idea we just dispelled above. Moreover, Eq. (4.21) may be applied to isolated systems as well as to those in contact with a thermal reservoir. For isolated systems, the sum is limited to the particular value of energy that the isolated system has, and the visiting frequency of each microstate is $f_i = 1/\Gamma$, and therefore $\ln f_i = -\ln \Gamma$. Substituting these values into Eq. (4.21), we find $S = k_B \ln \Gamma$.

According to Eq. (4.21), entropy is the average of $\ln f(E_i)$. What does $\ln f(E_i)$ mean? The answer may be found if we insist that we arrive at Eq. (4.21) following the same definition for entropy that we had for isolated thermal systems, namely, that it should be the logarithm of the accessible number of microstates. Since the present system visits many, many more microstates but with different frequencies, that number Γ can only be an appropriately chosen effective number. How should we determine it? Let me repeat that, since the energy of the system constantly changes, what we call their energy is the average energy, $U = \bar{E}$. Since the exponential function and logarithmic function are not a linear function, the average of $\exp(E)$ is not equal to $\exp(\bar{E})$, and the average of $\ln(E)$ is not equal to $\ln(\bar{E})$. Forgetting this simple matter can cause many confusions. Let us begin with $\ln f(E_i)$ given by Eq. (4.16):

$$\ln f(E_i) = -\alpha - \beta E_i \qquad (4.23)$$

where $\alpha = \ln Z$ and $\beta = 1/k_B T$. By taking logarithm, we have reduced the exponential function to a linear function, and this is a welcome feature because energy is an additive quantity. We know how all the energies of the visited states are averaged to yield $\bar{E}$, and while we do just that, whatever comes out on the left hand side, that is the average of $\ln f(E_i)$ and we should be safe from the confusion warned above. The Boltzmann factor should be limited only to the energy levels (N-particle microstate energy E), but because our system is so large that the spacings between successive energy levels are very small, we may write $f(\bar{E})$ for the average energy although there may not be a true N-particle microstate corresponding to $\bar{E}$. If the nearest energy level to $\bar{E}$ is E_k, by $f(\bar{E})$ we mean $f(E_k)$. With this proviso, let me ask: what is the probability for the system to visit a state corresponding to $\bar{E}$? The answer is: $f(\bar{E})$, which is only a very small fraction of the total probability 1. So let us include two more, one located to the left and one to the right side of $\bar{E}$. What is the probability for the system to visit any one of the three states? Since the two added states are so closely located to the central one, the probability of each is nearly identical to that of the central one. So, the answer is: $3 \times f(\bar{E})$, which is still much smaller than the total probability 1. We keep on adding more and more states until we find the probability of finding the system visiting any one of the $\Delta\Gamma$ states around $E = \bar{E}$ is unity. That number $\Delta\Gamma$ must be the effective number of microstates that we are after. So we may write

$$\Delta\Gamma f(\bar{E}) = 1 \qquad (4.24)$$

or

$$\Delta\Gamma = \frac{1}{f(\bar{E})}. \qquad (4.25)$$

The corresponding entropy is:

$$\begin{aligned}
S &= k_B \ln \Delta\Gamma \\
&= -k_B \ln f(\overline{E}) \\
&= -k_B(\alpha + \beta\bar{E}) \\
&= -k_B\langle \alpha + \beta E_i \rangle \\
&= -k_B\langle \ln f(E_i) \rangle
\end{aligned} \qquad (4.26)$$

in agreement with Eq. (4.21); the angular brackets mean taking average. Thus the number of microstates is still at the root of entropy, but the number to be used for thermal systems in contact with a thermal reservoir is the effective number. Notice that the effective number is not the weighted count of the microstates, and we now know why. We shall return to this in the next chapter.

4.5   Response Functions and Fluctuations

When a system is put in contact with a thermal reservoir, its energy fluctuates. The system seems to know how much energy to claim from the reservoir as the average energy is exactly the amount needed to minimize the free energy. How about the fluctuation itself? Is it just a set of insignificant ripples to be averaged out? Is it possible to measure it?

In the human world, we are accustomed to thinking of fluctuations as something bad. If your grades fluctuate widely between A and D, we assume that you are not a steady reliable worker, which is considered bad. As a scientist, we first learn the word "fluctuation" in an introductory laboratory course to learn how to analyze error. This leaves us with a long-lasting firm impression that fluctuation has something to do with error. Nature abounds with fluctuations and that is not a result of anybody's error. Actually nature's fluctuations carry useful signatures which we are about to learn how to read.

4.5.1 *Energy*

Let us start with energy. The fluctuation is defined, following Eq. (2.64), as

$$\delta U^2 = <(U - <U>)^2>$$
$$= <U^2> - <U>^2. \tag{4.27}$$

Since

$$<U> = \sum_i E_i \exp[-\beta E_i]/Z, \tag{4.28}$$

and

$$<U^2> = \sum_i E_i^2 \exp[-\beta E_i]/Z, \tag{4.29}$$

it follows that

$$\delta U^2 = \frac{1}{Z}\sum_i E_i^2 \exp(-\beta E_i) - \left(\frac{1}{Z}\sum_i E_i \exp[-\beta E_i]\right)^2$$

$$= \frac{\partial}{\partial(-\beta)}\left(\frac{1}{Z}\sum_i E_i \exp(-\beta E_i)\right)$$

$$= k_B T^2 \left(\frac{\partial <U>}{\partial T}\right)_V = k_B T^2 C_V, \tag{4.30}$$

where C_V is the heat capacity measured with the volume held fixed. The volume was held fixed during the differentiation because the differentiation did not alter E_i. Notice that we have related the fluctuation to the heat capacity, but without actually calculating the partition function and the two averages. Thus the relationship is general as it is valid for any set of microstate energies.

As Eq. (4.30) shows, the fluctuation is related to the specific heat which is a response function. The specific heat tells us how the system changes its energy U in response when the reservoir changes the temperature T.

The specific heat $C_V = (\partial U/\partial T)_V$ is a useful thermodynamic quantity because a great deal can be learned from it about the inner workings of thermal systems. One would then wish to calculate it. With that in mind, examine the first line in Eq. (4.30). Although the definition of C_V requires to change the temperature and then to calculate the consequent change of energy, but, thanks to Eq. (4.30), we can calculate it without changing the temperature at all. Keep the temperature where it is, and just watch how the energy fluctuates. In fact, this is how C_V is measured when one performs computer simulations. Because energy and temperature enter into the Boltzmann factor as $\exp(-E_i/k_B T)$, the temperature T dictates how much energy the system should have. When the dictator changes its message, the dictatee has to respond. Is there something that the dictator can read before changing its message to find out how favorably the dictatee will respond? Yes. It is the fluctuation of the dictated quantity U.

There is something very similar to this in the human world. Generals are supposed to defend their government, but that is not always what they do. There are some generals who wait for a chance to topple their own government. Is there a sign in the government that such a general can read to find out how favorable the result of his action will be if and when he attempts it? Yes: fluctuations in the government. The more fluctuations there are in the stability of the government, the more vulnerable is the government to a coup.

4.5.2 *Magnetization*

Next turn to magnets and see how the magnetization fluctuates. Each N-spin microstate may be specified with the spin configuration $(s_1, s_2, s_3, \cdots, s_N)$ which will be abbreviated as $\{s\}$. The magnetization is given by

$$M\{s\} = m(s_1 + s_2 + s_3 + \cdots + s_N). \qquad (4.31)$$

The fluctuation may be calculated from

$$\delta M^2 = <M^2> - <M>^2, \qquad (4.32)$$

where

$$<M> = \sum_{\{s\}} M \exp(\beta H M)/Z, \qquad (4.33)$$

$$<M^2> = \sum_{\{s\}} M^2 \exp(\beta H M)/Z \,, \tag{4.34}$$

and

$$Z = \sum_{\{s\}} \exp(\beta H M)/Z \,, \tag{4.35}$$

is the partition function.

It follows that

$$\delta M^2 = \frac{1}{Z} \sum_{\{s\}} M^2 \exp(\beta H M) - \left(\sum_{\{s\}} M \exp(H M) \right)^2$$

$$= k_B T \frac{\partial}{\partial H} \left(\frac{1}{Z} \sum_{\{s\}} M \exp(H M) \right)$$

$$= k_B T \frac{\partial M}{\partial H} = k_B T \chi \tag{4.36}$$

where χ is the susceptibility. It is a response function. Suppose we apply more magnetic field. In response the spins will further align along the direction of the magnetic field, resulting in more magnetization. The susceptibility measures the extent of the response. As will be shown in Chapter 13, the susceptibility, in turn, measures the tendency of spins to act collectively, which is the dominating factor responsible for all the fascinating thermal properties at low temperatures. So, the fluctuations are by no means insignificant ripples. Again notice that we have established the relationship without calculating the partition function, and therefore it is a general relationship.

Equations (4.30) and (4.36) are two examples of what is known as the fluctuation-dissipation theorem. It describes a relationship between the fluctuation of a quantity and its response against changes in the "field" that dictate the quantity. We have seen twice already how such a relationship arises. To generalize, if the Boltzmann factor has a term like $\exp(Aa)$ or $\exp(A/a)$, the fluctuation of A is a measure of how A will respond if a changes. Think of a as a field which tries to dictate A. By the way, the word "dissipation" in the fluctuation-dissipation theorem is somewhat misleading in the present context. The theorem actually covers more than what we have discussed. The name "fluctuation-dissipation" comes from the non-equilibrium statistical mechanics where the time-dependent fluctuations are related to the transport coefficients of dissipating quantities.

4.6   Canonical Ensemble

In Chapter 3, we described the equilibrium properties of thermally isolated systems in terms of microcanonical ensemble. The formalism that we have just presented for systems embedded in a heat reservoir is called the canonical ensemble. Each ensemble member represents the same system but frozen in different microstates. If we wish to find the average of, say, A, we take a survey over all members to find out what A is in each. Then we take a weighted average where each A_i is weighed according to its Boltzmann factor. The equilibrium average is

$$<A> = \frac{\sum_i A_i e^{-E_i/k_B T}}{Z} \qquad (4.37)$$

where

$$Z = \sum_i e^{-E_i/k_B T} \qquad (4.38)$$

normalizes the distribution given by the Boltzmann factors. In this way, We have found a way to handle thermal systems exposed to atmosphere or a similar heat reservoir. Since Z is based on microstates, one takes this route when one wishes to explore thermal systems from the very basic microscopic point of view. Such an endeavor is called Statistical Mechanics or Statistical Physics. By its virtue of $Z = \exp(-F/k_B T)$, it also provides a bridge to the more macroscopic thermal world of thermodynamics as the free energy holds all the keys to the macroscopic properties of thermal systems.

Let us practice this with as many examples as we can in the remainder of this chapter. To this end, we will revisit the systems that we studied in the previous chapter but we will regard them as being exposed to a heat reservoir which imposes a fixed temperature. In the previous chapter devoted to isolated thermal systems, you were somewhat uneasy with the notion of calculating temperature from the energy given. We are more accustomed to think that temperature is a kind of environment to which systems are exposed. Now we take the latter view, but this view is still based on the same fundamental postulate and therefore we should check that the two views bring the same results. After that, we will arrive at the same results using the free energy minimum principle, i.e. the modified fundamental postulate.

4.6.1 *Ising Spins*

We calculated the average energy and average magnetization for Ising spins in Chapter 3 treating them as a system isolated from the rest of the universe, i.e. using the method of microcanonical ensemble. There, the temperature is determined by the amount of energy the system is given. In the present canonical ensemble method, the temperature is imposed on the system by the reservoir. If the two temperatures are the same, and if the two methods are both correct, the two methods should yield the same results. Let us check this out first with ideal Ising spins in an external field.

The microstate energies may be written as

$$E\{s_1, s_2, s_3, \cdots s_N\} = -(s_1 + s_2 + s_3 + \cdots + s_N)mH, \qquad (4.39)$$

and the partition function as

$$
\begin{aligned}
Z &= \sum_{s_1, s_2, s_3, \cdots} e^{(s_1 + s_2 + s_3 + \cdots)\beta mH} \\
&= \left(e^{\beta mB} + e^{-\beta mH}\right) \sum_{s_2, s_3, \cdots} e^{(s_2 + s_3 + \cdots)\beta mH} \\
&= \left(e^{\beta mH} + e^{-\beta mH}\right)^N .
\end{aligned}
\qquad (4.40)
$$

With Z in hand, the average magnetization is given by

$$
\begin{aligned}
<M> &= \frac{1}{Z} \sum_{s_1, s_2, s_3, \cdots} m(s_1 + s_2 + s_3 + \cdots)e^{m(s_1 + s_2 + s_3 + \cdots)\beta mH} \\
&= \frac{1}{Z}\frac{\partial Z}{\partial(\beta H)} = \frac{\partial}{\partial(\beta H)} \ln Z \\
&= Nm \tanh(\beta mH)
\end{aligned}
\qquad (4.41)
$$

and the average energy by

$$
\begin{aligned}
<U> &= \frac{1}{Z} \sum_{s_1, s_2, s_3, \cdots} [-(s_1 + s_2 + s_3 + \cdots)mH]\, e^{(s_1 + s_2 + s_3 + \cdots)\beta mH} \\
&= \frac{-1}{Z}\frac{\partial Z}{\partial \beta} \\
&= -NmH \tanh(\beta mH),
\end{aligned}
\qquad (4.42)
$$

in agreement with Eq. (3.111) and shows that the results based on the present canonical ensemble method are the same as that based on the earlier microcanonical ensemble.

Unfortunately for real world problems, it is formidably difficult to calculate the partition function. For example, the Ising spin model must include an interaction term to represent the real world ferromagnets, which brings in terms like $s_1 \times s_2$ in the exponent. Then the summation over s_1 in Eq. (4.40) can no longer be performed without involving s_2, and similarly the summation over s_2 cannot be performed without involving s_3, and so on; all the spin coordinates get tangled up, making the task of carrying out the summations very difficult. Nevertheless, such a model has been solved for two dimensional magnets more than a half century ago by Lars Onsager, but it has frustrated many, many brilliant mathematical minds who wanted to extend the solution to three dimensions. But thermal physics still has advanced far more since then by taking other routes including the free energy minimum principle. To take this route, however, one has to have a deep intuition to the general form of the free energy. If one can write the free energy involving a certain number of parameters or in a certain form of function, the rest will be taken care of by the minimum principle and it often does not take any more than the back side of a typical envelope.[6] So let me illustrate how it works with our model of ideal Ising spin as it provides an excellent but simple example.

Can we write $F = U - TS$ in a parametric form with the net spin $s = M/m$ as a parameter? That is easy because we know that $U = -smH$, and Eq. (3.60) in Chapter 3 provides $S = k_B T \ln \Gamma$ as function of U and thus as a function of s. Thus we may start from

$$F = U - TS = -msH$$
$$- k_B T \left\{ N \ln N - \frac{N+s}{2} \ln \left(\frac{N+s}{2} \right) - \frac{N-s}{2} \ln \left(\frac{N-s}{2} \right) \right\}. \quad (4.43)$$

Upon demanding that s be so as to minimize F, i.e. $\partial F / \partial s = 0$, we obtain

$$mH = \frac{k_B T}{2} \ln \left(\frac{N+s}{N-s} \right) \quad (4.44)$$

and then inverting it for s, we arrive at

$$s = N \tanh(mH/k_B T) \quad (4.45)$$

---

[6]Let me add a personal opinion here. The 'deep' intuition does not come out of nowhere just by a stroke of hand as it often seems. It comes after many, many failures.

which we have seen twice before. A much better example is the Landau free energy which is, however, reserved for Chapter 14.

4.6.2 Schottky Defects

Let us repeat what we did above for the Schottky defects. The Ising model can be modified to suit the present situation. The state of each lattice site may be represented with an Ising spin. The spin s_i is zero if the site is still occupied by an atom or 1 if it is vacant. The single spin energy is then given by ϵs_i where ϵ represents the amount of energy needed to remove an atom from its site. The total energy for a particular N-spin microstate by

$$E\{s_1, s_2, s_3, \ldots, s_N\} = \epsilon(s_1 + s_2 + s_3 + \cdots s_N) \tag{4.46}$$

where ϵ represents the amount of energy needed to vacate a site. The partition function is then given by

$$Z = \sum_{s_1, s_2, s_3, \ldots} e^{-\epsilon(s_1 + s_2 + s_3 + \cdots s_N)/k_B T}$$

$$= \left[1 + e^{-\epsilon/k_B T}\right] \sum_{s_2, s_3, s_4, \ldots} e^{-\epsilon(s_2 + s_3 + \cdots s_N)/k_B T}$$

$$= \left[1 + e^{-\epsilon/k_B T}\right]^N . \tag{4.47}$$

Now let us calculate the average number of vacant sites, or the average net spin,

$$<s> = \frac{1}{Z} \sum_{s_1, s_2, s_3, \ldots} (s_1 + s_2 + s_3 + \cdots + s_N) e^{-\epsilon(s_1 + s_2 + \cdots)\beta}$$

$$= \frac{1}{Z} \frac{1}{\epsilon} \frac{\partial Z}{\partial(-\beta)} = \frac{1}{\epsilon} \frac{\partial \ln Z}{\partial(-\beta)}$$

$$= \frac{N}{e^{\epsilon/k_B T} + 1} \tag{4.48}$$

which we saw in Chapter 4.

Now it is time to sing the free energy song! Let us suppose that we have laboratory data which provide some clue on the free energy. We tried many, many different forms of free energy to match with the data. We failed but all indications pointed to the form of

$$F = U - TS = -\epsilon s - k_B T[N \ln N - (N - s) \ln(N - s) - s \ln s]. \tag{4.49}$$

If this is true, what should the vacancy number s be at temperature T? The minimum occurs at

$$s = \frac{N}{e^{\epsilon/k_B T} + 1} \tag{4.50}$$

which agrees with the previous calculations. Of course the reader knows where the above 'trial' form came from, but that is the best I can do at the moment. I did once heard an author whose work perfectly fit in with the above scenario, but I fear that I cannot locate it. The author was a chemical engineer. The topic did not particularly interest me, but as his talk progressed, the way he arrived at his final trial form after many, many failures was most inspiring. What I can locate is the so-called Landau free energy, but it will be reserved for Chapters 13 and 14.

4.6.3   Ideal Gas

Most probably this is what you have been waiting for. This time the calculation is more involved than in the above two examples, but let us see if we can be equally successful. Write the single particle microstate energy in the form of

$$\mathcal{E}\{n_x, n_y, n_z\} = \alpha^2 n^2, \tag{4.51}$$

where $n^2 = n_x^2 + n_y^2 + n_z^2$, and $n_x = 1, 2, \ldots$, etc. and

$$\alpha^2 = \left(\frac{\hbar^2}{2m}\right)\left(\frac{\pi}{V^{1/3}}\right)^2. \tag{4.52}$$

The N-particle microstate energies take the form

$$E\{n_{1x}, n_{1y}, n_{1z}, n_{2x}, n_{2y}, n_{2z}, \ldots\} = \alpha^2(n_1^2 + n_2^2 + n_3^2 + \cdots + n_N^2), \tag{4.53}$$

where $n_i^2 = n_{ix}^2 + n_{iy}^2 + n_{iz}^2$.

We should begin just with one molecule. The partition function is given by

$$Z_1 = \sum_{n_x, n_y, n_z} \exp(-\mathcal{E}/k_B T) = \sum_{n_x, n_y, n_z} \exp[-\alpha^2(n_x^2 + n_y^2 + n_z^2)/k_B T]. \tag{4.54}$$

Let us assume that $k_B T$ is much larger than α^2. Then the Boltzmann factor hardly changes when the quantum numbers change by unity, and therefore we may safely replace the three sums with three Gaussian integrals. Thus, assuming

$$\boxed{n_c \equiv \sqrt{k_B T/\alpha^2} \gg 1}.$$ (4.55)

We find that

$$Z_1 = \int_0^\infty dn_x \int_0^\infty dn_y \int_0^\infty dn_z \exp[-\alpha^2(n_x^2 + n_y^2 + n_z^2)/k_B T]$$

$$= \left(\frac{1}{2}\right)^3 \left(\frac{\pi}{\alpha^2/k_B T}\right)^{3/2}$$

$$= V \left(\frac{2\pi m k_B T}{h^2}\right)^{3/2}.$$ (4.56)

Two comments are in order. First, examine how the Boltzmann factor weigh each microstate. All single particle orbitals enter into the sum for the partition function, but the Boltzmann factor cuts the spectrum off at the upper end starting approximately from the cutoff n_c. Thus, the larger n_c is, the more states are made available for occupation. Second, notice that the contributions from the three integrals are factorized out. In general, contributions from different and unrelated modes of energy are partitioned into products.

Now let us consider N identical ideal molecules. If we tag the molecules and treat them for a moment as if they were distinguishable from one another, then the N-particle partition function would be a product of N single particle partition functions,

$$Z = Z^{(1)} \times Z^{(2)} \times Z^{(3)} \cdots$$ (4.57)

where $Z^{(1)} = Z^{(2)} = Z^{(3)} = \cdots = Z_1$. Unfortunately the particles are not distinguishable and a large over counting error has been committed. We obtained the wrong result at a lightening speed, but we are going to to take plenty of time to correct it.

Let me rewrite this in a more pedestrian style like

$$\{\exp(-\mathcal{E}_1/k_B T) + \exp(-\mathcal{E}_2/k_B T) + \exp(-\mathcal{E}_3/k_B T) + \cdots\}$$

$$\times \{\exp(-\mathcal{E}_1/k_B T) + \exp(-\mathcal{E}_2/k_B T) + \exp(-\mathcal{E}_3/k_B T) + \cdots\}$$

$$\times \{\exp(-\mathcal{E}_1/k_B T) + \exp(-\mathcal{E}_2/k_B T) + \exp(-\mathcal{E}_3/k_B T) + \cdots\}$$

$$\times \cdots$$ (4.58)

where $\mathcal{E}_1$ means the first microstate (orbital) energy, $\mathcal{E}_2$ the second, and so on. The first bracket stands for the first molecule, the second one for the second molecule, and so on. All the single-particle energies appearing here are assumed to have survived the cuts by the Boltzmann factor. When the brackets are opened up, Eq. (4.58) consists of many terms, each of which looks like

$$A = \exp(-\mathcal{E}_{46}/k_B T)\exp(-\mathcal{E}_3/k_B T)\exp(-\mathcal{E}_{17}/k_B T),\dots$$
$$= (46, 3, 17, \dots), \tag{4.59}$$

which says that the first molecule is in the 46th orbital, the second molecule in the 3rd orbital, the third in the 17th orbital, and so on. This is one of many N-particle microstates counted by Eq. (4.57). In addition to A, Eq. (4.57) also counts microstates such as

$$B = (3, 17, 46, \dots), \tag{4.60}$$

which is the same as A except that the first three registries have been swapped around while keeping the rest as they are in A. As we may see by writing B like the first line of Eq. (4.59), B does not constitute a microstate different from A for identical particles. Counting such terms in addition to A constitutes an overcounting error. How many times does Eq. (4.57) overcount A in this way? That depends on the registries in A. If they are all different from each other, then A is overcounted $N!$ times. As another example, if the other registries not explicitly shown in A and B are all, say, 8, then A is overcounted $3!$ times. Thus the overcounting error is different depending on whether all the registries in A are different or include some orbitals more than once. Under these circumstances, the best we can do is to assume that the overwhelming majority of the N-particle microstates fall into the former category. Under this scenario, the overcounting error may be corrected by simply diving Eq. (4.57) by $N!$.

The assumption that the majority of the occupied single-particle states are all different from one another may be justified only if

There are far more states than there are particles to populate them.

Under this condition, the chances of any single-particle orbital appearing in more than one bracket would be very slim, and in order to insure this scenario, the cutoff point should indeed be very large. So, the two boxed statements are consistent with each other. We needed the first assumption to calculate the single particle partition function Z_1, and the second assumption to find an easy way to correct the overcounting error. As it has turned out, the two assumptions are in fact the same. Such luck! We have boxed the two statements to keep a record of all the assumptions made and to make it sure that they are all consistent with each other.

Thus we have

$$Z = \frac{Z_1^N}{N!} \qquad (4.61)$$

and, upon substituting the result for Z_1, we find

$$Z = \frac{1}{N!} V^N \left(\frac{2\pi m k_B T}{h^2} \right)^{3N/2} . \qquad (4.62)$$

Taking logarithm, we now have the free energy

$$F = -k_B T N \ln \left[\frac{V}{N} \left(\frac{2\pi m k_B T}{h^2} \right)^{3/2} \right] - k_B T N . \qquad (4.63)$$

All other quantities of interest are obtained by taking the appropriate partial differentiations on F. The results are:

$$P = \frac{N k_B T}{V} , \qquad (4.64)$$

$$S = N \left\{ \ln \left[\frac{V}{N} \left(\frac{m k_B T}{2\pi \hbar^2} \right)^{3/2} \right] + \frac{5}{2} \right\} , \qquad (4.65)$$

$$U = 3N \frac{k_B T}{2} , \qquad (4.66)$$

$$\mu = k_B T \ln \left\{ \frac{N}{V} \left(\frac{2\pi \hbar^2}{m k_B T} \right)^{3/2} \right\} . \qquad (4.67)$$

These results are the same as those we obtained earlier for isolated systems. All four quantities exhibit the same temperature dependence as they do in isolated systems!

Now we wish to recapture the same results using the minimum free energy principle. This time, let us treat energy U as the unknown parameter. For a fixed V, T, and V, the free energy is given by by $F = U - TS(U)$. The free energy depends on U through two terms, U and TS, but the two are competing against each other. Since $S(U)$ is a monotonic function of U, the entropy term favors a large value of U while the energy term favors a small value. The temperature will tilt the balance, more in favor of the entropy term when the temperature is high and more in favor of the energy term when the temperature is low. What would be the outcome of the balancing act? We now have a chance to test the modified fundamental postulate.

This is again easy because we know from Eq. (3.60) of Chapter 3 how to write the entropy term as a function of U. The trial function is of the form

$$F = U - 3Nk_BT \left\{ \ln\left(\frac{V}{N}\right) + \frac{1}{2}\ln\left(\frac{2\pi mU}{h^2}\right) + \frac{5}{2} \right\}. \tag{4.68}$$

Demanding that U shall be so as to minimize F,

$$\frac{\partial F}{\partial U} = 1 - 3Nk_B\frac{1}{2}\frac{1}{U} = 0 \tag{4.69}$$

which yields Eq. (4.66) as a result of the balancing act.

A unique length scale has emerged again:

$$\Lambda_{th} = \frac{h}{\sqrt{2\pi mk_BT}}. \tag{4.70}$$

It is the thermal de Broglie wavelength which appeared first in Chapter 3. It has reappeared in

$$Z_1 = \frac{V}{\Lambda_{th}^3}, \tag{4.71}$$

and

$$Z = \frac{1}{N!}\left(\frac{V}{\Lambda_{th}^3}\right)^N. \tag{4.72}$$

The cutoff quantum radius n_c may also be written in terms of it:

$$n_c = \frac{\sqrt{k_BT}}{\alpha^2} = \frac{2}{\sqrt{\pi}}\frac{L}{\Lambda_{th}}. \tag{4.73}$$

As emphasized amply in the previous chapter, what matters is the ratio

$$\frac{(V/N)^{1/3}}{\Lambda_{th}} \equiv \frac{l}{\Lambda_{th}}, \tag{4.74}$$

where l is the linear size of the volume per particle or a crude measure of the inter-particle distance. A large n_C means a small Λ_{th}, which in turn means high temperature. Both Z_1 and Z show, however, that it is not just Λ_{th}; it is the ratio l/Λ_{th}. Thus the condition for the classical regime is the low density and high temperature. More specifically, the temperature should be high enough to insure that Λ_{th} is less than the inter-particle distance:

$$\Lambda_{th} \ll l. \tag{4.75}$$

As we were progressing in the previous chapter and in the present chapter, we made several assumptions at several points, but they are all identical assumptions; some are more precise than others. The above condition may replace all of them and gives the condition necessary for ideal gases to be in their classical regime. For example, in the interior of dying stars called white dwarf, the temperature is in excess of millions of degrees, but because the matter there is so compressed that the temperature is not high enough to put the system in its classical regime; the interior of a white dwarf is a quantum fluid.

4.6.4 *Equipartition Theorem*

In the formalism of microcanonical ensemble, the energy is given first and then the temperature and all the other thermodynamic quantities are determined according to it. In the present canonical ensemble method, the temperature is imposed by the reservoir and dictates how much energy the system may have. It is notable that the dictated energy bears no resemblance to the form of energy of the particles. All those factors entering into α, namely, m, h, and V do affect the microstate energies, but they disappear in the final expression for the average energy! This may be attributed to the fact that the mode of energy is quadratic in n, and n can be anything from zero to infinity. To see this, consider just one such a mode. Its average energy is given by

$$<ax^2> = \frac{1}{Z} \int_0^\infty \exp(-\beta ax^2)ax^2 dx \tag{4.76}$$

where

$$Z = \int_0^\infty \exp(-\beta ax^2)dx = \frac{1}{2}\sqrt{\frac{\pi}{\beta a}}. \tag{4.77}$$

It follows that

$$<ax^2> = \frac{1}{Z}\frac{\partial}{\partial(-\beta)}Z = \frac{\partial}{\partial(-\beta)}\ln Z$$

$$= \frac{\partial}{\partial(-\beta)}\left(\ln a^{-1/2} + \ln \beta^{-1/2} + \cdots\right)$$

$$= \frac{1}{2}k_B T. \tag{4.78}$$

In Eq. (4.53), there are $3N$ such modes (indexed with n_1, n_2, etcetra) and there is no coupling between them. Thus each Gaussian integral independently results in $k_B T/2$, whence Eq. (4.66).

This-is called the equipartition theorem. What if the molecules interact with each other? As we will discuss in Chapter 7, there is no quadratic pattern in the interaction energy and therefore the equipartition theorem does not apply to the interaction energy, but the theorem still applies to the kinetic energy part. Thus we may write as a general statement,

$$\left\langle \frac{mv^2}{2} \right\rangle = \frac{3}{2}k_B T, \tag{4.79}$$

which provides us with an idea of how temperature affects molecular motions, but unfortunately many people take it as the definition of temperature which is of course quite inappropriate.

As another example, consider an RC circuit with no voltage applied. Although there is no externally applied voltage across the capacitor, there is still fluctuating charge on the capacitor due to thermal fluctuations. What would be the average energy due to this fluctuating charge? When a capacitor is charged to Q, its energy is $Q^2/2C$, a quadratic function of Q and Q can be anything from zero to infinity. The equipartition theorem gives $<Q^2/2C> = k_B T/2$. Similarly, there will be a current through an inductor even when there is no external driving voltage. The thermal energy that the reservoir provides for the thermal current is $<LI^2/2> = k_B T/2$.

4.7   Maxwell–Boltzmann Distribution

When we use the N-particle microstate energies as we did throughout, convenience dictates us to tag the identical particles, which unfortunately results in an overcounting error. The task of correcting this error is simple if the temperature is sufficiently high, but it becomes formidable if the temperature is not high enough. The best way to avoid this dilemma is to avoid using the N-particle microstate energies E and to use instead the single particle microstate energies $\mathcal{E}$'s together with their occupation numbers. In this way, we never need to ask the identity of the particles; all we need in order to specify the N-particle microstates is a set of occupation numbers. We would say that so many are occupying the first orbital and so many are occupying the second orbital and so on. We will return to this issue later for full details but the idea can actually be adapted even for the present high temperature classical regime as well.

Consider a single molecule. The probability that it will be in a particular microstate $\mathcal{E}_j$ is given by

$$f_j = \frac{1}{Z_1} e^{-\beta \mathcal{E}_j} \tag{4.80}$$

where $Z_1 = V \left(2\pi m k_B T / h^2 \right)^{3/2}$. If we have N molecules, the expected number of molecules occupying the state will be

$$n_j = N f_j = \frac{N}{V} \frac{h^2}{2\pi m k_B T} e^{-\beta \mathcal{E}_j} . \tag{4.81}$$

From Eq. (4.67), it may also be written simply as

$$n_j = e^{\mu / k_B T} e^{-\beta \mathcal{E}_j} . \tag{4.82}$$

This is called the Maxwell-Boltzmann distribution function. Although it is called a distribution function, it is not a distribution function of probability. It distributes the given N particles among the infinitely many single particle microstates. It is valid only in the classical high temperature regime and will be replaced by the Fermi-Dirac and Bose-Einstein distribution functions in the quantum regime at low temperatures.

In this description, each of the N-particle microstates is specified with the set of occupation numbers $(n_1, n_2, n_3, \cdots)$, and the total number of particles N and the N-particle energy with

$$N = \sum_j n_j , \tag{4.83}$$

$$U = \sum_j \mathcal{E}_j n_j . \tag{4.84}$$

The Maxwell–Boltzmann distribution function is clearly at the microscopic level as it addresses the issue of how the microstates are populated by particles. In a continuing effort, let us see if we may arrive at it at a more macroscopic level using the minimum free energy principle. To this end, we divide the entire range of $\mathcal{E}$ from zero to ∞ into groups, group j representing the region from $j\Delta\mathcal{E}$ to $j\Delta\mathcal{E} + \Delta\mathcal{E}$. Here $\Delta\mathcal{E}$ is large enough to contain G_j single-particle microstates but small enough to make it safe to assume that all those G_j microstate energies are the same; all the microstate energies in each group will be represented by one arbitrarily chosen representative state energy to be written as $\bar{\mathcal{E}}_j$.

Let us suppose that a total of N_1 molecules are occupying some of the microstates in the first group and N_2 molecules in the second group, and so on. In this macroscopic description, a macrostate is specified with the set of occupation numbers $(N_1, N_2, N_3, \ldots)$. Given the set, the total number of particles N and their total energy U are respectively given by

$$N = \sum_j N_j, \tag{4.85}$$

and

$$U = \sum_j \bar{\mathcal{E}}_j N_j. \tag{4.86}$$

You see, in order to describe a thermal system at a macroscopic level, we have to make it clear what options the system has, and we just did that. How many ways are there for N_j particles to occupy some of the G_j microstates? Assuming as we did before that there are more microstates than there are particles to populate them in each group, the multiplicity is given by

$$\Delta\Gamma_j = \frac{G_j!}{(G_j - N_j)!N_j!} \approx \frac{G_j^{N_j}}{N_j!} \tag{4.87}$$

where because N_j is so much smaller than G_j, $G_j!/(G_j - N_j)! \approx G_j^{N_j}$. The total multiplicity Γ is given by

$$\Gamma = \prod \Delta\Gamma_j. \tag{4.88}$$

Although N_j is smaller than G_j, it is a large number by itself and thus it is safe to apply the Stirling approximation. After a short algebra, the total entropy comes out to be

$$S = k_B \ln\Gamma = k_B \sum_j N_j \ln\left(\frac{eG_j}{N_j}\right) \tag{4.89}$$

where $\ln e = 1$.

In search for the equilibrium occupation number distribution, we should minimize the free energy with respect to the group occupation numbers, but the group occupation numbers are not entirely independent of each other because they must add up to the given value of N, namely, the first of Eq. (4.85). We may handle this situation using the method of Lagrange multiplier,[7]

$$\frac{\partial}{\partial N_j}(U - TS + \lambda N) = 0 \qquad (4.90)$$

where λ is the undetermined Lagrange multiplier. Carrying out the partial differentiation using Eqs. (4.85), (4.86) and (4.89), we find

$$N_j = G_j e^{-\bar{\mathcal{E}}_j} e^{-\lambda/k_B T} . \qquad (4.91)$$

Now we need to determine the parameters λ to ensure that the constraining condition is indeed satisfied. Choosing $\lambda = -\mu$, the chemical potential, we have

$$\sum_j G_j e^{-\bar{\mathcal{E}}_j} e^{\mu/k_B T} = \sum_i e^{-\mathcal{E}_i} e^{\mu/k_B T} = \sum_i n_i = N \qquad (4.92)$$

where we converted the crude macroscopic expression with G_j and $\bar{\mathcal{E}}_j$ back into the corresponding microscopic expression Eq. (4.83), which completes the desired proof.

4.8 Maxwell Velocity Distribution

Let me rewrite the Maxwell–Boltzmann distribution function with the chemical potential written out as a function of temperature and density,

$$\bar{n}(p) = \frac{N}{V}\left(\frac{h^2}{2\pi m k_B T}\right)^{3/2} e^{-p^2/2m k_B T} . \qquad (4.93)$$

This gives the number of molecules with momentum p. We wish to convert this into a probability distribution $P(p)$ so that the probability of finding a particle with momentum anywhere in the range of p and $p + dp$ is given by $P(p)dp$. It is given by

$$P(p)dp = \frac{4\pi p^2 dp}{h^3}\frac{\bar{n}(p)}{N} \qquad (4.94)$$

---

[7]See Chapter 2.

where $n(p)/N$ converts the occupation number $n(p)$ into a probability and $4\pi p^2 dp/h^3$ gives the number of microstates in the targeted range of momentum. Putting all together, we find

$$P(p)dp = \frac{4\pi p^2 \, dp}{(2\pi m k_B T)^{3/2}} e^{-p^2/2mk_B T} . \qquad (4.95)$$

It may be put more conveniently by substituting $p = mv$ to obtain the velocity distribution function given by

$$P(v)dv = 4\pi v^2 \, dv \left(\frac{m}{2\pi m k_B T}\right)^{3/2} e^{-mv^2/2k_B T} \qquad (4.96)$$

which is most likely the best known distribution function to the general public. The probability is zero at $v = 0$ and $v = \infty$, and it has a peak at

$$v_{mx} = \left(\frac{2k_B T}{m}\right)^{1/2} . \qquad (4.97)$$

Another useful quantity is the root-mean-square (rms) value

$$v_{rms} = \sqrt{<v^2>} = \frac{2}{\sqrt{\pi}} v_{mx} . \qquad (4.98)$$

Finally, if the molecules are subject to an external potential such as the gravitational potential energy mgh, it may be added to the kinetic energy term: For example, the Maxwell distribution function will take the form

$$P(p)dp = \frac{4\pi p^2 \, dp}{(2\pi m k_B T)^{3/2}} e^{-(p^2/2m+mgh)/k_B T} \qquad (4.99)$$

which shows that the probability diminishes quite rapidly with the altitude in the atmosphere.

Exercise. *Consider a system consisting of just one particle which can exist only in one of two states, one of energy $-\epsilon/2$ and one of energy $\epsilon/2$, where ϵ is a constant. Calculate the average energy U and the heat capacity $C = \partial U/\partial T$. Show that the heat capacity vanishes both at the high and low temperature limits with a hump in the mid range of temperature. This is called a Schottky anomaly. The vanishing heat capacity at the low temperature limit is due to the gap between the lowest (ground) state energy and the next one while the vanishing heat capacity at the high temperature limit is due to the absence of sufficiently high energy states comparable to*

$k_B T \gg \epsilon$; *the reservoir provides plenty of energy, but the system cannot take it in the two limits. We will see similar examples in Secs. 4.1 and 5.4. As this example shows, we can learn a great deal about the atomic and molecular details of a system by measuring its heat capacity.*

Exercise. *Consider a paramagnet consisting of non-interacting magnetic dipole moments in external field $\vec{H}$. Unlike in the uniaxial ferromagnets, the dipole moments can point in any direction. The energy of a single moment $\vec{m}$ is given by $\epsilon = -\vec{m} \cdot \vec{H}$. Since there is no coupling between different moments, each moment may be treated as an independent system. The total magnetization is given by $M = Nm < \cos(\theta) >$ where θ is the angle between $\vec{m}$ and $\vec{H}$. Calculate $< \cos(\theta) >$ and the susceptibility $\chi = \partial M / \partial H$ in the high temperature limit. Observe that the susceptibility goes like $1/T$. This is called the Curie Law.*

Exercise. *Consider an oscillator in equilibrium with a thermal reservoir at temperature T. The energy of the oscillator is given by $U = mv^2/2 + Kr^2/2$. What is the average energy at temperature T?*

4.9   Specific Heat and Einstein's Solid

There was another challenging puzzle during the early part of the 20th century. That was the vanishing pattern of the heat capacity of solids near absolute zero,

$$C_V = \alpha T^3 + \gamma T. \tag{4.100}$$

For gases at room temperatures, $\dot{U} = 3N k_B T/2$ which gives $C_V = 3N k_B/2$, a constant. Recognizing $3N$ as the total number of degrees of freedom of N particles, it says that each degree of freedom contributes $k_B T/2$, which we described earlier as the equipartition theorem. In fact, if the temperature is further raised, diatomic molecules such as N_2 and O_2 begin to exhibit additional degrees of freedom, the rotational degree of freedom and the vibrational degree of freedom; the latter requires higher temperature than the former to be unlocked. If the temperature is lowered, on the other hand, so long as the atoms and molecules execute any kind of motion, the classical statistical mechanics predicts that the heat capacity should remain constant, but experiments have shown that it does not. Einstein took the first step to resolve this discrepancy.

In a solid phase, atoms are arranged on lattice sites, and each atom oscillates around its site. We would then have $3N$ coupled oscillators. In the Einstein model, we disregard the coupling between the oscillators and treat each atom as executing an independent simple harmonic motion around its lattice site. According to quantum mechanics, an oscillator can only have the following amount of energy:

$$\epsilon(n) = \left(\frac{1}{2} + n\right)\hbar\omega, \tag{4.101}$$

where $\omega = (K/m)^{1/2}$, and $n = 0, 1, 2, 3, \dots$. Since the oscillators are localized on the lattice sites, we may tag them and treat them as distinguishable objects with no concern for overcounting errors. Each microstate of N oscillators may then be indexed with a set of quantum numbers $(n_1, n_2, n_3, \dots, n_N)$, where the subscripts refer to the oscillator tag numbers, and the corresponding microstate energy is given by

$$E(n_1, n_2, n_3, \dots, n_N) = \epsilon(n_1) + \epsilon(n_2) + \cdots + \epsilon(n_N). \tag{4.102}$$

There is no coupling between oscillators; there are no terms like $n_1 n_5$. The partition function for the N oscillators partitions therefore into N single-oscillator partition functions,

$$Z = \sum_{n_1, n_2, n_3, \cdots} \exp[\epsilon(n_1) + \epsilon(n_2) + \cdots \epsilon(n_N)] = Z_1^N, \tag{4.103}$$

where the single-oscillator partition function z_1 is given by

$$Z_1 = \sum_{n_1=0}^{\infty} \exp[-\epsilon(n_1)] = \frac{\exp(-\beta\hbar\omega/2)}{1 - \exp(-\beta\hbar\omega)}. \tag{4.104}$$

The average oscillator energy is

$$<\epsilon> = -\frac{1}{Z_1}\frac{\partial Z1}{\partial \beta} = \frac{\hbar\omega}{2} + \frac{\hbar\omega}{e^{\beta\hbar\omega} - 1}. \tag{4.105}$$

The total energy in the radiation is $U = 3N<\epsilon>$ which gives the heat capacity

$$C_V = 3Nk_B \left(\frac{\hbar\omega}{k_B T}\right)^2 \frac{e^{\hbar/k_B T}}{\left(e^{\hbar\omega/k_B T} - 1\right)^2}. \tag{4.106}$$

Why does it say that the volume was held fixed? The frequency ω is given by $\omega = \sqrt{k_s/m}$ and the restoring force constant k_s depends on how tightly the atoms are packed, or on the interatomic distance which in turn depends on the volume of the solid, and we kept the frequency fixed during the differentiation. As T approaches zero,

$$C_V \approx \frac{e^{-\beta\hbar\omega}}{T^2} \tag{4.107}$$

and therefore it approaches zero faster than can be described with any power law. For diamond, Einstein's result compares with measured results very favorably down to about 30 K, but there is a noticeable deviation thereafter.

• Let us obtain the same result using the minimum free energy principle. To start with, assume that all the oscillators are on their ground state. They are now proposing to receive from the thermal reservoir $M = n_1 + n_2 + n_3 + \cdots + n_N$ units of quanta for a total energy of

$$U = N\frac{\hbar\omega}{2} + M\hbar\omega. \tag{4.108}$$

Then what is the multiplicity? A total of $M = (n_1 + n_2 + n_3 + \cdots + n_N)$ quanta are going to be distributed amongst N oscillators. How many ways are there to do so? The answer is (see Chapter 2),

$$\Gamma = \frac{(N-1+M)!}{(N-1)!M!}. \tag{4.109}$$

The entropy of the oscillators with that much energy would be, assuming that N and M are both very large and thus using the Sterling approximation,

$$S/k_B = (N-1)\left\{\frac{N-1+M}{N-1}\ln\left(\frac{N-1+M}{N-1}\right) - \frac{M}{N-1}\ln\left(\frac{M}{N-1}\right)\right\}$$
$$= N\left\{\left(1+\frac{M}{N}\right)\ln\left(1+\frac{M}{N}\right) - \frac{M}{N}\ln\left(\frac{M}{N}\right)\right\} \tag{4.110}$$

where -1 has been dropped for the last line. The trial free energy to be minimized is

$$F = M\hbar\omega + N\frac{\hbar\omega}{2} - k_B T N_i\left\{\left(1+\frac{M}{N}\right)\ln\left(1+\frac{M}{N}\right) - \frac{M}{N}\ln\left(\frac{M}{N}\right)\right\}. \tag{4.111}$$

By minimizing it with respect to them, we have

$$\frac{\partial F}{\partial M} = \hbar\omega - k_B T\left\{\ln\left(1+\frac{M_i}{N_i}\right) - \ln\left(\frac{M}{N}\right)\right\} = 0 \tag{4.112}$$

which may be solved to give

$$\frac{M}{N} = \frac{1}{e^{\hbar\omega/k_B T} - 1} \qquad (4.113)$$

which is the optimum value for the average number of quanta for each oscillator to receive, hence the average energy of each oscillator is

$$\frac{U}{N} = \frac{\hbar\omega}{2} + \frac{\hbar\omega}{e^{\hbar\omega/k_B T} - 1}. \qquad (4.114)$$

Exercise. *Calculate the free energy and the entropy from the partition function.*

Exercise. *We treated the Einstein's solid as a thermal system in contact with a reservoir. Now treat the system as an isolated thermal system. The necessary multiplicity factor can be obtained from the "fund raising" exercise that we did at the end of Chapter 3.*

4.10   Specific Heat and Debye's Solid

Debye followed the lead of Einstein and explained the T^3 term of Eq. (4.100). The atoms in solids are making very complicated collective motions; one atom does not vibrate independent of others as assumed in the Einstein model. The collective motion is the sound wave, and it is quantized in the same manner as the electromagnetic wave is quantized. There are some differences, but none requires any serious alteration. The quanta of the acoustic wave is called the phonons. Whereas the electromagnetic wave has two directions of polarization, the sound wave has three, one longitudinal (compression) oscillation and two transverse (sheer) oscillations. There is, however, one important difference. In the case of electromagnetic wave, there is no limit to the available number of frequency modes, but the sound wave is a collective motion of N atoms which collectively have $3N$ degrees of freedom, and therefore the total number of modes must be limited to $3N$. These modes correspond to what is called in mechanics the normal modes. When the atomic motions are viewed in the atomic coordinates, all the coordinates are tangled to each other. But when viewed in the normal coordinates, the same motions turn into $3N$ clean independent untangled motions.

The rest proceeds in the same way as in the Planck theory of radiation. In particular we adopt the same type of linear dispersion relationship

$$\omega = \frac{n\pi v}{L} \tag{4.115}$$

where $n = (n_x^2 + n_y^2 + n_z^2)^{1/2}$ and v is the average speed of the transverse and longitudinal modes. To ensure that the total number of allowed modes is $3N$, n must be limited to $n < n_{mx}$ and n_{mx} is given by

$$\frac{3}{8} \int_0^{n_{mx}} 4\pi n^2 dn = 3N \tag{4.116}$$

which yields

$$n_{mx} = \left(\frac{6N}{\pi}\right)^{1/3}. \tag{4.117}$$

The average number of phonons with frequency $\omega(n)$ is, as before,

$$<s(n)> = \frac{1}{e^{\hbar\omega(n)/k_B T} - 1}. \tag{4.118}$$

The average total energy of the phonons is therefore

$$U = \frac{3}{8} \int_0^{n_{mx}} 4\pi n^2 dn \hbar\omega(n) \frac{1}{e^{\hbar\omega(n)/k_B T} - 1}$$
$$= \frac{3\pi}{2} \left(\frac{L}{\pi c}\right)^3 \hbar \left(\frac{k_B T}{\hbar}\right)^4 \int_0^{x_{mx}} \frac{x^3 dx}{e^x - 1} \tag{4.119}$$

where $x_{mx} = \hbar\omega_{mx}(n)/k_B T = \hbar\pi v n_{mx}/k_B TL$. The temperature dependency in the exponent of the denominator has been pushed out in part to the front and in part to the integral limit. Because of the latter, it is only for very low temperatures where $x_{mx} \approx \infty$ that we can represent the temperature dependency as a single power law. The resultant integral in the above is $\pi^4/15$, which gives

$$U = \frac{3\pi^4 N k_B}{5\theta^4} T^4 \tag{4.120}$$

where

$$\theta = \frac{\hbar\pi v}{k_B L} \left(\frac{6N}{\pi}\right)^{1/3} \tag{4.121}$$

is called the Debye temperature. Finally the specific heat is then given by

$$C_V = \frac{12}{5}\pi^4 N k_B \left(\frac{T}{\theta}\right)^3 . \tag{4.122}$$

Debye's results agree with measured results impressively well. For insulators, the agreement extends near to the absolute zero, but for conductors, the two begin to deviate approximately starting from 10K. What Debye's result misses is the extra contribution from the linear term in $C_V = \alpha T^3 + \gamma T$, and the linear term comes from electrons and electrons at such low temperatures need to be treated on the basis of quantum statistical mechanics as we will learn in a later chapter.

As this and earlier examples have shown, by studying the specific heat, we can probe what the constituents of thermal systems do. The constituents could be mobile particles like molecules at room temperatures or products of quantum mechanics like photons and phonons, but the specific heat reveals a clue as to what the appropriate constituents are and what they are doing. We have not seen it all. We will return to this later for the last remaining leg of the story when we study electrons near the absolute zero temperature.

4.11   Solids and Gas in Equilibrium

We can put together what have been covered in the last four sections for a very simple description of solids and gas in equilibrium. We may treat the solid with the Eisntein model, the evaporation with the Schottky model, and the gas with the ideal gas model. The evaporation process of atoms and their vibration while still in the solid are, however, not two independent processes. The evaporation happens when an oscillator reaches a quantum state whose energy is comparable to the binding energy, and therefore we only need the Einstein model for both. And we have of course a very reliable model for gas, the ideal gas model. We should make it sure that the energy levels of the two models, the energy levels of the atoms in gas and those of the oscillating atoms, are both measured with respect to the same reference level. In the ideal gas, the zero energy means no kinetic energy and thus the energy levels of the oscillators are written as

$$\epsilon(n) = -\epsilon_0 + \left(\frac{1}{2} + n\right)\hbar\omega , \tag{4.123}$$

where ϵ_0 is the binding energy. The corresponding single atom partition function is given by

$$\zeta_1 = e^{\beta\epsilon_0}e^{-\beta\hbar\omega/2}\sum_{n=0}^{\infty}e^{-\beta n\hbar\omega} = \frac{e^{\beta\epsilon_0}e^{-\beta\hbar\omega/2}}{1-e^{-\beta\hbar\omega}}. \tag{4.124}$$

If there are N_S atoms still left in the solid, their partition function is given by

$$Z_S = \zeta_1^{3N_S}. \tag{4.125}$$

There are now N_G atoms in the vapor phase and $N_G = N - N_S$. Their partition function Z_G is given by

$$Z_G = \frac{Z_1^{N_G}}{N_G!}. \tag{4.126}$$

The free energy is then given by

$$F = -k_BT\left[-3N_S(-\epsilon_0 + \hbar\omega/2)\beta - 3N_S\ln\left(1 - e^{-\hbar\omega\beta}\right)\right]$$
$$- k_BT\left[N_G\ln Z_1 - N_G\ln N_G + N_G\right]. \tag{4.127}$$

We will let the minimum free energy principle determine the equilibrium value of N_G. By solving $\partial F/\partial N_G = 0$ for N_G, we find

$$N_G = Z_1 e^{-3\epsilon_0\beta}\left(e^{\hbar\omega\beta/2} - e^{-\hbar\omega\beta/2}\right)^3. \tag{4.128}$$

Substituting $N_G = PV_G/k_BT$ and writing out Z_1, we finally obtain

$$P = (2\pi mk_B/h)^{3/2}(k_BT)^{5/2}e^{-3\epsilon_0\beta}\left(e^{\hbar\omega\beta/2} - e^{-\hbar\omega\beta/2}\right)^3. \tag{4.129}$$

The frequency ω here is the Einstein frequency and is dependent on the restoring force constant and the atomic mass. On the other hand, ϵ_0 is the binding energy. If the atom breaks away when its quantum number reaches n_{bk}, then $\epsilon_0 = \hbar\omega(1/2 + n_{bk})$. The ϵ_0-dependent exponential and the ω-dependent term compete, but it should be safe to assume that $\epsilon_0 \gg \omega\hbar$. Then the former overpowers the latter, and therefore, when the atoms are more tightly bound the gas pressure drops, as it should.

Exercise. *Improve the above with Debye's model for the solid.*

4.12   Black Body Radiation

All bodies radiate when heated. We all have observed some of it. When a piece of metallic rod is heated, we first see something simmering. This is due to a radiation in the region of infrared. When it is further heated to a very high temperature, it begins to glow. The dominant radiation is now in the visible light range. When this kind of thermal behavior of radiation is studied in detail, it turned out to be the most challenging problem which baffled physicists starting from the last part of the 19th century and the turn of the 20th century. This is a period after the triumphant discovery of Maxwell that oscillating electric charges radiate electromagnetic waves and that electromagnetic waves carry energy. Thus, when a body is subject to a radiation field, depending on how freely the electrons in the body can be driven by the impinging field, different bodies react differently to the radiation. For example, in a shiny metallic body there are nearly free electrons which can oscillate nearly as driven by the field; thus nearly all the impinging energy is reradiated off right away with only small portion being used to cause a lattice vibration, namely heat. By contrast black bodies like soot are such that the accelerating electrons constantly collide with the atoms causing most of the impinging energy to be absorbed to the atoms as a heat energy. So a black body is a perfect absorber and thus gets heated more rapidly but heat being accelerating charges it is emitted away sooner or later. Thus a good absorber also has to be a good emitter; after all it cannot just keep on receiving energy from the radiation field without violating the second law.

To make a nearly perfect black body, make a cavity of dimension $L \times L \times L$ using any material and keep it at temperature T. Make a small hole on one of the walls. When a light enters in the cavity through it, it will be bounced around with little or no chance to come out right away like a light reflected by a mirror, and thus the interior would look dark like a black body. When the cavity reaches its thermal equilibrium with the radiation fields in it, several prominent properties were discovered. (a) the intensity of the radiation is different for different frequencies. The intensity diminishes at the extreme low and high frequencies with a maximum peak in the middle. This pattern is only depends on the temperature and independent of the make-up of the walls. (b) The frequency of the most intense radiation increases with increasing temperature.

The radiation is an ideal quantum system as we will study in Chapter 8, but we need not wait till then. We may introduce the quantum nature which provides a unique set of microstate energies. Once we have the microstate energies, we are all prepared to handle the rest. Moreover, the quantum nature may be introduced very easily.

The key to solving the puzzle was a powerful idea of Planck's, that, in order to account for the thermal behavior of the radiation in the cavity, the radiation in the cavity must be regarded as a collection of particles called photons. A photon in the mode of angular frequency ω carries energy $\hbar\omega$. There can only be an integer number of photons in each mode, not, e.g., 0.5 photon or 126.7 photons. High intensity simply means that there is a large number of photons. Let us study the radiation as a thermal system with the cavity walls serving as a reservoir. The reservoir exchanges energy with the system by emitting photons into the cavity or by absorbing the photons from the cavity. Thus the number of photons does not remain fixed.

Depending on the volume of the cavity, there are many different modes of radiation each of which is represented by different angular frequency. To find out the spectrum of the allowed angular frequencies, regard the radiation as a wave. The electromagnetic waves exist in the cavity as various standing waves. In order to form a standing wave, the wavelength must fit in just right with the cavity dimensions. To sort out the conditions for a standing wave, proceed just as you did in introductory physics for a standing wave on a string of length L, but at the end, declare that the integer n is actually $(n_x^2 + n_y^2 + n_z^2)^{1/2}$ and n_x, n_y, n_z can only be positive integers. Thus,

$$\lambda = \frac{2L}{n} \qquad (4.130)$$

where $n = (n_x^2 + n_y^2 + n_z^2)^{1/2}$. The corresponding angular frequency is then given by

$$\omega(n) = \frac{n\pi c}{L} \qquad (4.131)$$

where c is the speed of light. This dispersion relationship relates the photon energy $\hbar\omega$ to the photon momentum $\hbar k = h/\lambda$. It is the same relationship as that for mobile particles except that it is a linear relationship. To see it, square both sides:

$$\omega(n)^2 = \left(\frac{\pi c}{L}\right)^2 (n_x^2 + n_y^2 + n_z^2) \qquad (4.132)$$

where the triplet quantum numbers n_x, n_y, n_z play the same role as in Eq. (3.24) for single particle microstates. The only difference is that, while the particle energy is proportional to the square of the matter wave frequency, the photon energy is proportional to the frequency of the electromagnetic wave.

The microstate energies are then given by

$$E\{s_1, s_2, s_3, \dots\} = s_1 \hbar\omega_1 + s_2 \hbar\omega_2 + s_3 \hbar\omega_3 + \cdots, \qquad (4.133)$$

where $s_1, s_2, s_3, \dots$ are the photon numbers of the modes $\omega_1, \omega_2, \dots$, respectively, and each photon numbers can be independently $0, 1, 2, 3, \dots, \infty$. The indices 1,2,3, etc. are a short hand for (n_x, n_y, n_z).

The partition function is given by adding the Boltzmann factors for all microstates, i.e. by adding the Boltzmann factors over all $\{s_1, s_2, \cdots\}$,

$$
\begin{aligned}
Z &= \sum_{\{s_1, s_2, s_3, \cdots\}} \exp(-\beta s_1 \hbar\omega_1 - \beta s_2 \hbar\omega_2 - \cdots) \\
&= \sum_{s_1} \exp(-s_1 \beta\omega_1) \sum_{s_2} \exp(-s_2 \beta\omega_2) \sum_{s_3} \exp(-s_3 \beta\omega_3) \cdots \\
&= \prod_i \frac{1}{1 - \exp(-\beta\hbar\omega_i)} \\
&\equiv \prod_i Z_i .
\end{aligned}
\qquad (4.134)
$$

The average energy is then given by

$$U = \frac{1}{Z}\frac{\partial Z}{\partial(-\beta)} = \frac{\partial(\ln Z)}{\partial(-\beta)} = \sum_i \frac{\partial(\ln Z_i)}{\partial(-\beta)} = \sum_i \frac{\hbar\omega_i}{\exp(\beta\hbar\omega_i) - 1}, \quad (4.135)$$

which says that the average photon number is

$$<s_i> = \frac{1}{\exp(\beta\hbar\omega_i) - 1} . \qquad (4.136)$$

Let me rewrite the last line of Eq. (4.135) so that we may carry out the summation:

$$U = 2 \sum_{n_x, n_y, n_z} \frac{\hbar\omega(n)}{\exp[\beta\hbar\omega(n)] - 1}, \qquad (4.137)$$

where the factor 2 has been added to take into account the fact that each mode can be polarized in two different directions. Strictly speaking, n is limited to the grid points in the (n_x, n_y, n_z) space, but since $\omega(n)$ is multiplied by $\hbar$ which is very small, the discrete sum may be converted into an integral. We should remain in the first quadrant only, but since it is symmetric, we cover all 8 quadrants and then divide the result by 8. The result for U is

$$U = \frac{2}{8} \int_0^\infty dn \, 4\pi n^2 \frac{\hbar\omega(n)}{\exp[\hbar\omega(n)/k_B T] - 1} \, . \tag{4.138}$$

How does the total energy or the total intensity change with the temperature? Let $x = \hbar\omega/k_B T = \hbar(\pi c/L)n/k_B T$, and write all n's and ω's appearing in the integral in terms of x. Then the integral may be written entirely in terms of x (the integral converges), and all the things that get pushed out of the integral include 4 factors of T, two from n^2, one from dn and one from the $\hbar\omega$ in the numerator.

$$U = \pi k_B T \left(\frac{k_B T L}{\hbar \pi c} \right)^3 \int_0^\infty \frac{x^3 dx}{e^x - 1} \, . \tag{4.139}$$

The integration over x gives $\pi^4/15$. Thus the total energy per volume of the cavity is given by

$$\frac{U}{V} = \frac{\pi^2}{15\hbar^3 c^3} (k_B T)^4 \tag{4.140}$$

where emphasis should be placed on the T^4 behavior and is called the Stefan–Boltzmann's law.

The heat capacity therefore goes like T^3 and vanishes at the low temperature limit. The vanishing heat capacity is a result of the gap between the lowest microstate energy and the next one, a phenomenon which we saw in an exercise problem in Sec. 4.7; we will see more of it later. The angular frequency ω may be thought of as the degree of freedom for the radiation in cavity, but since the Boltzmann factor is not a quadratic function of ω, there is no reason to expect the equi-partition theorem to hold, and it does not.

Next go back to Eq. (4.138) and this time convert the integration over n to one over ω to represent U/V as a sum of light intensities over all ranges of frequency,

$$\frac{U}{V} = \int d\omega u(\omega, T) \tag{4.141}$$

where

$$u(\omega, T) = \frac{\hbar}{\pi^2 c^3} \frac{\omega^3}{\exp[\hbar\omega/k_B T] - 1} \, . \tag{4.142}$$

This is the famous Planck's radiation law. The light intensity diminishes both in the the low and high frequency limits. It was this pattern that frustrated many physicists before Planck. It has had an enormous impact in all branches of physics, most notably in astrophysics.

According to Eq. (4.142), the intensity reaches a maximum at

$$\frac{d}{dx}\left(\frac{x^3}{e^x - 1}\right) = 0, \quad \text{or} \quad 3(1 - e^{-x}) = x \tag{4.143}$$

where $x = \hbar\omega/k_B T$. This can be solved graphically with the result

$$\hbar\omega_{mx}/k_B T \approx 2.82. \tag{4.144}$$

If the intensity is measured over a reasonably wide range of ω and ω_{mx} is measured, see what Eq. (4.144) gives us! The temperature of the cavity! The best known example of such a measurement is that of the cosmic background radiation, namely, the remnant radiation from the Big Bang. The cavity in this case, namely the Universe, has cooled down to approximately 2.9 K. It is remarkable that the temperature of the universe may be measured while sitting on the earth.

Next let us calculate the free energy and all the related thermodynamic quantities. From Eq. (4.134),

$$F = -k_B T \ln Z = 2k_B T \sum_{n_x, n_y, n_z} \ln[1 - e^{-\beta\hbar\omega(n)}] \tag{4.145}$$

where the factor 2 is again to count two different polarizations. The entropy is then given by

$$S/2k_B = -\frac{\partial F}{\partial T} = -\sum_{n_x, n_y, n_z} \ln[1 - e^{-\beta\hbar\omega(n)}] + \sum_{n_x, n_y, n_z} \frac{\hbar\omega(n)/k_B T}{\exp(\beta\hbar\omega(n)) - 1}. \tag{4.146}$$

Next we shall calculate the pressure, the pressure that the lights exert on the cavity wall. If you have read this book this far, this is a reward for you: we can calculate the pressure of light. If a person were to try to calculate this without any knowledge of thermal physics, he would have to ask: how much momentum does a photon carry? What atoms make up the cavity wall? How does a photon interact with the atoms? We do not have to ask any of these questions. According to Eq. (4.8), all we have to ask is: does the free energy depend on the volume of the space in which the photons are confined? Equation (4.131) says that the answer is yes; notice the presence of $L = V^{1/3}$. It then takes only one stroke of differentiation to find out the pressure. The result is:

$$P = -\frac{\partial F}{\partial V} = 2 \sum_{n_x,n_y,n_z} \frac{\hbar\omega(n)}{\exp[\beta\hbar\omega(n)] - 1} \left(\frac{1}{3V}\right) = \frac{U}{3V}. \qquad (4.147)$$

Compare this with the gas pressure which may be written as $P = 2U/3V$. Suppose that photons and molecules have the same amount of energy per volume. The light exerts half as much pressure as the molecules do. Considering the fact that photons do not give us the sensation of collision that molecules do on a windy day, I would say that half is a surprisingly high fraction.

• Now that we have worked out the above results via the route of the statistical mechanics, it is time to recapture the same results via the route of the free energy minimum principle. Actually Planck did not work out the puzzle as in the above. He treated the system as an isolated system and calculated the entropy as a function of energy. We shall now extend this microcanonical ensemble version to a cannonical ensemble version.

Since the cavity walls are in thermal equilibrium with the radiation in it, the equilibrium distribution of energy in the radiation should be the same as the energy distribution in the cavity walls. Take the cavity walls as consisting of oscillators or resonators. The oscillators possess energy just like the photons in bundles or quanta to be written simply as ϵ. The oscillators come in groups, some with a strong spring constant (thus with a large resonant frequency) while some with a very weak spring constant (thus with a very small resonant frequency), and every thing in between; we shall refer to these groups with index i. Consider a particular group of oscillators whose energy comes in integer multiples of ϵ_i. Assume that there are N_i of them. Suppose that this group is to possess energy $U_i = M_i\epsilon$. What is the multiplicity? How many ways are there to distribute M_i pieces of identical candies amongst N_i children? The answer is (see Chapter 2):

$$\Gamma_i = \frac{(N_i - 1 + M_i)!}{(N_i - 1)!M_i!}. \qquad (4.148)$$

Assume that N_i and M_i are both large enough to approximate the factorials using Sterling's formula. Then the entropy of this group of oscillators is given by

$$S_i/k_B = (N_i - 1)\left\{\frac{N_i - 1 + M_i}{N_i - 1}\ln\left(\frac{N_i - 1 + M_i}{N_i - 1}\right) - \frac{M_i}{N_i - 1}\ln\left(\frac{M_i}{N_i - 1}\right)\right\}$$

$$= N_i\left\{\left(1 + \frac{M_i}{N_i}\right)\ln\left(1 + \frac{M_i}{N_i}\right) - \frac{M_i}{N_i}\ln\left(\frac{M_i}{N_i}\right)\right\} \qquad (4.149)$$

where -1 has been dropped for the last line.

In the mean time, the energy in this group of oscillators is

$$U_i = M_i\epsilon_i. \qquad (4.150)$$

Putting these together for all groups, we have for the trial free energy

$$F = \sum_i M_i\epsilon_i - k_BT\sum_i N_i\left\{\left(1 + \frac{M_i}{N_i}\right)\ln\left(1 + \frac{M_i}{N_i}\right) - \frac{M_i}{N_i}\ln\left(\frac{M_i}{N_i}\right)\right\}$$

$$\qquad (4.151)$$

where M_i's are the trial parameters. By minimizing F with respect to them, we have

$$\frac{\partial F}{\partial M_i} = \epsilon_i - k_BT\left\{\ln\left(1 + \frac{M_i}{N_i}\right) - \ln\left(\frac{M_i}{N_i}\right)\right\} = 0, \qquad (4.152)$$

which may be solved to give

$$\frac{M_i}{N_i} = \frac{1}{e^{\epsilon_i/k_BT} - 1} \qquad (4.153)$$

and

$$\frac{U_i}{N_i} = \frac{M_i\epsilon_i}{N_i} = \frac{\epsilon_i}{e^{\epsilon_i/k_BT} - 1}. \qquad (4.154)$$

Identifying ϵ_i with $\hbar\omega_i$, the average number of quanta in each group has come out in agreement with the average photon number, and the average total energy in each group in agreement with the total photon energy of angular mode ω_i.

Chapter 5

Entropy as a Measure of Disorder

5.1 Introduction

Can we describe entropy in any other terms more accessible by the general public? Such descriptions do exist and are the subject of this chapter. Generalize Eq. (4.21) to all situations where the the state of matter is constantly changing amongst $1, 2, 3, \ldots$ with frequency (probability) $p_1, p_2, p_3, \ldots$, respectively. The functional

$$D\{p_1, p_2, \cdots\} = -C \sum_i p_i \ln p_i \qquad (5.1)$$

is regarded as a measure of 'disorder'. Here the Boltzmann constant has been replaced with a constant. This has been widely practiced in physics, and for a good reason. It is important to understand in what sense the term 'disorder' is used. Shannon arrived at the same equation for his information theory. There are those who wish to develop thermal physics based on the information theory and call entropy 'uncertainty' or 'missing information'. It is therefore equally important to understand in what sense they use the term 'uncertainty'.

5.2 Order and Disorder

According to this description, entropy is a measure of **disorder** and nature tends to change irreversibly from an ordered state to a more disordered state. The usual examples given include the way unruly children mess up neatly ordered chairs in their classroom and the way monkeys might mess up a well-ordered library catalog. This description associates disorder with a state of affairs which can be realized in many different ways, that is, a large number of microstates, and it associates order with a state of affairs which can be realized only in a small number of ways, that is, a small number of microstates. According to the film "Amadeus", Mozart effectively said that there is in fact only one way to compose an opera based on the story of The Marriage of Figaro, namely, the way in which he composed it. In other words, there is only one microstate which achieves the maximum musical order possible for the story. To the emperor who ordered Mozart to change the composition, he said, "Your Majesty, it cannot be changed because it is perfect." Likewise, there are fewer ways to arrange chairs in straight lines than there are to arrange them in crooked lines or with no geometrical pattern whatsoever.

This is not, however, how the average person thinks of disorder. Most people assume that if things are left to themselves, they will tend towards a state of confusion and chaos, and therefore for the worse. This is totally inappropriate for molecules and spins. Recall the "orderly" manner in which molecules direct their traffic of energy flow, for example, when we disrupt them by adding an excessive amount of energy to a certain region. When the same type of circumstances arise for humans with, say, gridlocked cars in one area, we need traffic-control officers, but molecules do not. Even for human affairs, who is to say that crooked lines are necessarily worse than straight lines, or that the absence of a prominently visible geometric pattern necessarily means artistic disorder? In the case of the monkey and the library cards, I would challenge the perception even more firmly. Suppose that the monkey mixed the cards for thermal physics books with those for sociology. Perhaps the monkey saw a connection between thermal physics and sociology and wanted to say that sociology is a highly sophisticated kind of thermal physics.

This is typical of the confusion which arises when a scientific term meant for a specific meaning is taken in the context of words used in our daily life. This creates some trouble because what is order to one person can easily be disorder to another, which is why we just poked fun at it. When a word is adopted from a living language to describe a scientific concept, there is always a reason although it is never perfect. Let me attempt to explain this point using a half dozen examples taken from our daily life. Our definition of order and disorder is straightforward. If there is only a very limited number of ways an event can occur, we shall associate that event with order. If it can occur in a large number of ways, we shall associate it with a disorder. There is no moral or social implication.

Consider driving on a busy city street. Most of the ordinary orderly drivers do about the same thing under the same condition. This is because they all follow the rules which do not leave much room for us to do anything else. You drive straight on your lane at the suggested speed, change lane only if there is room for you and give a turn signal before your attempt, slow down when you see a yellow light, and stop when you see a red light, and so on. So, can we agree that an orderly driving can be done only in a very limited number of ways? Contrast that with what you find when you are in the middle of a bunch of drunken drivers. They seem to think that they have infinite number of ways to drive the same street under the same condition. It is next to impossible to predict when the cars on your next lane will try to get in front of you, and you have no idea when you should speed up so as not to be hit by the car behind you, and so on. So, can we agree that there are zillion (1000 trillion) ways to be a 'disorderly' driver? Wait! Change the number to one zillion plus one. I have just read that a star football player from none other than my alma mater has just been arrested for drunk driving. Guess what he did? He stopped at a busy traffic light and slept! That is how limitless the number of ways drunken drivers can drive. The word 'disorder' should be taken in that meaning.

Return to the statement about school children messing up their room. Strictly speaking, in terms of geometry, it is an agreeable statement. Put two points on the floor. There is only one way to connect the two points with a straight line. In that sense, the straight line has the perfect order. Tie the two points with a rope slightly longer than the shortest distance. The rope cannot be straight everywhere, and there are zillions of different ways the rope can take its shape. In that sense we may state that crooked lines are 'disorderly'.

A boy is told by his mom to stay only in his room until she returns. If mom can count on him to do so, namely, if the boy knows only one way to be home while his mom is away, namely to be in his own room, then we will all agree to call him an orderly boy. But if there is no telling to which room he will dash to as soon as his mom leaves, and he could be found in any one of the dozen rooms in the house at any moment, we will call him a disorderly boy.

When we say that a person's room is orderly or disorderly, we know what we mean. In an orderly room, things are put together in an orderly fashion, all T shirts together in one drawer, all socks are together in another drawer, and therefore the belongings occupy only a small portion of the room space leaving the rest empty. To get a sense of the number of the microstates of the owner's belongings, give him one more T shirt. We know exactly where it will go. Contrast this with a disorderly room where all belongings are scattered everywhere. Give him one more T shirt. Would you like to predict where it will go? It could land somewhere on the floor, somewhere in a drawer, or it could be hanging on a lamp hat or on a door nob. Without doing any enumeration, this example should suffice to conclude that there are more ways for a disorderly room owner to keep his belongings than there are for an orderly room owner. The number of collective microstates is very large in the case of a disorderly room while it is limited in the case of an orderly room.

Do you have a spectacular marching band at your school? Watch them when they perform. They move together maintaining a coordination as if there were some kind of invisible bonds between them. Good bands tend to choose a difficult routine which allows absolutely no room of error for any one of them. There is only one way to do what they do. If one member tries anything else, the whole performance will break down and we all know immediately. When Ohio State band writes the letter 'Ohio' on the field, although everyone is moving, we see the steady 'hand-written' letters 'Ohio'. You can easily imagine how hard this must be. There are two places where the lines cross. If just one person strays or moves too slowly or too fast, we would immediately see a broken pattern. I saw them performing several times on TV. Each did what each was supposed to do and nothing else. Judging from the pattern they make, it is totally safe to state that there is only one way to accomplish what they accomplish. Of course, to appreciate a truly spectacular band, you have to see the Golden Eagle Marching band of my school.

Now let me describe a truly disorderly band. I saw one. With some exaggeration, they were like this. Some were walking while others were sitting and chatting. Some were playing their instruments, but I could not recognize any tune because everybody was playing different pieces. Some were wandering around while others were just staring at the stand. One was staring at the sky, but I did not see anything in the sky, and so, we should assume that it was a choreographed act. In short, I could not tell whether they were performing or taking a break, but believe it or not, I was told that they actually performed. They were proud of themselves because they did not make a mistake. They said that they had about one zillion repertoire and that they could perform any of them at any time without' making a single error. Most likely, their band director, if they had one, told them to go out and do whatever each of them wanted to do, and of course the outcome can be any of zillions of possibilities. Since each can do what each wants independent of others, the total multiplicity is $\Gamma = \Gamma_1\Gamma_2\Gamma_3\ldots$. Since each can do anything he wants, Γ_i is infinity for all members, whence Γ is infinity. If you see them, they do look disorderly.

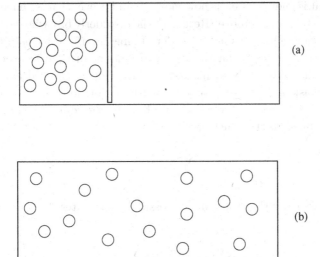

Fig. 5.1 (a) The middle wall keeps the particles in the left end. (b) Without the middle wall or 'Mom'.

These examples are from our daily life. Now let us take an example from the thermal world. Our gas molecules are no more orderly than the disorderly boy. They are just a bunch of disorderly boys. This is actually true in spite of all those passionate words that I uttered earlier in their defense. Figure 5.1 is a snapshot of a gas in a container. The middle wall keeps the gas molecules in the left part leaving the right part empty. Imagine that the middle wall is removed. No sooner does the wall leave than the molecules begin to rush into the vacuum part, just for the pleasure of doing what the middle wall did not let them until then. The middle wall did 'order' the molecules to stay only in the left part until it returns, but it did not do any good. We know that the entropy is less in (a) than in (b). Do molecules appear more ordered in (a) than in (b)? Yes, molecules are more gathered close to each other in (a) then in (b). Consult with the Sakur-Tetrode equation. If a system is limited to a lesser volume, the entropy is less, and hence the multiplicity is less. If all the air molecules in your room suddenly gather themselves into the left half of your room leaving the right half empty of air, would you not consider that an orderly act? Of course that does not happen spontaneously, which is exactly what the second law says. Just like the disorderly boy, they exhibit such an order only in the presence of their 'Mom', the middle wall.

OK that is enough. Let us now move on and prove what we have implied again and again in the above. Return to the disorderly boy who dashes from room to room against his mom's order. Using Eq. (5.1), let us prove that the disorder is largest if he spends equal amount of time in every room, and zero if he stays only in his room. This can be checked easily. The question to ask is: for what probability distribution does Disorder reach its maximum? To answer this question, we have to regard p_i as a variable, but they are subject to the constraint

$$\sum_{i=1}^{n} p_i = 1 \tag{5.2}$$

where we assumed that there are n possible microstates. So what we need to maximize is

$$Q = D\{p_i\} + \lambda \sum_i p_i$$

$$= \sum_i p_i(-\ln p_i + \lambda) \tag{5.3}$$

where λ is the undetermined Lagrange multiplier (see Chapter 2).

Demanding the extremum condition,

$$\frac{\partial Q}{\partial p_i} = 0 \tag{5.4}$$

we find

$$p_i = \exp(\lambda + 1). \tag{5.5}$$

Substituting it into Eq. (5.2), we find that λ is given by

$$\lambda = -\ln n - 1 \tag{5.6}$$

which gives the final answer for p_i,

$$p_i = 1/n. \tag{5.7}$$

As we expected, Disorder is at its maximum when the probability distribution is completely uniform. The other opposite extreme is easy to check. Let $p_1 = 1, p_2 = p_3 = \cdots = p_n = 0$. Clearly $p_1 \ln p_1 = 0$. Since $x \ln x$ approaches zero as x approaches zero, it follows that $p_i \ln p_i = 0$ for all the other terms as well. Thus the measure of disorder is zero when the system has only one accessible microstate.

Our disorderly room owner can become an orderly room owner and sort out all the T shirts together at one place and all those socks together at another place, and so on. So can our disorderly atoms and molecules in their gas phase! Certain atoms and molecules prefer to be mixed at high temperatures, but as the temperature falls they decide to phase separate, one species together at one place and another species together at another place. Our gas molecules act individually at high temperatures, but they tend to gather together at low temperatures to form a liquid with a much higher density and lower entropy. We can even see a short-range order in the liquid phase if we examine their way of positioning themselves around each other. This partial ordering turns into a long range order with further lowered entropy when they go into the solid phase. Just as it is spectacular when an orderly band performs a highly ordered routine, so are our molecules when they dance together at very low temperatures. That is the wonder of the low temperature physics! We will see all these in later chapters. Our theme will gradually change from disorder to order in the following chapters.

5.3   Uncertainty and Missing Information: Shannon's Entropy and Information Theory

Boltzmann obtained Eq. (5.1) for entropy. Shannon also obtained the same equation for his information theory, and the equation is often referred to as Shannon's entropy. The purpose of the two are, however, different and they address different issues. Thus, in spite of the same equation, they speak in different terms.

Suppose that we have to guess the outcome of an event based on several pieces of information we have been given. When the information is incomplete, as is often the case, it would be helpful if the uncertainty or poverty of information could be quantified. The purpose of Shannon's information theory is to find a mathematical entity that can serve as a measure of the uncertainty.

Suppose that a man is trapped in a burning two-story house and the firefighter is told that there are n rooms in the house and the probability of finding him in each room is $(p_1, p_2, p_3, \cdots p_n)$. Shannon demands that the measure of uncertainty should be constructed so as to satisfy the following conditions:

(a) The measure should be a single valued, continuous and differentiable function of all p_i's.

(b) The measure should reach its maximum when the probability distribution is completely uniform, i.e. all p_i's are equal to $1/n$, and the maximum should increase monotonically with increasing n.

(c) The measure should remain unchanged if the probabilities are grouped into subsets, each with its own measure of uncertainty. This means that we can measure the uncertainty in choosing a floor first and then the uncertainty of choosing a room on the chosen floor next. Shannon demands that the measure should not depend on such a strategy.

It appears that Shannon tried all conceivable forms for the measure and was led at the end to the following form:

$$H(p_1, p_2, p_3, \cdots) = \sum_i^n p_i I(p_i) \tag{5.8}$$

and

$$I(p_i) = \log_2 \left[\frac{1}{p_i} \right] \tag{5.9}$$

where $\log_2$ means logarithm based on 2 and $I(p)$ is called the missing information. We will find out shortly how Shannon finds that $I(p)$ as given above may be called the missing information. We have already seen that H as given by Eq. (5.1) satisfy (a) and (b). We will check (c) after we learn more about the equation in terms of information.

In Shannon's information theory, entropy is regarded as the average of missing informations. The idea of missing information is the key, which is why Eq. (5.8) emphasizes its role. No one can possibly object the choice of the words 'uncertainty' and 'missing information' because it is an urgent matter to guess in which room the man is trapped. Unlike Boltzmann's thermal systems and our disorderly boy, the man does not jump from one room to another. He is trapped in one of the rooms and we just do not know in which room he is now. There are n different possibilities with probabilities $(p_1, p_2, \ldots, p_n)$. Shall we guess that he is room 1 or 2, or what? Shannon is not actually just trying to guide us how to guess. If that is all what he wants, the choice would be the room for which the probability is maximum. He wants more. Suppose that there are 8 rooms, and $p_1 = 0.9, p_2 = p_3 = \cdots = p_{10} = 0.1/9$. If the firefighter says that he will first try Room 1, how uneasy would you be with his decision? Our answer would be that he was given a pretty good amount information to make that guess, but we are still slightly uneasy because the man could be in any other room. If he says that he will first try Room 2, how uneasy would you be with his decision? Our answer would be that a probability as small as $0.1/9$ is hardly sufficient to make that guess. There is a good bit of missing information to make that decision. We are quite uneasy. If someone said that he saw the man entering room 2 but the previous day, then even that small piece of information would help and make us less uneasy. Such an answer is not good enough for Shannon. He wants us to quantify the degree of uneasiness for each room. I made up the word 'uneasiness'. Call it 'uncertainty' or 'missing information(MI)' $I(p)$.

Are you not tempted to say $I(p) = 1 - p$? No! Shannon pursues the answer in the following way. Assume that someone knows the correct room number, but he is not going to tell us the room number straightaway. We have to ask him a question, but his answer will only be either YES or NO. So we have to ask him binary questions, and by the way, he charges for each question. The larger the degree of uncertainty is, the more questions we will have to ask. The minimum number of questions (MNQ) that we have to ask (to find out the correct room number) is a good measure of the missing information. That is what Shannon wants.

Shannon's entropy is then the average of MI, but this is no ordinary average because the missing pieces of information to weigh for the average are dependent on the very weighing factor with which they are weighed. Let us figure out how $I(p)$ should depend on p. Since $I(p)$ depends only on p, we will figure it out with an easy example, and the easiest example that we can work out is when the probability is the same for every room, $p = 1/8$ for all rooms. We are ready to interrogate the man to find out the exact room number. It is an integer between 1 and 8. Shall we ask him: is it Room 1? and if he says no, then ask him if it is Room 2, and so on. If we take this strategy, we may get away with one question or we may end up asking as many as 7 questions. We have to do better. We should ask questions in such a way that we can eliminate as many room numbers as possible after each question. So our first question could be: is it greater than 4? If we hear NO, the next question could be: Is it greater than 2? Whatever we hear next, we only need to ask one more question. So the

MNQ is 3. Notice that here $p = 1/8 = 1/2^3$ and MNQ is 3. For any 2^n-room house, $p = 1/2^n$ and the minimum number is n. To see why, divide all the candidate room numbers into 2 groups, and eliminate one with one question. Repeat this process again and again until each group consists of one candidate. So how many times do we have to repeat the process? The answer is: n times or $\log_2 2^n = \log_2 \frac{1}{p}$, where $\log_2$ means logarithm based on 2. So we arrive at

$$I(p_i) = \log_2 \left[\frac{1}{p_i} \right] . \tag{5.10}$$

So, the lower the probability is, the higher the uncertainty is.

We figured out $I(p)$ but using the easiest example. How good is our answer in general? Consider an unfair die which has 4 faces and the probability of landing on each face is given by $(p_1 = 1/2, p_2 = 1/4, p_3 = 1/8, p_4 = 1/8)$. Substituting these numbers into the Shannon entropy equation, we find $H = 1.75$. It says that it will take sometimes more questions and sometimes less because MI is different for different face numbers, and that on average it will take 1.75 questions. It is cumbersome to check this out with a brute force, but let us do it. To minimize the number of questions, we have to be cautious. Since face 1 has half of the total probability, we may get away with just one question by guessing it first. So our first question would be: Is it face 1? There is 0.5 chance of hearing YES, in which case the minimum number is 1, but there is also 0.5 chance of hearing NO, in which case we have to ask one more question and try face 2. Proceeding in this way, we find

$$\text{Av MNQ} = 1 \times 0.5 + 2 \times (1 - 0.5) \times \frac{0.25}{1 - 0.5}$$

$$+ 3 \times (1 - 0.5) \times \left(1 - \frac{0.25}{1 - 0.5} \right) \times 1 = 1.75 . \tag{5.11}$$

Here the second term shows how often we can get away with two questions which happens when we hear NO for the first question but YES for the second question. If we hear NO for the second question, we are forced to ask one more question, but we need not be right for this guess, whence $\times 1$; if the answer is NO again, we know that it is face 4. So average MNQ=1.75, in agreement with Shannon's entropy, but it is so because we chose just the right numbers. In general, it turns out that Shannon's H can be off but not by more than one question.

$$H \leq \text{Average MNQ} \leq H + 1 \tag{5.12}$$

which is still quite good. For a proof, see the book by Cover and Thomas. So H stands for something which can be obtained in an operational way.

There is one important virtue of the logarithmic function. Suppose that what we have to guess is actually a result of two independent events. Thus $p = q_1 \times q_2$. An example would be like: Throw a fair coin first, and then a 4-faced fair die. We are betting that the coin will land on its head and the 4-faced fair die will land on face 4. Then notice that $\log_2(p) = \log_2(q_1 q_2) = \log_2 q_1 + \log_2 q_2$, namely that the missing informations add as they should. Imagine what it would be like if we choose any other function like, say, a trigonometric function. The uncertainty would change with p like a capricious wind!

When the probability distribution is completely uniform providing absolutely no clue, we have to deal with the largest amount of MI. In the opposite extreme, namely when p_i is unity for one face and the rest are all zero, MI is zero. There is no missing information and there is no need to ask any question! Next let us check if H works as demanded in item (c). For this issue, the firefighter is a better example to work with than dice. The house has four rooms, two on the first floor and two on the second floor. Call the two rooms on the first floor 1 and 2, and those two on the second floor 3 and 4. Suppose that the probability of the man being in each room is as follows: $p_1 = 1/2, p_2 = 1/4, p_3 = 1/8, p_4 = 1/8$. We already know that $H = 1.75$ if we count MNQ without grouping.

We will now ask the questions in two groups. We will first guess which floor the man is on. The probability distribution for this issue (the first floor versus the second floor) is given by $(0.75, 0.25)$, for which $H_{floor} = 0.811$. Now we address the room issue. Let H_{room}^1 represent H for the first floor and H_{room}^2 for the second floor. The probability distribution for the former is $(0.5/0.75, 0.25/0.75)$ and that for the latter is $(0.125/0.25, 0.125/0.25)$. Calculating H for each, the final result is:

$$H\{p_i\} = H_{floor} + 0.75 H_{room}^1 + 0.25 H_{room}^2$$

$$= 0.811 + 0.75 \times 0.9182 + 0.25 \times 1.0)$$

$$= 1.75 . \tag{5.13}$$

Thus grouping does not make any difference. This is another remarkable virtue due to the logarithmic function.

In many other fields, the idea of analyzing complexity using the information theory appears most promising. Above all, it is exciting to know that we have a science based on ignorance. One day it may help cure all the ills of human society. There are also physicists who develop their thermal physics based on the information theory. My only reservation is that one has to make fuzzy an important thermodynamic quantity such as temperature as a mechanical quantity, which limits the depth. Their formulation tends to be more mathematical than physical. If that is your inclination, there are two excellent books listed in the reference section at the end of this book.

Exercise. *Repeat the above example with* ($f_1 = 0.6, f_2 = 0.3, f_3 = 0.06,$ $f_4 = 0.04$).

To summarize, Shannon's entropy allows us to quantify the degree of uncertainty that we face when we have to guess on the basis of a given a probability distribution. In the language of information theory, entropy is the average uncertainty or the amount of missing information. It measures how difficult it is to guess a random variable. As a such, it has proven useful not only in the originally intended field of communication, but also in other fields as well. The idea of analyzing complexity using the information theory appears most promising. Above all, it is exciting to know that we have a science based on ignorance.

Here is a story which tells us how the difficulties of making decisions on the basis of limited information are dealt with in fairy-tale worlds. This is a story about water nymphs.

When the nymphs wish for something badly, they go to their master spirit who gives them a quiz. They have to pass the quiz, and to pass the quiz they have to do lots of guessing. Sometimes they are given some clue and sometimes not. Only a very lucky few pass the quiz. Luckily, if they really really wish for something, they have Jezibaba who has the power to make them as lucky as they need to be to pass the quiz. So they go to Jezibaba and tell her what they wish and why. Jezibaba then knows what they will have to guess. If they agree with the condition that she imposes, she can do a miracle for them. The condition that she imposes can be deadly, and so they have to be careful what to wish for because they will get it. When Jezibaba decides to perform her magic, she performs some mysterious calculations and on the basis of those calculations she gives the nymphs some potion. She calculates the right amount of potion based on—(ssshhhee! she learned how to calculate H!). Once they drink it, they become lucky enough, not needlessly too lucky, not insufficiently lucky, but just lucky enough to pass the quiz. *The most famous nymph whom Jezibaba helped out is Rusalka, who wished to become a human because she saw a prince deep down in their pond one day and fell in love with him, and she did become a beautiful lady. Unfortunately her conversion ended in a tragedy which needs not be told here.*

5.4 Disorder Versus Uncertainty

Entropy is the heart and soul of thermal physics, but it is also a source of grief to many who find the idea too abstract or mysterious. Those seem to find Shannon's interpretation of entropy as a measure of 'uncertainty' comforting. I have several comments on this claim.

To start with, let me briefly explain how the notion of entropy evolved. Thermal physics started as a science of heat engine. The most important milestone of heat engine is the discovery of Carnot who showed that there is an upper limit to the efficiency of any heat engine. We enjoy the luxury of learning this using the second law, but Carnot and his contemporaries did not know anything about entropy yet. It was Clausius who saw on the basis of Carnot's discovery the need for entropy, but Clausius did not know the physical meaning of what he wanted to introduce. So, in an effort to prevent any possible misconception and to keep everybody on the same page, he chose the Greek word 'entropy' as it does not mean anything. This was a thoughtful and agreeable intention, but it became something to cry about ever since. According to a legend or rumor, when Shannon worked out his information theory, he consulted with John von Neumann and asked him how he should call what we now call Shannon's entropy. Von Neumann apparently noticed the identity between Eqs. (5.8) and (5.1) and suggested the word 'entropy', saying something to the effect that people doing thermal physics did not know the meaning of entropy and he had something to offer for them.[8] There are even those who claim that entropy is a misnomer and should be replaced with 'uncertainty' or 'missing information'.

The issues that Boltzmann and Shannon addressed and their premises are different. The uncertainty addressed in the information theory is genuine and there is an urgent need to quantify it. What more is there to say about the urgency required for the firefighter to know the room number?

---

[8]Either my understanding of this episode is incorrect or there was some misunderstanding between the two. Shannon's work took place in 1940s. As early as the latter part of 1800s, the physical meaning of entropy was established thanks to Boltzmann and Gibbs.

At stake is whether or not the man will be rescued. Moreover, this is a well-defined uncertainty because the trapped man is not hopping through ceilings or walls from room to room like a thermal system would. He is in one of the rooms; we just do not know in which room he is. By contrast there is no such issue in thermal physics. Thermal systems never get trapped or stuck in one microstate like the man trapped in the burning house. Amorphous materials do get stuck to turn into a glassy state but the general thermal systems that we have studied do not. Instead, they move around from one microstate to another constantly, and there is no compelling need for us to know the exact microstate that the system is visiting at any given instant of time. It is more like iPod. When we have a large collection of songs, we tend to randomize it. We let it play the songs in a random sequence. We do so because we have no compelling reason to know in advance which one will be played at any instant. Yes, it is true that we do not know what song is going to be played next, but we do not want to know and have effectively said 'surprise me'. Shannon did not have the iPod. If he had, his iPod story would go like this: one of these songs brought a big reward to the singer. Can you guess which one did?

Nevertheless, let us regard the situation as being uncertain anyway. The pertinent question would then be whether or not there is any physically meaningful quantity that we cannot predict due to that uncertainty. We followed the rules of thermal physics based on this very 'uncertainty', but everything turned out to be in accord with the central limit theorem and predictable. Extensive variables such as energy and magnetization do fluctuate, but not only their mean values are precisely predictable, but even the fluctuations are also predictable. Thermal physics is a perfectly predictable science, and there is no uncertainty, let alone an urgent uncertainty.

Chapter 6

Open Systems Exposed to Heat and Particle Reservoir

6.1 Introduction

So far we have considered closed systems with a fixed number of molecules in a fixed volume. In this chapter we will study open systems in a fixed volume but with a variable number of molecules. To give an example, designate somewhere in your room a volume of space of 1 cm × 1 cm × 1 cm. That designated volume of space is an open system. The space is wide open and molecules can enter or leave as they please. The rest of the room may be regarded as a reservoir as it is ready to exchange both energy and molecules with the designated space. We will study such open systems in Secs. 5.2 through 5.5. Systems consisting of moving particles, gases, liquids and the free electrons in solids may all be studied as an open system, but not magnets because the atoms cannot move in and out freely. Let me warn that the words 'open' is used here in the sense that particles may enter or leave. The same words are often used in literature, however, to mean that energy may enter or leave.

6.2 Open Systems

Consider an open system in a fixed volume but consisting of a variable number of molecules. The system is in contact with a reservoir which exchanges with the system not only energy but also molecules. What is the appropriate energy-like function? Since N is not fixed, we wish to remove it from F. Since $\partial F/\partial N = \mu$, the variable to replace N with is μ, the chemical potential. The energy-like function is

$$\Omega = F - \mu N = U - TS - \mu N \,, \tag{6.1}$$

155

which some call the Landau potential while others call it the grand potential. Substituting Eq. (3.86) for U, it follows that

$$\Omega = -PV. \tag{6.2}$$

The reservoir and the system are to share N_R molecules and U_R energy; both N_R and U_R are very large. Suppose that the system claims N_s molecules and energy U_s. The system is only a small part of the combined system, and the reservoir is large enough to guarantee $N_s \ll N_R$ and $U_s \ll U_R$. In order to maximize the total entropy of the combined system, how many molecules and how much energy should the system claim? The decision should be made on the basis of the total entropy,

$$S_{R+s}(U_R, N_R, U_s, N_s) = S_R(U_R - U_s, N_R - N_s) + S_s(U_s, N_s). \tag{6.3}$$

Since $N_s \ll N_R$ and $U_s \ll U_R$, we may expand S_R to first order in U_s and N_s to obtain

$$S_{R+s}(U_R, N_R, U_s, N_s) = S_R(U_R, N_R) - \frac{1}{T}U_s + \frac{\mu}{T}N_s + S_s$$

$$= S_R(U_R, N_R) - \frac{1}{T}(U_s - TS_s - \mu N_s) \tag{6.4}$$

where we recognize that what is in the parenthesis is Ω. Two observations should be made here. First, both T and μ are determined by the reservoir and do not change with U_s and N_s because $U_R \gg U_s$ and $N_R \gg N_s$. Second, the three extensive quantities appearing in Ω, U_s, N_s and S_s, are all those of the system. How should N_s and U_s be chosen? Eq. (7.9) says that the total entropy of the combined system may be maximized if U_s and N_s are chosen so as to minimize Ω. We do not have to know anything about the reservoir except that it imposes a fixed temperature T and a fixed chemical potential μ on the system. Once again everything has come out the way we wanted.

Examine where the three terms of Ω/T come from. The first two terms represent the entropy cost that the reservoir pays for allowing the system to have N_s molecules and energy U_s while the third term is what the system brings to the total entropy in return. Therefore Ω/T is the net entropy cost. The reservoir and the system conspire to minimize it.

Taking the differential for Ω, we obtain

$$d\Omega = -SdT - PdV - Nd\mu, \tag{6.5}$$

and

$$S = -\left(\frac{\partial \Omega}{\partial T}\right)_{V,\mu}, \tag{6.6}$$

$$P = -\left(\frac{\partial \Omega}{\partial V}\right)_{T,\mu}, \tag{6.7}$$

$$N = -\left(\frac{\partial \Omega}{\partial \mu}\right)_{T,V}. \tag{6.8}$$

The appropriate variables of Ω are T, V and μ, as we wanted. Given Ω as a function of T, V and μ, we may calculate S, P, and N via the above equations.

Let us find out what Ω is for ideal gases. Rewriting Eq. (4.67) in the form of

$$N = V \left(\frac{mk_BT}{2\pi\hbar^2}\right)^{3/2} e^{\mu/k_BT} \tag{6.9}$$

and using Eq. (4.63), it follows that

$$\Omega = F - \mu N = -k_BTN = -k_BTV \left(\frac{mk_BT}{2\pi\hbar^2}\right)^{3/2} e^{\mu/k_BT}. \tag{6.10}$$

6.3 Grand Canonical Ensemble Formalism

The corresponding statistical formalism called the grand canonical ensemble formalism proceeds like this. What is the probability that N molecules will come into the open system to occupy the microstate $E_i(N)$? Because the combined system visits all of its microstates with equal frequency, the probability in question is proportional to the number of ways the reservoir can accommodate what the N molecules want to do:

$$P[E_i(N), N] \sim \Gamma_R[U_R - E_i(N), N_R - N]$$
$$= \exp\left\{S_R\left[U_R - E_i(N), N_R - N\right]/k_B\right\}. \tag{6.11}$$

Upon expanding S_R to first order in E_i and N, we may write

$$S_R = S_R(U_R, N_R) - \frac{1}{T}E_i(N) + \frac{\mu}{T}N. \tag{6.12}$$

It then follows that

$$P[E_i(N), N] \sim \exp[-(E_i(N) - \mu N)/k_B T], \tag{6.13}$$

which is called the Gibbs factor. To normalize it for all microstates of all number of molecules, the normalization factor takes the form of

$$\Xi = \sum_N \sum_i \exp[-(E_i(N) - \mu N)/k_B T] \tag{6.14}$$

which is called the grand partition function. Let us calculate it for the ideal gas. The summation over all microstates of a particular N-particle system is just the partition function we have calculated earlier. So, we may write

$$\Xi = \sum_{N=0}^{N=\infty} e^{\beta \mu N} e^N \left[\frac{V}{N} \left(\frac{2\pi m k_B T}{h^2} \right)^{3/2} \right]^N$$

$$= \sum_{N=0}^{N=\infty} \frac{e^{\beta \mu N} V^N \left(2\pi m k_B T / h^2 \right)^N}{N!}$$

$$= e^{V\zeta} \tag{6.15}$$

where

$$\zeta = e^{\beta \mu} \left(\frac{2\pi m k_B T}{h^2} \right)^{3/2}. \tag{6.16}$$

In the first line, what follows $\exp(\beta \mu N)$ is $Z = e^{-\beta F}$ with Eq. (4.63) substituted for F. In the second line, the N-dependent factors $e^N (1/N)^N$ in the first line has been reduced to $N!$ by using the Stirling approximation in reverse. Since $\Xi = e^{-\beta \Omega}$, we have for the grand potential

$$\Omega = -k_B T V e^{\beta \mu} \left(\frac{2\pi m k_B T}{h^2} \right)^{3/2} \tag{6.17}$$

in agreement with our earlier thermodynamic prediction in Eq. (6.10).

Just like its canonical-ensemble counterpart Z, Ξ becomes the "master" because everything that can be calculated using the Gibbs factor can be obtained directly from it. To see why, notice that

$$<U - \mu N> = \frac{1}{\Xi} \sum_i \sum_N (E_i(N) - N\mu) \exp[-(E_i - \mu N)/k_B T]$$

$$= kT^2 \frac{\partial \ln \Xi}{\partial T}. \tag{6.18}$$

From Eqs. (6.1) and (6.5), we may also write

$$<U - \mu N> = \Omega + TS$$

$$= \Omega - T \left(\frac{\partial \Omega}{\partial T} \right)_{V,\mu}$$

$$= T^2 \frac{\partial}{\partial T} \left(-\frac{\Omega}{T} \right)_{V,\mu}, \qquad (6.19)$$

which, by comparing with Eq. (6.18), gives us

$$\Xi = \exp \left(-\frac{\Omega}{k_B T} \right). \qquad (6.20)$$

Everything we said earlier for Z and F can be reiterated here for Ξ and Ω.

Since the number of molecules N is not fixed, it is interesting to see how it is determined and how it fluctuates. To this end, rewrite Ξ as follows

$$\Xi = \sum_i \sum_N \exp[\mu N / k_B T] \exp[-E_i(N)/k_B T]. \qquad (6.21)$$

It follows that

$$<N> = \sum_i \sum_N N \exp[\mu N / k_B T] \exp[-E_i(N)/k_B T]/\Xi$$

$$= k_B T \frac{1}{\Xi} \frac{\partial \Xi}{\partial \mu}. \qquad (6.22)$$

Thus $<N>$ is determined by the chemical potential that the reservoir imposes on the system. This is also true in the human world. The city of Hattiesburg is open to anyone who wishes to move in or out. There is no guard who controls the population count. However, the city can control its population by controlling the "unlivability" of the city. If they wish to accelerate the growth of population, they can help make it more livable in many different ways, which they do quite well. If they decide to slow down the growth, on the other hand, they have many ways of doing so. For example, they could make it very difficult to obtain a building permit.

In an open system, the number of molecules N fluctuates. To study it, let us first calculate

$$<N^2> = \sum_i \sum_N N^2 \exp[\mu N / k_B T] \exp[-E_i(N)/k_B T]$$

$$= (k_B T)^2 \frac{1}{\Xi} \frac{\partial^2 \Xi}{\partial \mu^2}. \qquad (6.23)$$

It then follows that

$$\delta N^2 = <N^2> - <N>^2$$

$$= k_B T \left(\frac{\partial <N>}{\partial \mu} \right)_{V,T}. \tag{6.24}$$

This is another example of the fluctuation-dissipation theorem. Because N and μ are coupled in the Gibbs factor, the fluctuation of N is related to the response of N to a change in μ.

The response function in Eq. (6.24) is difficult to measure directly in the laboratory. To find a way to measure it more easily, we need to do some juggling. Invert the partial derivative and write it as $(\partial \mu / \partial N)_{V,T}$, where the volume V and T should be left untouched during the differentiation. Now invoke the Gibbs-Duhem relationship, Eq. (3.90). Since T must be held fixed, μ can only change via p. Thus

$$\left(\frac{\partial \mu}{\partial N} \right)_{V,T} = \frac{V}{N} \left(\frac{\partial P}{\partial N} \right)_{V,T}$$

$$= \frac{1}{N} \left(\frac{\partial p}{\partial n} \right)_{V,T}, \tag{6.25}$$

where $n = N/V$ represents the concentration. This calls for changing n and finding out the consequent change in P while holding V fixed, but P is blind and cannot tell whether n changes through V or through N. Thus we may replace $(\partial n)_{V,T}$ with $(-N/V^2)(\partial V)_{N,T}$ and write

$$\frac{1}{N} \left(\frac{\partial P}{\partial n} \right)_{V,T} = -\frac{V^2}{N^2} \left(\frac{\partial p}{\partial V} \right)_{N,T}. \tag{6.26}$$

Thus we finally arrive at

$$\delta N^2 = k_B T \left(\frac{\partial N}{\partial \mu} \right)_{V,T}$$

$$= k_B T N n K_T, \tag{6.27}$$

where

$$K_T = -\frac{1}{V} \left(\frac{\partial V}{\partial P} \right)_{N,T} \tag{6.28}$$

is called the isothermal compressibility and is a very important thermodynamic coefficient. A large isothermal compressibility means a large fluctuation in the number of particles. Suppose that the total population of a large country is widely fluctuating. How could that happen? It cannot happen if each region is randomly different in terms of livability. The entire country must be becoming livable or unlivable and therefore every part of the country is bringing population from outside or chase them out in unison as if they conspired to do so together. That turns out to be also the case for thermal systems. A large isothermal compressibility means, as we will find out later, that different parts of the system act in unison over a long length scale, namely that there is a correlation over a long length scale.

6.4   Hemoglobin and Myoglobin

The grand canonical ensemble formulation is most convenient in that it offers a way of handling identical particles without tagging them, and this advantage will be fully exploited in Chapter 8. In the mean time, we have here a very simple thermal system which is well suited for a much simpler grand canonical ensemble formalism. Our blood stream contains protein molecules called hemoglobin and myoglobin which serve the role of collecting and distributing oxygen molecules to tissues. In these molecules, there are oxygen collecting sites (called heme) and parts that seem to serve to protect the collected oxygen while in transportation. Between the two protein molecules, myoglobin is much simpler because it has only one heme which can carry at most one oxygen molecule or remain empty. So, our thermal system consists of oxygen molecules and myoglobin molecules (Mb). Focus on one myoglobin molecule. It is like a flower in garden waiting for a bee to come and visit or a bench in the park waiting for someone to come and sit on. It can remain empty with zero energy or take one oxygen molecule with energy ϵ_0. The oxygen gas serves the role of particle reservoir. In this open system, the particle number can only be zero or unity. There are only two microstates to consider, one with zero energy and zero number of particles and one with energy ϵ_0 and one particle.

Thus the grand partition function consists of only two terms:

$$\Xi = 1 + e^{-(\epsilon_0 - \mu)/k_B T} , \qquad (6.29)$$

where the first term is the not-yet normalized probability to be empty of oxygen and the second to be occupied by one oxygen. The probability of the myoblobin to have one oxygen molecule is then

$$f = \frac{e^{-(\epsilon_0 - \mu)/k_B T}}{1 + e^{-(\epsilon_0 - \mu)/k_B T}} . \qquad (6.30)$$

Equivalently, this may be taken as the fraction of the total oxygenated myglobin molecules to the total sum of oxygenated and non-oxygenated myglobins. For the chemical potential, we know from Chapter 3:

$$e^{\mu/k_B T} = \frac{N}{V}\left(\frac{h^2}{2\pi m k_B T}\right)^{3/2}$$

$$= \frac{P}{k_B T}\left(\frac{h^2}{2\pi m k_B T}\right)^{3/2} \tag{6.31}$$

where the ideal gas equation of state is used for the second line. The result may further be simplified to the form

$$f = \frac{P}{P + P_0} \tag{6.32}$$

where

$$P_0 = \frac{1}{k_B T}\left(\frac{h^2}{2\pi m k_B T}\right)^{-3/2} e^{\epsilon_0/k_B T}. \tag{6.33}$$

So, the fraction changes with the gas pressure of the oxygen following a pattern of hyperbola. When $P = P_0$, the saturation is half full. This thermal system may be constructed and studied in laboratory and the results of measurement for the fraction is in excellent agreement with the prediction. Written in this form, it is called the Langmuir adsorption isotherm. Its general purpose is to give the fraction of solid surface covered by an adsorbate as a function of its gas pressure.

Chapter 7

Flexible Systems: Systems in Contact with Heat and Volume Reservoir

7.1 Introduction

Systems which are closed for N but open for V, like a gas in a balloon, are our subject in this chapter. The reservoir imposes a fixed temperature and pressure on the system. We will study a gas confined to a cylinder equipped with a mobile piston which is designed to exert a constant pressure on the gas. The cylinder is embedded in a heat bath. If the gas pressure increases (or decreases) for any reason, the gas will expand (or shrink) until its pressure matches the fixed external pressure.

7.2 Gibbs Free Energy

We would like to work with a thermodynamic function for which T, P, and N are independent variables. Since $\partial F / \partial V = -P$, the appropriate energy-like function is

$$G = F + PV$$
$$= U - TS + PV, \tag{7.1}$$

which is called the Gibbs free energy. Substituting Eq. (3.86) into Eq. (7.1) for U, it follows that

$$G = \mu N. \tag{7.2}$$

For ideal gases, the Gibbs free energy is given by

$$G = N k_B T \ln \left(\frac{P}{k_B T \Lambda_{th}^3} \right) \tag{7.3}$$

where $\Lambda_{th} = h/\sqrt{2\pi m k_B T}$. Taking differentials on Eq. (7.1), we find

$$dG = -SdT + VdP + \mu dN \tag{7.4}$$

which means that the independent variables are indeed T, P and N, as intended, and that

$$S = -\left(\frac{\partial G}{\partial T}\right)_{P,N}, \tag{7.5}$$

$$V = \left(\frac{\partial G}{\partial P}\right)_{T,N}, \tag{7.6}$$

$$\mu = \left(\frac{\partial G}{\partial N}\right)_{T,P}. \tag{7.7}$$

Let us see once more how the maximum entropy principle is converted into a minimum free energy principle. A system is brought to a reservoir which can exchange energy and volume with the system. Thus a total energy U_T and a total volume V_T are to be shared between the two. Suppose that the system claims energy U_s and volume V_s. How does the entropy change as a result of this sharing?

$$S_{R+s}(U_T, V_T, U_s, V_s) = S_R(U_T - U_s, V_T - V_s) + S_s(U_s, V_s). \tag{7.8}$$

Since the system is only a very small part of the combined system, $U_l \ll U_T$ and $V_s \ll V_T$, which allows us to Taylor-expand the S_R term, resulting in

$$S_{R+s}(U_R, N_R, U_s, V_s) = S_R(U_T, V_R) - \frac{1}{T}U_s - \frac{P}{T}V_s + S_s$$

$$= S_R(U_R, V_R) - \frac{1}{T}(U_s - TS_s + PV_s) \tag{7.9}$$

which shows that the entropy cost to be paid for the proposed sharing is G/T. How much energy and volume should then the system claim? In order to maximize the total entropy of the combined system, the system's energy and volume should be so as to minimize the cost, namely, to minimize the Gibbs free energy. That shall be the thermodynamic principle governing the system.

7.3 Constant Pressure Ensemble Formalism

Now let us construct the appropriate machinery of statistical mechanics. Since the energy and volume will fluctuate, we may start with the question: What is the probability of the system having volume V and falling into the microstate $E_i(V)$? Because the combined system visits all of its microstates with equal frequency, the probability in question is proportional to the number of ways the reservoir can accommodate what the system is proposing to do:

$$P[E_i(V), V] \sim \Gamma_R[U_T - E_i(V), V_T - V]$$

$$= \exp\{S_R[U_T - E_i(V), V_T - V]/k_B\}. \qquad (7.10)$$

Upon expanding S_R to first order,

$$S_R = S_R(U_T, V_T) - \frac{1}{T}E_i(V) + \frac{P}{T}V. \qquad (7.11)$$

It then follows that

$$P[E_i(V), V] \sim \exp[-(E_i(V) + PV)/k_B T], \qquad (7.12)$$

which is called the Gibbs factor. To normalize it, allow all microstates of all possible volumes:

$$\Psi = \sum_V \sum_i \exp[-(E_i(V) + PV)/k_B T] \qquad (7.13)$$

which is called the constant pressure partition function. Let us calculate it for the ideal gas.

At a glance, it appears very similar to the grand partition function except that we have to sum over volume V instead of the number of particles N. That is problematic, however, because whereas N comes in integer values only but volume is a continuous variable. There are two known ways to proceed. In version I, we replace the sum over V with an integral. The partition function is a untiles number, but if we follow this version, it will then carry the unit of volume. We may repair this problem by dividing it with an unknown volume V_0 and hope that it will not remain in any pertinent thermodynamic variable, and it does not. In version II, we quantize volume V to discrete set of V_n without introducing any arbitrary element. So, this is much more attractive, and indeed it yields the correct result for the Gibbs potential as given by Eq. (7.3) and hence correct thermodynamics.

But unfortunately it fails on tasks related to fluctuations. Nevertheless, I will go through it because it offers a nice ground where we can practice several tools that we have learned.

7.3.1   *Constant Pressure Ensemble Formalism I*

In this version, we replace the sum over V with an integral. To keep the partition function unitless, we then divide the result with an unknown volume V_0.

$$\Psi = \sum_V \sum_i \exp[-(E_i(V) + PV)/k_B T]$$

$$= \frac{1}{V_0} \int_0^\infty \exp(-PV/k_B T) Z(T, V, N) dV$$

$$= \left(\frac{2\pi m k_B T}{h^2}\right)^{3N/2} \frac{1}{N!} \frac{1}{V_0} \int_0^\infty \exp(-PV/k_B T) V^N dV$$

$$= \left[\left(\frac{2\pi m k_B T}{h^2}\right)^{3/2} \frac{k_B T}{P}\right]^N \frac{k_B T}{P} \frac{1}{V_0}. \tag{7.14}$$

It then follows that

$$G = -k_B T \ln \Psi = k_B T (N+1) \ln\left[\frac{P\Lambda_{th}^3}{k_B T}\right] - \ln V_0. \tag{7.15}$$

In the thermodynamic limit of $N \to \infty$, the unity in $(N+1)$ in the first term may be ignored. The third term which involves V_0 is not only negligible in the limit but it vanishes anyway whenever differentiated with respect to pertinent thermodynamic variables and therefore its arbitrary introduction does no harm. The remaining first term is in agreement with the correct Gibbs free energy given by Eq. (7.3).

Let us check the result more with ideal gases. Easiest to check is the average

$$<E + PV> = \frac{\partial \ln \Psi}{\partial(-\beta)} = \frac{3N k_B T}{2} + (N+1)k_B T. \tag{7.16}$$

Since $<E> = 3N k_B T/2$, it follows that

$$P<V> = (N+1)k_B T \tag{7.17}$$

where the unity is negligible in the thermodynamic limit.

Now let us calculate how the volume fluctuates. We find that

$$<V> = k_B T \frac{1}{\Psi} \frac{\partial \Psi}{\partial P} \tag{7.18}$$

and

$$<V^2> = (k_B T)^2 \frac{1}{\Psi} \frac{\partial^2 \Psi}{\partial P^2} . \tag{7.19}$$

Thus,

$$\delta V^2 = <V^2> - <V>^2 = (k_B T)^2 \frac{1}{\Psi} \frac{\partial^2 \Psi}{\partial P^2} - (k_B T)^2 \left(\frac{1}{\Psi} \frac{\partial \Psi}{\partial P} \right)^2$$

$$= -k_B T \frac{\partial}{\partial P} <V>$$

$$= k_B T V K_T \tag{7.20}$$

where not surprisingly we see again the isothermal compressibility.

7.3.2   *Constant Pressure Ensemble Formalism II*

The present formalism is much more attractive than version I as it does not require any arbitrary element, but it is only partly correct. Nevertheless, it provides us with a wonderful exercise ground where we can practice several useful tools that we have learned, and for that reason, I would like to go through it.

Returning back to Eq. (7.13), we may quantize volume V so that the summation over volume may be performed as it stands. Let me remind you that the micro state energy $E_n(V)$ is given by

$$E_n(V) = \frac{h^2 n^2}{8 m V^{2/3}} \tag{7.21}$$

where $n^2 = \sum_i (n_{ix}^2 + n_{iy}^2 + n_{iz}^2)$ and n_{ix}, n_{iy}, n_{iz} can each be any positive integer. Now consider the ensemble member frozen in the microstate of energy $E_n(V)$. The pressure in this member system is given by

$$P = -\frac{dE_n(V)}{dV} \tag{7.22}$$

which shows that the pressure is different depending on the microstate the system is in. The idea of Byers Brown was to quantize the volume so that

$$P = -\frac{dE_n(V_n)}{dV_n} = \frac{1}{12} \frac{h^2 n^2}{m V_n^{5/3}} \tag{7.23}$$

comes out to be the given external pressure regardless of n for all microstates. Solving the above equation for V_n, we find

$$V_n = \left(\frac{h^2 n^2}{12mP}\right)^{3/5}. \tag{7.24}$$

It then follows that

$$H_n = E_n + PV_n = \frac{5}{2}PV_n = \frac{5}{2}P^{2/5}\left(\frac{h^2 n^2}{12mP}\right)^{3/5} \equiv \alpha n^{6/5}. \tag{7.25}$$

The partition function is now reduced to a single summation,

$$\Psi = \sum_n e^{-H_n/k_B T}. \tag{7.26}$$

This calls for a weighted sum of the quantum number space. To remind you this space, stack up many simple cubes of size $1 \times 1 \times 1$ in the $3N$-dimensional space. The microstates we need to sum over are on the cubic lattice sites. Since the volume of each cube is unity, we may carry out the summation by converting it into

$$\Psi = \frac{1}{N!2^{3N}} \int_0^\infty e^{-H_n/k_B T} A_{3N}(n)dn \tag{7.27}$$

where the surface area $A_{3N}(n)$ is given by (see Chapter 2)

$$A_{3N}(n) = \frac{2\pi^{3N/2} n^{(3N-1)}}{(3N/2 - 1)!}. \tag{7.28}$$

The two factors $N!$ and 2^N are to correct the overcounting errors, one committed by tagging the identical particles and the other by extending the integral to all quadrants. Upon carrying out the integration, we arrive at

$$\Psi = \frac{\pi^{3N/2}(5N/2)!}{N!2^N(3N/2)!} \frac{5}{3} \frac{1}{(\alpha/k_B T)^{5N/2}}. \tag{7.29}$$

Upon taking logarithm and then going through some lengthy but simple algebra, we find that $G = -k_B T \ln \Psi$ turns out to be as given by into Eq. (7.3). So this version of the constant pressure ensemble works correctly for thermodynamics.

Now let us check the fluctuations in volume using Eq. (7.26). First, notice that

$$\frac{\partial H_n}{\partial P} = V_n \tag{7.30}$$

and therefore the average volume is given by

$$<V> = \frac{k_B T}{\Psi} \sum_n V_n e^{-H_n/k_B T} = \frac{k_B T}{\Psi} \frac{\partial \Psi}{\partial(-P)} \qquad (7.31)$$

as we would like to have it. Next we should calculate

$$<V^2> = \frac{1}{\Psi} \sum_n V_n^2 e^{-H_n/k_B T}$$

$$= \frac{k_B T}{\Psi} \frac{\partial}{\partial(-P)} \sum_n V_n e^{-H_n/k_B T}$$

$$- \frac{k_B T}{\Psi} \sum_n \frac{\partial V_n}{\partial(-P)} e^{-H_n/k_B T} \qquad (7.32)$$

not as we would like to have it. The presence of the second term breaks the usual pattern that leads to the fluctuation-dissipation theorem. So, the volume quantization accomplishes its intended purpose but at the expense of the fluctuation-dissipation theorem. The response function K_T and the fluctuation δV^2 are supposed to be brought into the fluctuation dissipation theorem by the factor $\exp(-PV/k_B T)$ in Eq. (7.13). This assumes that P and V are just a pair of conjugate variables, but that is no longer true because we interfered and fixed V according to P.

The present version assumes that each ensemble member is in mechanical equilibrium with the externally applied pressure. This gives correct results for average thermodynamic quantities, but their fluctuations turn out to be incorrect.

7.4 Weak Solutions and Osmosis

Because Gibbs free energy is the chemical potential per particle and chemical potential is the traffic controller for particle migration, it is ideally suited for studying liquid solutions. We first need to clarify on the issue of entropy of mixing for mixtures of liquids. Consider a dilute solution of sugar in water. We know how easily sugar melts in water, which means that sugar molecules would rather be surrounded by water molecules than by other sugar molecules. The hosting Water molecules are called the solvent (denoted with index 1) and the guest sugar molecules the solute (denoted with index 2). The argument we made in Chapter 3 for the additivity still apply for energy and volume. Thus, if the solution consists of N_1 water molecules and N_2 sugar molecules, we may write

$$U = \sum_{i=1}^{2} N_i u_i(T, P) \tag{7.33}$$

$$V = \sum_{i=1}^{2} N_i v_i(T, P) \tag{7.34}$$

where lower case symbols are used to mean 'per particle'.

We learned near the end of Chapter 3 the entropy of mixing for solid mixtures and gas mixtures. Let us see how we may argue for liquid solutions. We may start from the first law,

$$dS = \sum_{i} N_i \frac{1}{T} \left\{ du_i(P, T) + P dv_i(P, T) \right\} . \tag{7.35}$$

If you would just imagine that we carry out an indefinite integration over this differential, the end result must have the structure of

$$S = \sum_{i} N_i s_i(P, T) + C(N_1, N_2) \tag{7.36}$$

where the first term is the obvious outcome of the integration over du and dv which depend on P and T, and the second term $C(N_1, N_2, \cdots)$ is a result of all possible constants of integration, and thus it is only a function of N_1 and N_2 which appear as a common multiplier to du and dv. Physically speaking, s_i in the first term is what the entropy would be if all we have for the system were just N_1 molecules of water or N_2 molecules of sugar. The second term says that as far as the total entropy is concerned the system cannot be regarded as such a sum of pure systems and need an additional term to account for the fact that water and sugar coexist in the solution. This gives a good reason to suspect that it is the entropy of mixing that we are after, but we do not know at the moment what the function $C(N_1, N_2)$ is.

Now imagine that we raise the temperature of the solution that this equation is meant to represent. The solution will turn into a mixture of two gases. The corresponding entropy equation should then gradually turn from what is given above into what we know for gas mixtures. For gas mixtures, the entropy has to be the sum of those in the non-mixed states plus the entropy of mixing. Thus the second term should turn into $C = -\sum_{i} k_B N_i \ln x_i$. But wait! Since the second term is independent of P and

T, there was no change for C to make while we boiled the solution. If it is as given above in the gas phase, that was what it was in the liquid phase. The entropy of mixing takes the same for all phases and is given by

$$S = \sum_i N_i s_i(P,T) - \sum_i k_B N_i \ln x_i \qquad (7.37)$$

where x_1 and x_2 are the fractions of the solvent and solute, respectively. The free energy then takes the form of

$$G = \sum_i N_i g_i(P,T) + \sum_i k_B T N_i \ln x_i . \qquad (7.38)$$

Since $G = \mu N$, the chemical potential of the solvent water is given by

$$\mu_1 = g_1(P,T) + k_B T \ln x_1 . \qquad (7.39)$$

With this in hand, we are ready to make good the promise that we made earlier in Chapter 3 that we will study in more detail the fascinating phenomenon of osmosis.

Figure 7.1 shows the osmosis. The glass tube is open ended at both ends, but the bottom end is closed with semipermeable membrane which allows water to pass through but not the solute molecules. To begin the show, the water level in the cup outside the tube is made to be the same as that inside the tube. So the temperature and pressure are the same everywhere. As time progresses, however, more and more water enters into the tube and the osmosis stops only when the resultant extra pressure due to the raised water column in the tube is large enough to stop the migration of water.

Sweetened water is a good example. We know how easily sugar melts in water, which means that sugar molecules would rather be surrounded by water molecules than by other sugar molecules. The hosting Water molecules are called the solvent and the guest sugar molecules the solute. We learned near the end of Chapter 3 the entropy of mixing. When two species A and B are mixed at the rate $N_B/(N_A + N_B) = x$, the entropy of mixing is $S_{mixing} = -N_A k_B \ln(1 - x) - N_B k_B \ln x$. Here $N_A \gg N_B$ and therefore $x \ll 1$. The first term of S_{mixing} is the contribution of the solvent and the second the solute. Each solvent molecule therefore picks up extra entropy in the amount of $-k_B \ln(1-x)$. Since the chemical potential is the Gibbs free energy per particle and since $G = U - TS + PV$, the chemical potential of the solvent is given by

$$\mu(P,T,1-x) = \mu_0(P,T) + k_B T \ln(1-x) \qquad (7.40)$$

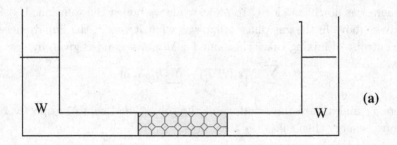

(a)

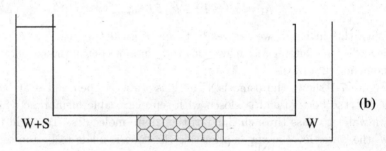

(b)

Fig. 7.1   The general pattern of the specific heat of diatomic gas molecules. See the text for the jumps, Δ_1, Δ_2 and Δ_3.

where μ_0 is the chemical potential of pure solvent in its non-mixed state. So it is clear that the chemical potential of water is lower inside the tube than outside, and therefore water will migrate into the tube, which explains the osmosis. Now let us calculate the osmotic pressure ΔP. When the pressure difference ΔP is large enough to stop the migration, the water chemical potential is the same outside and inside, which gives

$$\mu_0(P,T) = \mu_0(P + \Delta P, T) + k_B T \ln(1 - x). \qquad (7.41)$$

Since x is very small, $\ln(1 - x) \approx -x$. Thus we have

$$\mu_0(P + \Delta P, T) - \mu_0(P, T) \approx k_B T x. \qquad (7.42)$$

In the meantime,

$$\mu_0(P + \Delta P, T) - \mu_0(P, T) = \left(\frac{\partial g}{\partial P} \right)_T = v \Delta P. \qquad (7.43)$$

where g is the Gibbs free energy per particle and v is the volume per particle in the pure solvent.

Putting the two together, we have

$$\Delta P = k_B T \frac{x}{v} = k_B T \frac{N_s}{N_s + N_w} \frac{N_w}{V_{tube}} \approx k_B T \frac{N_s}{V_{tube}}, \qquad (7.44)$$

or $\Delta P.V_{tube} = k_B T$ which resembles the ideal gas equation for the solute molecules in the tube although what is in the tube is hardly a mixture of ideal gases. This equation is called van 't Hoff's formula. It has been well verified for a wide variety of solvent-solute systems. Judging from how fascinating the phenomenon is but how little we had to put in to derive it, I would rate it as one of the most delightful triumphs of thermal physics.

We worked for the water, but we ended up with a result which resembles an ideal gas equation of state for the sugar, not for the water. How is it that the sugar molecules behave like an ideal gas? First, keep in mind that they do so in the background of water. Second, each sugar molecule is surrounded by water molecules. So the sugar molecules should not be regarded as bare molecules but rather as 'dressed' molecules, dressed by a coat of water. These quasi molecules are wandering around in the vast background of water, and therefor they do not collide with each other; if they do, no more frequently than the ideal gas molecules collide with each other. So, the ideal-gas-like outcome is not totally surprising.

7.5 External Chemical Potential

Return to Eq. (3.9) and recall how the chemical potential was obtained for ideal gases. It was obtained via $\mu = \partial U/\partial N$ and for U we only took into account the total kinetic energy of the particles. If the particles are subject to a potential energy due to an external source, it also should contribute to the chemical potential. We may call this addition an external chemical potential and what is already in Eq. (3.9) the internal chemical potential. The total chemical potential may then be written as

$$\mu_{total} = \mu_{int} + \mu_{external}. \qquad (7.45)$$

A good example illustrating this is the air molecules in atmosphere. Let us treat the air as consisting of just one species of ideal gas molecules of mass m. Then the chemical potential at altitude h is

$$\mu(h) = k_B T \ln \left[n(h) \left(\frac{h^2}{2\pi m k_B T} \right)^{3/2} \right] + mgh \qquad (7.46)$$

where $n(h)$ is the air concentration N/V. We may use this equation to calculate how the concentration changes with altitude. Unfortunately changing with altitude is not just the concentration; the air temperature also changes. But to simplify the matter, we will assume that air temperature is a constant, which is a good approximation up to about $h = 30$ km.

Assume that the air is calm and is in a diffusive equilibrium, namely, that there is no upward or downward preferential migration. It means that the unlivability is the same near the ground and up at altitude h, and therefore we may state that $\mu(h) = \mu(h = 0)$ which gives

$$k_B T \ln n(0) = k_B T \ln n(h) + mgh \qquad (7.47)$$

which in turn gives

$$n(h) = n(0) e^{-mgh/k_B T}. \qquad (7.48)$$

Since $P = n k_B T$, we arrive at

$$P(h) = P(0) e^{-mgh/k_B T} \qquad (7.49)$$

which is called the barometric pressure equation. With $m = 48 \times 10^{-21}$ kg, the mass of an N_2 molecule, and $T = 217$ K, the result of the barometric equation agrees with actual measurements remarkably well up to about $h = 50$ km.[9]

Now that we know how to handle systems subject to an external potential energy, let us return to the topic of solution and ask how the gravity affects the concentrations. So, consider a solution in a vertical tube. Let me rewrite here the chemical potential as a function of height h:

$$\mu_2(h) = g_2\left(P(h), T\right) + k_B T \ln x_2(h) + m_2 g h \qquad (7.50)$$

$$\mu_1(h) = g_1\left(P(h), T\right) + k_B T \ln\left(1 - x_2(h)\right) + m_1 g h \qquad (7.51)$$

where the solvent molecules are designated '1' and the solute molecules '2' as before.

---

[9] See, Kittel and Kroamer, p. 126. For a more recent data, consult with Digital Dutch.

Let us work on the solute molecules first. Differentiating μ_2 with respect to h,

$$\frac{\partial \mu_2}{\partial h} = \frac{dP}{dh}\frac{\partial g_2}{\partial P} + k_B T \frac{1}{x_2}\frac{dx_2}{dh} + m_2 g$$

$$= \frac{dP}{dh}v_2 + k_B T \frac{1}{x_2}\frac{dx_2}{dh} + m_2 g. \tag{7.52}$$

Once equilibrium is reached, the chemical potential should be the same at all heights, hence $\partial \mu_2 / \partial h = 0$, which gives

$$-\frac{dP}{dh}v_2 = k_B T \frac{1}{x_2}\frac{dx_2}{dh} + m_2 g. \tag{7.53}$$

Repeating the same for the solvent molecules, we arrive at

$$-\frac{dP}{dh}v_1 = k_B T \frac{1}{1-x_2}\frac{dx_2}{dh} + m_1 g. \tag{7.54}$$

Combining these two equations, we have

$$k_B T \frac{dx_2}{dh}\left[\frac{1}{1-x_2}\frac{v_2}{v_1} - \frac{1}{x_2}\right] = m_2 g - m_1 g \frac{v_2}{v_1}. \tag{7.55}$$

Compare the two terms inside the vertical bracket. Since $x_2 \ll 1$, the first term may safely be ignored. Then the rest may be integrated for x_2 with the result

$$x_2(h) = x_2(0)e^{-(m_2 g - m_1 g v_2 / v_1)h / k_B T}. \tag{7.56}$$

Look at the content of the bracket in the exponent. The first term is the weight of a sugar molecule. Notice that v_2/v_1 is the number of water molecules displaced by the sugar molecule, and $m_1 g v_2 / v_1$ is the weight of those displaced molecules, namely, the buoyant force acting on the sugar molecule! Did you expect to see Archimedes' principle in action here? This is very interesting, but we should not stretch it too far. If Archimedes' principle were the overriding principle, there should be no solution; the solute molecules should all sink to the bottom or float on the top. Not so because of the entropy of mixing. The principle of maximum entropy or minimum free energy overrides the mechanical principle.

Chapter 8

Two More Energy-Like Functions and Concluding Remarks

8.1 Introduction

In this short chapter we will introduce two more energy-like functions, the energy itself and the enthalpy. Then we will briefly revisit the issue of extensivity but for all of the energy-like functions. This is followed by the minimum work theorem which adds the notion of thermodynamic potential to the Helmholtz and Gibbs functions F and G. The meaning of thermodynamic potential is not as simple and clear as their mechanical counterparts, but it is important to be aware of the fact that these functions are not there just to be minimized; they have a physical meaning on their own.

8.2 Energy

We have examined so far three different energy-like functions. They all follow a minimum principle. It would be odd if energy itself did not behave as energy-like functions do. It does, but in the following mathematical sense.

For energy, the independent variables are S, V, and N. The main actor S remains, and in the presence of the main actor, we can only attempt to translate what the main actor says in terms of energy. Entropy says that in the equilibrium state energy should be distributed such that the entropy is maximized. To put this more quantitatively, divide an isolated system into two parts. part 1 consists of N_1 molecules occupying volume V_1 while part 2 consists of N_2 molecules occupying volume V_2. The total entropy of the two-part system differs depending on how the total energy is distributed between the two parts. Suppose that the entropy is maximized when part 1 holds energy $\hat{U}_1$ and part 2 holds $\hat{U}_2$. Should U_1 and U_2 deviate from $\hat{U}_1$

and $\hat{U}_2$ by ΔU, the entropy must decrease, i.e.

$$S(\hat{U}_1 + \Delta U, V_1, N_1) + S(\hat{U}_2 - \Delta U, V_2, N_2) < S(\hat{U}_1, V_1, N_1) + S(\hat{U}_2, V_2, N_2).$$
(8.1)

Let us invert this equation, allowing S and U to exchange their roles. The entropy must then be treated as a conserved quantity while the energy must follow whatever is necessary to invert the equation. Call the right-hand side of Eq. (8.1) $\hat{S}_1 + \hat{S}_2$. If we repartition this total entropy differently, how would the energy in each part have to change? Look at Eq. (8.1). Since the same amount of total entropy is going to be distributed into two parts, the total entropy on the left-hand side should be equal to that on the right-hand side. Our question is: how should the two energy arguments on the left-hand side change if we want to replace the inequality sign with an equality sign? Since S is a monotonically increasing function of U, and since what is on the left-hand side of the equation is not large enough to match what is on the right-hand side, the two energy arguments on the left-hand side have to be increased. Then the total of the energies to enter as arguments on the left-hand side of the equation must be larger than those on the right-hand side. Thus the inverted equation reads:

$$U(\hat{S}_1 + \Delta S, V_1, N_1) + U(\hat{S}_2 - \Delta S, V_2, N_2) > U(\hat{S}_1, V_1, N_1) + U(\hat{S}_2, V_2, N_2).$$
(8.2)

This equation says that the entropy should be distributed such that the energy is minimized. It is a useful mathematical property which we will use in Sec. 5.7.2 and in Appendix C.

8.3   Enthalpy

Finally, consider the Legendre transformation,

$$H = U + PV,$$
(8.3)

which is called the enthalpy or heat function. Taking a differential for H,

$$dH = dU + P\,dV + V\,dP = T\,dS + V\,dP.$$
(8.4)

Thus for all processes taking place at a constant pressure, $dH = T\,dS$. The total change in H will tell us the total amount of heat transfer, which makes H a useful thermodynamic function. For this reason, it is also called the heat function.

We may argue for a minimum principle for W in the following way. Consider a flexible system in contact with a pressure reservoir. The system can only exchange volume with the reservoir and is insulated from the rest of the universe. Since changing volume requires work, the reservoir exchanges energy with the system by means of work and work only. Thus we may write

$$U_{R+s}(V_R, V_s) = U_R(V_R - V_s) + U_s(V_s), \qquad (8.5)$$

where V_s is the volume of the system and V_R the total volume available for both. Since $V_R \gg V_s$, we may expand the reservoir term to first order in V_s to write

$$U_{R+s}(V_R, V_s) = U_R(V_R) + PV_s + U_s(V_s), \qquad (8.6)$$

where $PV_S + U_s = H$. The work exchange process is an isentropic process, and therefore we may apply the minimum energy principle to it. To minimize the total energy of the combined system U_{R+s}, H needs to be minimized. This is a physically meaningful minimum principle, but because the principle stipulates thermal insulation, it has little or no practical utility. If we allow a heat exchange as well, then the appropriate free energy is G, the Gibbs free energy.

With all these in mind, examine the Gibbs factor for flexible systems, $\exp\left[-(E_i + PV)/k_B T\right]$ and notice the presence of the heat function in the exponent. So, we may say that heat function is to Gibbs free energy what energy is to Helmholtz free energy. In other words, Gibbs free energy takes into account only the heat exchange with the reservoir, not the work exchange. This will prove useful when we discuss the maximum work theorem.

8.4 Extensivities of Free Energies

Since we now have covered all free energy functions and they are all extensive quantities, this is a good place to find out how they change with the particle number N. Recall that we argued in Chapter 3 that both energy U and entropy S are extensive variables and therefore they should depend on N following the form of homogeneous functions of the first order, which led to

$$U(S, V, N) = N f_1(S/N, V/N),$$
$$S(U, V, N) = N f_2(U/N, V/N). \qquad (8.7)$$

For ideal gases the two functions f_1 and f_2 may be obtained from the Sackur-Tetrode equation. As we will show, free energies are all thermodynamic potentials and therefore they should change with N as a hormones function of the first order, which leads to

$$F(T,V,N) = Nf_3(T,V/N),$$
$$G(T,P,N) = Nf_4(T,P) = N\mu(T,P). \qquad (8.8)$$

Remember that the chemical potential μ is a function of T and P. This is important when we deal with multi-component systems. If a system consists of N_1 of species 1 and N_2 of species 2,

$$G(P,T,N_1,N_2) = (N_1 + N_2)\mu\left[P,T,N_2/(N_1 + N_2)\right]. \qquad (8.9)$$

For Landau potential Ω, N is not an independent variable, and therefore we cannot put its extensivity in the same format. The particle number inside the designated volume V can change by changing μ. Thus the average number $<N>$ depends on the size of the designated volume V and the chemical potential μ. Let us see if we can infer how Ω is related to V and $<N>$ from what we already know for ideal gases.

From Eq. (6.8), we find

$$<N> = -\frac{\partial\Omega}{\partial\mu} = -\frac{\Omega}{k_BT} \quad \text{or} \quad \Omega = -k_BT<N> \qquad (8.10)$$

The additivity is there, but playing the role of N is $<N>$, as it should.

8.5 Do Free Energies Have a Physical Meaning on Their Own? Maximum (Minimum) Work Theorem and Thermodynamic Potential

We know how the minimum free energy principle works and how valuable it is. Having practiced that important working principle a number of times, we may now pose for a moment and ask if there is any physical meaning to the free energies on their own. The answer is yes but in a very limited sense. What is presented below is an answer from the point of view of thermodynamics. The issue is very complex and difficult to comprehend. An easier answer will be presented from the point of view of statistical mechanics in Chapter 11, but it is limited to a special case of the Helmholtz free energy.

Consider a mechanical object connected to a mechanical work source. If the mechanical object has decreased its energy, we may state with certainty that it has performed the same amount of work to the work source. May we say the same for a thermal system in contact with a reservoir? The answer to this question is no longer straightforward. Let us specify the question in more detail. The system is in a certain state specified by the thermodynamic coordinates (U, S, V, N) and is to change by $\Delta U, \Delta S$ and ΔV to transfer to another state. The system is in contact with a heat reservoir and a volume reservoir. Neither the initial state nor the final state are necessarily equilibrium states. As usual, we assume that the reservoir is so large that its temperature remains fixed at T_0 and its pressure at P_0 regardless of how much energy and how much volume the system may claim. The work source is fully mechanical and carries no entropy.

Focusing on the system, suppose that it receives heat ΔQ from the heat reservoir and receives work ΔW from the volume reservoir, and ΔW_{ws} from the work source. Assuming that the changes in the reservoirs took place reversibly, we have

$$\Delta Q = -T_0 \Delta S_0$$
$$\Delta W = P_0 \Delta V_0$$

$$(8.11)$$

where the zero subscripts refer to the reservoir; their absence elsewhere will refer to the system. The minus in the first line is because the heat gained by the system is equal to the heat lost by the reservoir. The first law then states for the system that:

$$\Delta U = -T_0 \Delta S_0 + P_0 \Delta V_0 + \Delta W_{ws} . \qquad (8.12)$$

Now let us write this in terms of the system variables. For the volume,

$$\Delta V_0 = -\Delta V \qquad (8.13)$$

and for the entropy, we bring in the second law

$$\Delta S_0 + \Delta S \geq 0, \quad \text{or} \quad -\Delta S_0 \leq \Delta S . \qquad (8.14)$$

Substituting these into the first law, we have

$$\Delta U \leq T_0 \Delta S - P_0 \Delta V) + \Delta W_{ws} \qquad (8.15)$$

which, since T_0 and P_0 are constant, may be rewritten in the form of

$$\Delta(U - T_0 S + P_0 V) \leq \Delta W_{ws} \qquad (8.16)$$

where the equality holds when the changes in the system take place via a reversible process.

In a similar way, for systems of fixed volume and fixed number in contact only with a heat reservoir

$$\Delta(U - T_0 S) \leq \Delta W_{ws}, \tag{8.17}$$

and likewise, for systems of fixed volume but in contact with a heat reservoir and a particle reservoir,

$$\Delta(U - T_0 S - \mu_0 N) \leq \Delta W_{ws}. \tag{8.18}$$

Equations (8.16), (8.17) and (8.18) are what we needed to answer the question we posed earlier. Let me bring all of them here so that we can see them at one glance:

$$\Delta(U - T_0 \Delta S + P_0 \Delta V) \leq \Delta W_{ws}$$

$$\Delta(U - T_0 \Delta S) \leq \Delta W_{ws}$$

$$\Delta(U - T_0 \Delta S - \mu_0 N) \leq \Delta W_{ws}.$$

What is on the left hand side in each equation is the change in the appropriate free energy corresponding to the proposed change in the thermodynamic coordinates. All three equations say unanimously that if the change takes place reversibly the change requires a minimum amount of work but if it has to occur irreversibly it will cost more work from the work source. We may call this the minimum work theorem. If, for example, the proposed change is to a new state far far away from equilibrium states, then the cost could be substantially higher than the minimum cost necessary for a change to an equilibrium state. That should be taken as being reasonable.

In this way, we have recovered a relationship between work and free energies, which resembles the work-energy relationship of mechanical systems. A notion of thermodynamic potential has emerged, but unfortunately, there are all those inequality signs in the above equations, which make the resemblance meaningful only for reversible processes.

From a practical point of view, we are not really interested in finding what it will cost to put a thermal system into a certain non-equilibrium state of high free energy. We are more interested in the reverse, namely, if a thermal system is already in a non-equilibrium state of very high free energy, say, by a violent act of nature, how much useful work can we get from it when it changes to another state of lower free energy. Pocket of such energy exists at all times all over the world. If only we could become an energy scavenger hunting such pockets for 'free' energy! To answer this question, put a negative sign to both sides of all three equations to obtain

$$-\Delta(U - T_0\Delta S + P_0\Delta V) \geq -\Delta W_{ws}$$

$$-\Delta(U - T_0\Delta S) \geq -\Delta W_{ws} \tag{8.19}$$

$$-\Delta(U - T_0\Delta S - \mu_0 N) \geq -\Delta W_{ws}.$$

These equations tell us how much useful work we may get when a thermal system decreases its free energy. In this sense, the free energies are said to give the available energy. The available energy is maximum if the changes take place reversibly.[10] If the the change takes place irreversibly we get less than the maximum. This is called the maximum work theorem.

So, ΔW_{ws} may be related to the change in the free energy appropriate for the system, but the relationship is not as straightforward as for a mechanical system. The relationship can be straightforward, but that requires that all the changes take place reversibly. This is a difficult proposition because if we leave the systems at the mercy of their reservoirs, since T_0, P_0 and μ_0 remain fixed, there is only one equilibrium state and there are no reversible changes to speak of. Either we have to force the issue by forcing those intensive variables to change slowly or let something happen so that those constant intensive reservoir variables alone can no longer determine the equilibrium state. The latter is the case when there occurs a slow chemical reaction. An example of the former would be changing the pressure slowly so that the system remains in equilibrium at every stage of the process. Then we can prove that the free energy versus work relationship becomes just like for mechanical systems, thus firmly establishing the notion of thermodynamic potential at least under this highly controlled condition. I would like you to sleep on this matter until we will return to it in Chapter 10.

Finally, remove the work source. Then $W_{ws} = 0$, and we rediscover the extremum principle

$$\Delta(U - T_0 S) = \Delta F \leq 0$$

$$\Delta(U - T_0 S + P_0 V) = \Delta G \leq 0 \tag{8.20}$$

$$\Delta(U - T_0 S - \mu_0 N) = \Delta \Omega \leq 0.$$

---

[10]Suppose that we have a set of numbers bound by a minimum like, $(5, 6, 9, 23, 258, \dots)$. If we put a minus to every one of them, we get $(-5, -6, -9, -23, -258, \dots)$ which is bound by a maximum.

There may have been a remark overdue. The free energies that we dealt with earlier have T, P and μ without the zero subscript. Although there is no subscript, these are always the intensive reservoir variables. If a thermal system is not in equilibrium, the local temperature would be different from the reservoir temperature, and likewise the local pressure and chemical potential would be different from their reservoir counterparts. Let these local variables be T, P and μ. Is the non-equilibrium Helmholtz free energy, for example, then $F = U - TS$ or $U - T_0 S$? It is the latter. The former has no meaning.

Chapter 9

Quantum Ideal Gas

9.1 Introduction

We have studied ideal gases so far in the classical regime where the thermal de Broglie wavelength is much shorter than the average inter-particle distance, i.e. $\Lambda_{th} \ll (V/N)^{1/3}$. In this regime, quantum effects are negligible. As the temperature is lowered or the density is made higher, Λ_{th} increases or $(V/N)^{1/3}$ decreases. Either way, the system enters into a regime where quantum effects begin to play roles. In Sec. 9.2, we will explore the region where the thermal wavelength is still shorter than, but very close to, the average inter particle distance. In this transient regime, the ideal gas is called 'slightly degenerate'. In the opposite regime where the thermal wavelength is longer than the inter particle distance, the gas becomes fully degenerate and behave quite differently, which will be our subject in Sec. 9.3.

I do not know the reason for the choice of the word 'degenerate'. It is strange to say the least and even offensive as it sounds like implicating someones moral standard. But this is a widely used terminology in physics and we simply mean with this word that the behavior is different from the ordinary or that quantum effects are dominating. Yes, the behavior is indeed radically different. How different? As the gases enter into the 'degenerate regime', their thermal behavior becomes more and more 'orderly'. In fact their entropy approaches zero as the temperature approaches zero. If we have to pick one from the classical ideal gas and the quantum ideal gas and call it 'degenerate', it is actually the classical ideal gas, not the quantum ideal gas. We can poke fun at it because we now understand the quantum effects, but during the early days when some strange behavior was noticed, I will dare guess that someone thought it was more of a threat than an exciting new regime to explore.

9.2   The Occupation Number Formalism and Slightly Degenerate Ideal Gas

We did statistical mechanics calculations with a fixed number of particles using the N-fixed microcanonical and canonical ensemble methods. In both methods, we were forced to tag the identical particles and ran into the over-counting error problem. It is quite simple to correct this error when the temperature is high, but not so when the temperature is low. To find an alternate way, we turn to the μ-fixed grand canonical ensemble formalism. Its virtue is that in this formalism the multiparticle microstates may be constructed involving any number of particles. So, rather than asking particles which single particle microstate they are in, we shift the focus to the single particle microstates (orbitals) and ask how many particles are in each orbital. Since we would then only count the number of occupants of each orbital, we need not ask the identity of the occupants and may avoid the overcounting problem. Denote the occupation number for orbital $\mathcal{E}_i$ with n_i. The multiparticle microstates are then represented by the set of occupation numbers $(n_1, n_2, n_3, \dots)$. The corresponding energy and the total number of particles are, respectively, given by

$$
\begin{aligned}
E(n_1, n_2, n_3, \dots) &= n_1\mathcal{E}_1 + n_2\mathcal{E}_2 + \cdots \\
N(n_1, n_2, n_3, \dots) &= n_1 + n_2 + n_3 + \cdots .
\end{aligned}
\tag{9.1}
$$

The total number of particles would then fluctuate, but we need not worry about whether the number is too large or too small. That is the job that the chemical potential takes care of. With μ, the unlivability has been declared. If a microstate enters into the partition function with an outrageously large number, the $\exp(\mu N)$ factor will accordingly weigh it. The sum over all microstates of all energies and of all number of particles is carried out by summing over all possible occupation numbers:

$$
\begin{aligned}
\Xi &= \sum_{n_1, n_2, \cdots} e^{-(n_1\mathcal{E}_1 + n_2\mathcal{E}_2 + \cdots)/k_B T + \mu(n_1 + n_2 + \cdots)/k_B T} \\
&= \sum_{n_1} e^{-n_1(\mathcal{E}_1 - \mu)/k_B T} \sum_{n_2} e^{-n_2(\mathcal{E}_2 - \mu)/k_B T} \cdots \\
&\equiv \Xi_1 \times \Xi_2 \times \Xi_3 \cdots
\end{aligned}
\tag{9.2}
$$

where Ξ_1 represents the partition function for orbital 1 and Ξ_2 for orbital 2, and so on. So, the particles are no longer the focus. The orbitals are.

Think of them as seats waiting for particles to visit, and as such, they may even all be empty.

All known particles — electron, proton, neutron, Hydrogen atom, water molecule — may be divided into two groups, Fermi–Dirac particles or fermions and Bose-Einstein particles or bosons. All those particles with a non-integer spin are fermions while all those with an integer spin are bosons. The spin of a particle is a resultant of the intrinsic angular momenta of its constituents. To find out the spin of an atom or molecule, one has to know how to add angular momenta, which requires more advanced quantum mechanics than this book presumes. But for our purpose, we can get by with the following simple guide. Electrons, protons, and neutrons are all fermions. Just count how many of these fermions are in the atom or molecule in question. If the number is even, it is a boson. If the number is odd, it is a fermion. For example, He^4 has two protons, two neutrons and two electrons. Thus it contains 6 fermions and therefore it is a boson. He^3, on the other hand, contains two protons, two electrons and one neutron, which add to 5 fermions and therefore it is a fermion.

Whether a system consists of fermions or bosons has a profound effect on their thermal behavior. According to quantum mechanics, the occupation number of each orbital (single particle microstate) is subject to a different rule depending on whether the system consists of fermions or bosons.

First, consider fermions. The rule says that the occupation number can only be 0 or 1. It then follows that

$$\Xi_i = \sum_{n_i} e^{-n_i(\mathcal{E}_i - \mu)/k_B T}$$

$$= 1 + e^{(\mu - \mathcal{E}_i)/k_B T}. \tag{9.3}$$

For bosons, the rule allows the occupation number to be any positive integer from zero to infinity, and therefore we have

$$\Xi_i = \sum_{n_i} e^{-n_i(\mathcal{E}_i - \mu)/k_B T}$$

$$= \frac{1}{1 - e^{(\mu - \mathcal{E}_i)/k_B T}}. \tag{9.4}$$

We may now calculate the grand potential Ω. Since $\Omega = -k_B T \ln \Xi$, and since Ξ factorizes into products of single orbital partition functions, the grand potential is given by a sum of single orbital grand potentials

$$\Omega = \sum_i \Omega_i = -k_B T \sum_i \ln \Xi_i$$

$$= \mp k_B T \sum_i \ln \left(1 \pm e^{(\mu - \mathcal{E}_i)/k_B T}\right) \qquad (9.5)$$

where the upper sign represents fermions and the bottom sign bosons.

We wish to convert the summation into an integral. This brings us back to the problem of counting the microstates. We will do it in the semi-classical way using the plane wave method as opposed to the standing wave method. The latter is necessary because the zero momentum state is not included in the standing wave method. Near $T = 0$, however, the ground state is no longer negligible for bosons. In fact at $T = 0$, the zero-momentum ground state will play the dominating role for bosons.

Corresponding to a given orbital, $\mathcal{E} = p^2/2m$, there are many degenerate states. The degeneracy has three sources. (a) Momentum $\vec{p}$ can take on many different directions with the same magnitude and thus the same energy. (b) The particle can be at different parts of the volume with the same amount of energy. (c) The third comes from the spin states of the constituent particles. If spin is s, the projection of the spin angular momentum on any direction can take on $g = 2s + 1$ different values, and therefore g gives the spin degeneracy. For example, for electrons, s is $1/2$, and therefore $g = 2$; the two spin states are called 'up' and 'down'. Thus the electrons can have the same amount of energy with spin up or spin down.

I hope the reader can remember how we made in Chapter 3 city blocks in the position space and momentum blocks in the momentum space. Count the number of these blocks anywhere in the position part of the configuration space but with the momentum anywhere between p and $p + dp$ in the momentum part of the space. The number of single particle microstates without counting spin is given by Eq. (3.14), namely, the volume of the corresponding phase space divided by h^3. This then has to be multiplied by g for the total degeneracy. Since $2p\,dp = 2m\,d\mathcal{E}$, we may write the result as

$$\Delta\Gamma = \frac{gV 4\pi p^2 dp}{h^3}$$

$$= gV \xi \mathcal{E}^{1/2} d\mathcal{E}, \qquad (9.6)$$

where

$$\xi = 2^{5/2} \pi m^{3/2}/h^3. \qquad (9.7)$$

This tells us how to convert the sum into an integral:

$$\sum_i \to \int gV\xi \mathcal{E}^{1/2}d\mathcal{E}. \tag{9.8}$$

Proceeding this way, we find

$$\Omega = \mp k_B T \sum_i \ln\left(1 \pm e^{(\mu-\mathcal{E}_i)/k_B T}\right)$$

$$= \mp k_B T g V \xi \int \ln\left(1 \mp e^{(\mu-\mathcal{E})/k_B T}\right)\mathcal{E}^{1/2}d\mathcal{E}$$

$$= -\frac{2}{3}gV\xi \int_0^\infty \mathcal{E}^{3/2}\frac{d\mathcal{E}}{e^{(\mathcal{E}-\mu)/k_B T} \pm 1}. \tag{9.9}$$

The integral depends on μ and $k_B T$. To make these dependencies as explicit as possible, change the integration variable $\mathcal{E}$ to z by $z = \mathcal{E}/k_B T$. Then we have

$$\Omega = -\frac{2}{3}gV\xi(k_B T)^{5/2}\int_0^\infty \frac{z^{3/2}dz}{e^{(z-\mu/k_B T)} \pm 1}. \tag{9.10}$$

Call the remaining integral

$$f(\mu/k_B T) = \int_0^\infty \frac{z^{3/2}dz}{e^{(z-\mu/k_B T)} \pm 1}. \tag{9.11}$$

This type of integral is encountered frequently in thermal physics. There is a simple guide which we can follow: bring the exponential function in the denominator to the numerator so that we do not have to deal with the trouble-some term ± 1 in the denominator. This guides us to write

$$\frac{1}{e^{(z-\mu/k_B T)} \pm 1} = \frac{1}{e^{z-\mu/k_B T}\left(1 \pm e^{-z+\mu/k_B T}\right)}. \tag{9.12}$$

Now look at what is being added or subtracted to 1 in the denominator. Since $z > 0$, $e^{-z} < 1$. Assume that $e^{\mu/k_B T} \ll 1$ because we can then reduce the trouble-some part of the integrand to an infinite series:

$$\frac{1}{e^{(z-\mu/k_B T)} \pm 1} = e^{-z+\mu/k_B T}(1 \mp e^{-z+\mu/k_B T} \pm e^{(-z+\mu/k_B T)2} \mp \cdots) \tag{9.13}$$

and the accuracy of our integration will be dependent on how many terms of this series we are willing to include. To the leading order, we have

$$f^{(0)}(\mu/k_B T) = e^{\mu/k_B T} \int_0^\infty z^{3/2} e^{-z} dz$$

$$= e^{\mu/k_B T} \frac{3}{2} \int_0^\infty z^{1/2} e^{-z} dz$$

$$= e^{\mu/k_B T} \frac{3}{2} \frac{\sqrt{\pi}}{2} \qquad (9.14)$$

where we performed an integration by part for the second line but the integrated parts vanished at both ends.

Next consider the first order correction,

$$f^{(1)}(\mu/k_B T) = \mp e^{2\mu/k_B T} \int_0^\infty z^{3/2} e^{-2z} dz$$

$$= \mp e^{2\mu/k_B T} \frac{1}{2^{5/2}} \frac{3}{2} \frac{\sqrt{\pi}}{2} \qquad (9.15)$$

where we have again performed the same integration by part and the integration variable was changed from z to $2z$ for the final integration. Adding the two, we have to the first order of correction

$$f(\mu/k_B T) = e^{\mu/k_B T} \frac{3\sqrt{\pi}}{4} \left(1 \mp \frac{1}{2^{5/2}} e^{\mu/k_B T}\right). \qquad (9.16)$$

Substituting this into Eq. (9.10), we now have the grand potential

$$\Omega = -g \frac{V}{\Lambda^3} k_B T e^{\mu/k_B T} \left(1 \mp \frac{1}{2^{5/2}} e^{\mu/k_B T}\right). \qquad (9.17)$$

Here $\Lambda = h/\sqrt{2\pi m k_B T}$ is again the thermal de Broglie wavelength. The chemical potential has served us well, but in the end we have to replace it with the expected (average) particle number $\bar{N}$, and $\bar{N}$ is determined according to the level of the 'unlivability' declared by μ. Let us determine $\bar{N}$ to the zeroth order using the first term in Eq. (9.17):

$$\bar{N}_0 = -\frac{\partial \Omega}{\partial \mu} = g \frac{V}{\Lambda^3} e^{\mu/k_B T}. \qquad (9.18)$$

This is only an approximate result good only to the zeroth order for the given chemical potential. It is the exact result for the corresponding chemical potential given by

$$\mu_0 = k_B T \ln \left(\frac{\bar{N}_0}{gV} \Lambda^3 \right). \tag{9.19}$$

Except the presence of g, this is the same relationship between the particle number and the chemical potential that we found earlier for the classical ideal gas. For all the atoms and molecules that the ideal gas was supposed to represent, the spin was assumed to be zero and thus $g = 1$. So, we have proved that the present formalism gives the correct classical result at the zeroth order. Moreover, eliminating μ_0 from Ω_0, which gives

$$\Omega_0 = -g\bar{N}_0 k_B T. \tag{9.20}$$

Since $\Omega = -PV$, it reproduces the classical ideal gas equation of state. Thus the leading order term in Eq. (9.17) is the classical grand potential $\Omega_{classic}$. The second term gives the leading quantum correction.

By repeating the above but with both the leading term and the first order correction term for Ω, we are now about to see how fermions and bosons are going to deviate from the classical behavior. We find

$$\bar{N}_1 = -\frac{\partial \Omega}{\partial \mu} = g\frac{V}{\Lambda^3} \left(e^{\mu/k_B T} \mp \frac{1}{2^{3/2}} e^{2\mu/k_B T} \right). \tag{9.21}$$

To solve this equation for μ, it may appear that we need to solve a quadratic equation. That is not necessary. Remember that we assumed for Eq. (9.12) that $e^{\mu/k_B T} \ll 1$. The leading zeroth order term is of this order while the next first order correction term is even smaller than this by one order. The sensible thing to do under this circumstance is to approximate the correction term using μ_0 and solve the above equation for the first term $e^{\mu/k_B T}$. (I know that many wonder why we do not do the other away around. So, let me give you a quick example. We need to solve $x^2 + 0.0200x = 1.2321$, but we do not know anything about the quadratic equation. We are told that x is greater than 1 and very close to 1. If we approximate the bigger x^2 term with $x = 1$ and solve for the remaining x, $x = 11.545$. If we approximate the small second term with $x = 1$ and solve the equation, $x = 1.1004$. The correct answer is $x = 1.1000$.) This gives

$$e^{\mu/k_B T} = \frac{\bar{N}_1}{gV} \Lambda^3 \pm \frac{1}{2^{3/2}} \left(\frac{\bar{N}_0}{gV} \right)^2 \Lambda^6 \tag{9.22}$$

where all the subscripts to $\bar{N}$ are shown to show where each term came from. Now that you have seen it, we will drop them. Substitute the above into Eq. (9.17) to obtain the final result

$$
\Omega = -k_B T \frac{gV}{\Lambda^3} \left\{ \left[\frac{\bar{N}}{gV} \Lambda^3 \pm \frac{1}{2^{3/2}} \left(\frac{\bar{N}}{gV} \right)^2 \Lambda^6 \right] \mp \frac{1}{2^{5/2}} \left(\frac{\bar{N}}{gV} \right)^2 \Lambda^6 \right\}
$$

$$
= -k_B T \bar{N} \mp \frac{k_B T}{2^{5/2}} \bar{N}^2 \frac{\Lambda^3}{gV} . \tag{9.23}
$$

Since $\Omega = -PV$, the pressure is given by

$$
P = k_B T \frac{\bar{N}}{V} \pm \frac{k_B T}{2^{5/2}} \frac{\bar{N}^2}{gV^2} \Lambda^3 . \tag{9.24}
$$

This is what we have been after. Let me briefly reflect on how we have arrived here. We took advantage the μ-fixed formalism and obtained the grand potential. But we had to return back to the N-fixed formalism so that we can see what we see in the above equation. The above equation tells us that, should we put N fermions or bosons in a box of fixed volume and keep it at a fixed temperature T, then the pressure will be such and such. Of course, our main motive was to find out how the quantum rules alter the classical ideal gas behavior, but we also wanted to learn how the grand canonical formalism works, and it was for the latter reason why we stayed on the course expending a great deal of effort to learn how to return back to the N-fixed formalism. We will have to go through this part several times more, but the task will not be as tedious as in the above. In any event, now that we have learned the hardest part, we may also learn that there is a shortcut. It is based on the theorem of small increments.

Write Eq. (9.17) as

$$
\Omega = \Omega_0 + \Delta\Omega . \tag{9.25}
$$

where the first term is the zeroth order term which gives the classical ideal gas behavior, and the second term a much smaller correction term due to quantum effects. If we knew the Helmholtz free energies corresponding to Ω_0 and $\Delta\Omega$, we could calculate the pressure right away. We know F_0, the counterpart of Ω_0. Do we know the counterpart of $\Delta\Omega$? According to the theorem of small increments, $(\Delta F)_{V,N,T} = (\Delta\Omega)_{V,T,\mu}$. So, ΔF

corresponding to $\Delta\Omega$ is

$$\Delta F = \pm \frac{gVk_BT}{\Lambda^3 2^{5/2}} e^{2\mu/k_BT}$$

$$= \pm \frac{gVk_BT}{\Lambda^3 2^{5/2}} \left(\frac{N\Lambda^3}{gV}\right)^2$$

$$= \pm \frac{N^2 k_BT\Lambda^3}{2^{5/2}gV} . \tag{9.26}$$

The change in pressure due to this is given by $P = -\partial\Delta F/\partial V$, which gives the second term in Eq. (9.24).

Now let us examine the final result and find out how the quantum effects alter the ideal gas behavior. Suppose that we put N fermions and N bosons in two separate boxes of identical volume and keep them at the same temperature T. The pressure is higher in the fermion box than in the classical ideal gas. For bosons, it is the opposite. The pressure is lower than would be if the gas behaved like the classical ideal gas. It appears as if fermions now repel each other while bosons attract each other. In fact that is effectively what they do. The effective force is called the exchange force. It is repulsive for fermions and attractive for bosons. The actual derivation of this exchange force may be seen in the thesis of Boris Khan.[11] You will enjoy reading it because it is very readable, but you will also have to read a very sad story about this promising but sadly fallen young theoretical physicist.

The enhanced pressure of fermion gas is called the degeneracy pressure and plays a vital role in astrophysics. To appreciate this fully, let me repeat again that the quantum effects play role when the thermal wavelength is larger than the interparticle distance, i.e.

$$\left(\frac{V}{N}\right)^{1/3} < \frac{h}{\sqrt{2\pi m k_BT}} . \tag{9.27}$$

In the interior of stars, the temperature is very high but because the density is so high to the extent that the above criterion can still be satisfied. In the early and prime time of stars, electrons are no more than just by-standers. But they play a surprising role when small stars have exhausted all their fuel. At this stage, if the enormous gravitational pressure trying to squeeze the star is to be held against in a balance, it is not the thermal

---

[11] It has been reproduced in a monogram edited by G. Uhlenbeck.

pressure. It is the degeneracy pressure that keeps them stable. In the late stages of massive stars, every thing turns into neutrons. When the degeneracy pressure of neutrons is strong enough to balance against the truly enormous gravitational force, we have neutron stars. When that is not the case, there is no other hope against the merciless gravitational pull, and the star turns into a black hole.

Exercise. *Find out how the entropy of quantum ideal gas deviates from that of classical gas.*

9.3 Fermi–Dirac and Bose-Einstein Distribution Functions

Let us calculate the average occupation number of each orbital,

$$\bar{n}_i = -\frac{\partial \Omega_i}{\partial \mu} = \frac{1}{e^{(\mathcal{E}_i - \mu)/k_B T} \pm 1} \qquad (9.28)$$

where we used

$$\Omega_i = \mp k_B T \ln(1 \pm e^{(\mu - \mathcal{E}_i)/k_B T} . \qquad (9.29)$$

With the occupation number in hand, the expected average energy is

$$U = \sum_i \bar{n}_i \mathcal{E}_i = -gV\xi' \int_0^\infty d\mathcal{E} \frac{\mathcal{E}^{3/2}}{e^{(\mathcal{E} - \mu)/k_B T} \pm 1}, \qquad (9.30)$$

where ξ is a shorthand given in Eq. (9.7) and the sum has been converted into the integral following Eq. (9.8). Compare this with the total grand potential

$$\Omega = \sum_i \Omega_i = -\frac{2}{3}gV\xi \int_0^\infty d\mathcal{E} \frac{\mathcal{E}^{3/2}}{e^{(\mathcal{E} - \mu)/k_B T} \pm 1} . \qquad (9.31)$$

Since $\Omega = -PV$, it follows that

$$PV = \frac{2}{3}U . \qquad (9.32)$$

This general relationship which we saw before in the classical regime also holds in the quantum regime both for fermions and bosons. Finally the expected total number of particles is given by

$$\bar{N} = \sum_i \bar{n}_i = gV\xi \int_0^\infty d\mathcal{E} \frac{\mathcal{E}^{1/2}}{e^{(\mathcal{E} - \mu)/k_B T} \pm 1} . \qquad (9.33)$$

All these results from Eq. (9.28) through Eq. (9.33) are valid not only in the slightly degenerate regime but also in the fully degenerate regime. We shall study these equations in the next two sections very near and at $T = 0$.

9.4   Degenerate Electron Gas Near and At $T = 0$

Now we shall apply the tools developed above to a very important fermion system. When atoms come together to form a metal, each atom loses one or two of its loosely bound outer electrons. To a good approximation, these electrons may be regarded as a degenerate fermion gas. Examine the fermi distribution function very near $T = 0$,

$$n_i = \frac{1}{e^{(\mathcal{E}_i - \mu)/k_B T} + 1},\tag{9.34}$$

where the average sign has been omitted. As T approaches zero, the exponential function gives either infinity or zero depending on whether $\mathcal{E}$ is greater than μ or less than μ. Hence, $n_i = 1$ for $\mathcal{E} < \mu$ and $n_i = 0$ for $\mathcal{E} > \mu$, i.e. it is a step function. The chemical potential μ is therefore equal to the energy of the highest occupied orbital, which is called $\mathcal{E}_F$, the Fermi energy. The identification $\mu = \mathcal{E}_F$ near $T = 0$ makes the task of eliminating μ in favor of N easy.

What should $\mu = \mathcal{E}_F$ be if the average number of particles is to be N?

$$N = gV\xi \int_0^{\mathcal{E}_F} \mathcal{E}^{1/2} d\mathcal{E}\tag{9.35}$$

which gives

$$\mathcal{E}_F = \frac{1}{2m}\left(\frac{3h^3 N}{4\pi gV}\right)^{2/3}.\tag{9.36}$$

The Fermi momentum p_F, the momentum of the highest orbital, is obtained from $\mathcal{E}_F = p_F^2/2m$, and the Fermi temperature T_F defined by $T_F = \mathcal{E}_F/k_B$:

$$p_F = \left(\frac{3h^3 N}{4\pi gV}\right)^{2/3}\tag{9.37}$$

$$T_F = \frac{1}{k_B}\left(\frac{3h^3 N}{4\pi gV}\right)^{2/3}.\tag{9.38}$$

The energy at $T = 0$ may be obtained similarly from Eq. (9.30) with the distribution function replaced by the step function,

$$U_0 = gV\xi \int_0^{\mathcal{E}_F} \mathcal{E}^{3/2}d\mathcal{E}$$

$$= gV\xi \frac{2}{5}\mathcal{E}_F^{5/2} \tag{9.39}$$

$$= \frac{3}{5}\mathcal{E}_F N . \tag{9.40}$$

Is there any degeneracy for this N-particle ground state? None! There is only one way for the system to be in the ground state. The particles must systematically occupy all the orbitals up to the Fermi level and none above it. There is no room to shuffle around. In other words, we have a perfect order! We will calculate the entropy slightly above the zero temperature shortly, but we may now predict that it will approach zero as as T approaches zero. We started our study with the ideal gas at very high temperature where the multiplicity was astronomical. What a change!

If the system is so ordered at $T = 0$, then perhaps they are so frozen that their pressure is zero. Or is it? Let us calculate the pressure again. Since $PV = 2U/3$,

$$P = \frac{2\mathcal{E}_F}{5}\frac{N}{V} . \tag{9.41}$$

Even at $T = 0$, the gas pressure is not zero! They are not frozen! Since the pressure is not zero, they require a container! Welcome to the wonderful world of low temperature physics. An order without being frozen is the hall mark. Liquid He^3 can be in a phase where its entropy is lower than that of solid He^3 at the same temperature. All these may sound bizarre, but not if viewed in the context of order and disorder as explained in Chapter 6. The particles can be moving, but its entropy can still be very low if they move in a highly coordinated fashion to the extent that they are actually more ordered than in the solid phase where the particles are wiggling around their lattice sites.

Now let us explore slightly above the zero temperature. To this end, return to Eq. (9.9) for Ω, and to Eqs. (9.30) and (9.33) for U and N, respectively. For Ω and U, we need the integral

$$I_{3/2} = \int_0^\infty \frac{\mathcal{E}^{3/2}d\mathcal{E}}{e^{(\mathcal{E}-\mu)/k_B T} + 1} \tag{9.42}$$

and for N, we need

$$I_{1/2} = \int_0^\infty \frac{\mathcal{E}^{1/2} d\mathcal{E}}{e^{(\mathcal{E}-\mu)/k_B T} + 1}. \tag{9.43}$$

Let us work out $I_{3/2}$. First change the integration variable to $z = (\mathcal{E} - \mu)/k_B T$, which gives

$$I_{3/2} = k_B T \left(\int_{-\mu/k_B T}^0 \frac{(k_B T z + \mu)^{3/2} dz}{e^z + 1} + \int_0^\infty \frac{(k_B T z + \mu)^{3/2} dz}{e^z + 1} \right). \tag{9.44}$$

Next, change the integration variable of the first integral to $-z$, and obtain

$$I_{3/2} = k_B T \left(\int_0^{\mu/k_B T} \frac{(-k_B T z + \mu)^{3/2} dz}{e^{-z} + 1} + \int_0^\infty \frac{(k_B T z + \mu)^{3/2} dz}{e^z + 1} \right). \tag{9.45}$$

We are trying to make the two integrals as alike as possible so that we may combine the two. Since T approaches zero, the two will have the same upper limit, but the denominator of the integrand has changed. To repair this setback, notice that

$$\frac{1}{e^{-z} + 1} = \frac{1}{e^{-z} + 1} - 1 + 1 = -\frac{1}{e^z + 1} + 1, \tag{9.46}$$

which gives

$$I_{3/2} = \int_0^\mu \mathcal{E}^{3/2} d\mathcal{E} + k_B T \int_0^\infty \frac{(\mu + k_B T z)^{3/2} - (\mu - k_B T z)^{3/2}}{e^z + 1} dz. \tag{9.47}$$

The first term comes from the unity in Eq. (9.46), and for this term, it is better if we change the integration variable z back to $\mathcal{E}$, as shown. Now focus on the numerator in the second integral. Since we are near $T = 0$, $k_B T z$ is very small compared to μ. Thus Taylor expand the two terms,

$$(\mu + k_B T)^{3/2} = \mu^{3/2} + \frac{3}{2}\mu^{1/2} k_B T z + \frac{3}{4}\mu^{-1/2}(k_B T z)^2 + \cdots \tag{9.48}$$

$$(\mu - k_B T)^{3/2} = \mu^{3/2} - \frac{3}{2}\mu^{1/2} k_B T z + \frac{3}{4}\mu^{-1/2}(k_B T z)^2 + \cdots$$

which finally gives

$$I_{3/2} = \int_0^\mu \mathcal{E}^{3/2} d\mathcal{E} + 3(k_B T)^2 \mu^{1/2} \int_0^\infty \frac{z}{e^z + 1} dz$$

$$= \frac{2}{5}\mu^{5/2} + (k_B T)^2 \mu^{1/2} \frac{\pi^2}{4}. \tag{9.49}$$

Following the same algebra, the result for $I_{1/2}$ is

$$I_{1/2} = \frac{2}{3}\mu^{3/2} + (k_B T)^2 \mu^{-1/2}\frac{\pi^2}{12}. \tag{9.50}$$

All preparations are done! We now ask: how bad the unlivability must be if the average population is only N? Substituting Eq. (9.50) into Eq. (9.33), μ determines N via

$$N = gV\xi\left(\frac{2}{3}\mu^{3/2} + (k_B T)^2\mu^{-1/2}\frac{\pi^2}{12}\right) \tag{9.51}$$

which we need to solve for μ. Notice that if we set $T = 0$ here, we recover the zero temperature chemical potential $\mu_0 = \mathcal{E}_F$. Since T is very small, the second term is of order $(k_B T)^2$ while the first term is two order higher. So it is appropriate to approximate the second term with with $\mu = \mathcal{E}_F$

$$N = gV\xi\frac{2}{3}\mu^{3/2} + gV\xi(k_B T)^2 \mathcal{E}_F^{-1/2}\frac{\pi^2}{12} \tag{9.52}$$

and then solve for the remaining μ.

$$\mu = \left(\mathcal{E}_F^{3/2} - \frac{3}{2}(k_B T)^2\mathcal{E}_F^{-1/2}\frac{\pi^2}{12}\right)^{2/3}. \tag{9.53}$$

Now Taylor expand the right hand side and keep only the first two terms to obtain

$$\mu = \mathcal{E}_F - \frac{\pi^2}{12\mathcal{E}_F}(k_B T)^2. \tag{9.54}$$

With the chemical potential in hand, we are now ready to see how U and Ω changes with the temperature very near $T = 0$. Let us first work on U. Substituting Eq. (9.49) into Eq. (9.30), we have

$$U = gV\xi\left\{\frac{2}{5}\mu^{5/2} + (k_B T)^2\mu^{1/2}\frac{\pi^2}{4}\right\}$$

$$= gV\xi\left\{\frac{2}{5}\left[\mathcal{E}_F - \frac{\pi^2}{12}\frac{(k_B T)^2}{\mathcal{E}_F}\right]^{5/2} + (k_B T)^2\mathcal{E}_F^{1/2}\frac{\pi^2}{4}\right\}$$

$$= gV\xi\left\{\frac{2}{5}\mathcal{E}_F^{5/2} + \frac{(k_B T)^2}{6}\mathcal{E}_F^{1/2}\right\}. \tag{9.55}$$

The leading term is the ground state energy U_0 at $T = 0$ given by Eq. (9.39). As the temperature approaches zero, the energy decreases like T^2. The heat capacity therefore is given by

$$C = \frac{dU}{dT} = gV\xi\frac{k_B^2\mathcal{E}_F^{1/2}}{3}T. \tag{9.56}$$

The linear behavior of the heat capacity provides the final answer to the heat capacity puzzle. Einstein started the search for answers with his theory based on individual free oscillators. His theory agrees with measurements quite well but begins to deviate at very low temperatures. Debye picked up the second leg. For insulators his T^3 prediction agrees with measurements for the whole temperature range, but not for metals. The deviation of metals from the Debye theory is due to the free electrons. Sommerfeld took this last leg with his theory which we have just studied. For metals, contributing to the heat capacity are both the lattice vibrations and the motions of the degenerate electron gas. When the T^3 behavior of lattice vibrations and the T behavior of free electrons are added, the combined predictions agree with measurements to a near perfection.

Contrast the predictions of all three theories with that of the classical ideal gas. The heat capacity of classical ideal gas is a constant which means that their hunger for energy continues down to the entire temperature range. The ideal oscillators of Einstein, by contrast, freezes dramatically as reflected by the exponential decay of the heat capacity. The Debye theory, on the other hand, shows that the lattice vibration mode of energy remains far more active than that of Einstein's oscillator mode at low temperatures. The reason is that in Debye's theory the ions on the lattice sites vibrate in concert as opposed to the individual uncoordinated random vibrations in Einstein's model. As we will discuss in Chapter 14, such modes of excitation can come with infinitely low energy compatible to the thermal energy. The free electron mode of energy remains even more active near absolute zero, but of course the degree of activity is not anywhere near that of the classical ideal gas. All these show that quantum systems are like picky eaters who do not eat much because they require 100 conditions to eat what is on their plate. No, make it 200 conditions.

What has made the electrons so picky? It is the quantum rule that each orbital can be occupied by no more than one fermion. At $T = 0$, all the orbitals are occupied up to the top Fermi level. At near $T = 0$, this puts those electrons deep down below the Fermi level in a distinctive disadvantage. If they are to take energy from the heat source, they have to move to a higher level, somewhere above the Fermi level. At near zero temperature, the available thermal energy is only of order $k_B T$ which is not enough to do so, and therefore they cannot take the thermal energy. Only those very near the Fermi level can. The fraction of these lucky electrons to the total is $k_B T/\mathcal{E}_F$. Since each of these electrons can absorb energy of the order of $k_B T$, the total thermal energy that they can absorb goes like T^2 and the heat capacity like T.

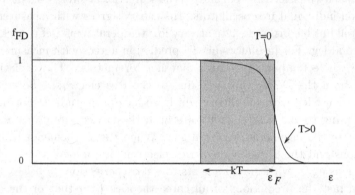

Fig. 9.1 Fermi gas near $T = 0$. Those fermions in the shaded area can absorb the thermal energy, but those outside cannot.

Next let us calculate the entropy. We may start from

$$\Omega = -\frac{2gV\xi}{3}I_{3/2} = -\frac{2gV\xi}{3}\left(\frac{2}{5}\mu^{5/2} + (k_B T)^2 \mu^{1/2}\frac{\pi^2}{4}\right), \qquad (9.57)$$

which immediately gives

$$S = -\left(\frac{\partial\Omega}{\partial\mu}\right)_{V,\mu} = gV\xi\frac{k_B^2\mathcal{E}_F^{1/2}}{3}T. \qquad (9.58)$$

Notice here again what we did to keep track of the orders in which the entropy depends on the temperature. The second term in Eq. (9.57) already is in the order of $(k_B T)^2$ and therefore the chemical potential that accompanies it was safely approximated with $\mathcal{E}_F$. The entropy shares the same pattern with the heat capacity. It vanishes at $T = 0$, which we anticipated, but we had to wait till $I_{3/2}$ was calculated to see its linear dependency on the temperature.

9.5 Degenerate Ideal Boson Gas At and Near $T = 0$

Bosons and Fermions appear similar except a difference in sign, but that difference is by no means trivial. I once heard about a colloquium speaker who apparently exploded to someone in his audience who asked about a sign in the equation he wrote down. He said, "Forget about that stupid sign. Just listen to what I say." No one can say so about the sign difference in boson and fermion distribution functions. The two can be vastly different when $\exp[(\mathcal{E}-\mu)/k_BT]$ is near unity. In fact there is little or no similarity in their thermal behavior which is why we discuss the two in separate sections.

Let us start from the Bose-Einstein distribution function, the distribution function

$$n_i = \frac{1}{e^{(\mathcal{E}_i-\mu)/k_BT} - 1}. \tag{9.59}$$

The presence of -1 in the denominator has a profound consequence. That is that μ cannot be positive. If it were, i.e. $\mu = |\mu|$, then for all orbitals below $|\mu|$, i.e. $\mathcal{E}_i < |\mu|$, the distribution would be negative! So,

$$\mu < 0. \tag{9.60}$$

Given $\mu = -|\mu|$, the average number of bosons is, according to Eq. (9.33), given by

$$\begin{aligned}
N &= gV\xi \int_0^\infty \frac{\mathcal{E}^{1/2}d\mathcal{E}}{e^{(\mathcal{E}-\mu)/k_BT} - 1} \\
&= gV\xi(k_BT)^{3/2} \int_0^\infty \frac{z^{1/2}dz}{e^{(z-\mu/k_BT)} - 1} \\
&= gV\xi(k_BT)^{3/2} \int_0^\infty \frac{z^{1/2}dz}{\exp(z)\exp(|\mu|/k_BT) - 1}.
\end{aligned} \tag{9.61}$$

Now we are going to lower the temperature closer and closer to zero and we are going to ask the chemical potential to be merciful enough to allow the population to remain fixed at the given value of N. This is a legitimate request. See what $|\mu|$ has to do in response. Since $\exp(z)\exp(|\mu|) > 1$ in the denominator, $|\mu|$ has to approach zero. Moreover, since the front factor $(k_BT)^{3/2}$ decreases N as T approaches zero, $|\mu|$ has to decrease faster than would be necessary to keep the integral unchanged. This is strange because in order to keep the population fixed it has to get meaner and meaner and makes the boson box more and more unlivable for the bosons. Well, then

$|\mu|$ may reach zero before T does. Suppose that this happens at $T = T_C$. Then since μ cannot be positive there will be no population left when $T = 0$! How absurd! It appears as if there were a miniature black hole gobbling up our bosons.

Something must be wrong with Eq. (9.61). What could it be? Since we obtained this equation by converting what was a sum into an integral, that conversion must be blamed. How could it be wrong? There is no restriction to the occupation numbers of bosons, and therefore, as the temperature is lowered the ground orbital, namely, the zero momentum state, can be occupied by a macroscopic number. When the original sum over the orbitals is converted into an integral, such a macroscopic occupation is not recognized under the integral sign. The particles that seemed to be disappearing simply transferred from a finite momentum state to the ground orbital. This massive occupation of the ground orbital is called the Bose-Einstein condensation.

What Eq. (9.61) told us was not the total population number but rather the number of those that remain above the ground orbital, i.e. the number of particles above the condensate. Let $\mu = 0$, i.e. $T \leq T_C$. The z integral is

$$\int_0^\infty \frac{z^{1/2} dz}{e^z - 1} = \frac{\pi^{1/2}}{2} 2.612 = 2.313 \,. \tag{9.62}$$

and Eq. (9.61) should read

$$N_{\mathcal{E}>0} = gV\xi(k_B T)^{3/2}(2.313) \,. \tag{9.63}$$

Exactly at $T = T_C$ where the condensation begins, $N_{\mathcal{E}>0} = N$, and hence

$$N = gV\xi(k_B T_C)^{3/2}(2.313) \,. \tag{9.64}$$

which determines T_C,

$$T_C = \left(\frac{N}{V}\right)^{2/3} \frac{h^2}{(\pi g)^{2/3} 2^{5/3} m k_B} \,. \tag{9.65}$$

Combining Eq. (9.63) and (9.64), the fraction of those not not in the condensate to the total is

$$\frac{N_{\mathcal{E}>0}}{N} = \left(\frac{T}{T_C}\right)^{3/2} \tag{9.66}$$

or the fraction of those in the condensate to the total is

$$\frac{N_0}{N} = 1 - \left(\frac{T}{T_C}\right)^{3/2}. \tag{9.67}$$

The Bose-Einstein condensation is complete at $T = 0$. All bosons are in the ground orbital, which constitutes again a perfect order. There is no multiplicity and we may predict that the entropy should be zero at $T = 0$.

Next let us calculate the energy. Since the condensation takes place on the zero energy orbital, converting the sum to the integral did no harm as far as the energy is concerned. Moreover, since $\mu = 0$ for $T < T_C$, we can calculate the energy without any worry about calculating μ. What a relief!

$$U = gV\xi \int_0^\infty \frac{\mathcal{E}^{3/2} d\mathcal{E}}{e^{\mathcal{E}/k_B T} - 1}$$

$$= gv\xi(k_B T)^{5/2} \int_0^\infty \frac{z^{3/2} dz}{e^z - 1}$$

$$= gv\xi(k_B T)^{5/2} 1.7826. \tag{9.68}$$

The z-integral is given by $\Gamma(3/2 + 1)\zeta(3/2 + 1) = \frac{3\sqrt{\pi}}{4} 1.341 = 1.7826$. Here ζ is the Riemann zeta function which is conveniently tabulated in any mathematics handbook including the popular one in the Schaum's series. The specific heat is then given by

$$C_V = \left(\frac{\partial U}{\partial T}\right)_V = gv\xi k_B^{5/2} T^{3/2} 4.4565. \tag{9.69}$$

It vanishes as T approaches zero faster than that of fermions, which is because more and more of them condense into the condensate carrying no energy.

Since $PV = (2/3)U$, from Eq. (9.68)

$$P = g\xi(k_B T)^{5/2} 1.7826 \times 2/3. \tag{9.70}$$

Notice that the pressure is independent of volume, which means that the isothermal compressibility is infinite.

$$K_T = -\frac{1}{V}\left(\frac{\partial V}{\partial P}\right)_T = \infty. \tag{9.71}$$

It is like when you play a game of tug of war. You pull a little harder and so does your opponent and nothing happens. You try even harder.

Your opponent cannot pull any harder and is now being pulled. Similar instability occurs when Bose particles begin to condense into the zero momentum state. This is because those in the condensate have no energy and make no contribution to the pressure.

Finally, let us calculate the entropy. We can start from the grand potential which is identical to U except a factor of $2/3$. We will take the same but a shortcut. Since $\Omega = -PV$ and $PV = (2/3)U$,

$$
\begin{aligned}
S = -\left(\frac{\partial \Omega}{\partial T}\right)_{V,\mu} &= \frac{2}{3}\left(\frac{\partial U}{\partial T}\right)_{V,\mu} \\
&= \frac{2}{3}gV\xi k_B^{5/2}T^{3/2}4.4565 \\
&= gV\xi k_B^{5/2}T^{3/2}2.971 .
\end{aligned}
\tag{9.72}
$$

As we predicted earlier, the entropy goes to zero as T approaches zero. The order in Bosons at $T = 0$ is truly spectacular. The wave function of the ground orbital is a constant. The wave length is infinity. This means that they are everywhere just like a ghost and the wave functions add coherently everywhere. They would look like one macroscopically large atom. For bosons in the degenerate regime, the notion of spatial entropy loses its meaning, which is why the pressure is independent of volume.

For long time, it was uncertain whether or not similar condensation occurs in real world systems of interacting atoms or molecules. That has changed ever since the Bose-Einstein condensation in interacting atoms was realized about a decade ago. Now it appears that the interactions can even be tuned, which opens all kinds of exciting possibilities as we will discuss in Chapter 14.

9.6 The Classical Regime Revisited

Whether a particle is a boson or fermion is an intrinsic property which cannot change with its environment. Then question arises as to why the distinction disappears in the high temperature classical regime. In addition, what is the occupation number or distribution function in the classical regime? Let us clear up these questions. Return to Eq. (9.28), the Fermi–Dirac and Bose-Einstein distribution functions. Bosons and fermions differ by the $+1$ and -1 factors in the denominator of their respective distribution functions. If this difference has disappeared, we can only blame $\exp(-\mu/k_B T)$.

That is indeed true. In the classical regime, $\exp(\mu/k_BT) \ll 1$, and hence $\exp(-\mu/k_BT) \gg 1$, and the $+1$ and -1 factors may be ignored. Then the F-D and B-E distribution functions converge to

$$n_i = \exp(\mu/k_BT)\exp(-\mathcal{E}_i/k_BT). \tag{9.73}$$

We are in the μ-fixed formalism. To go to the N-fixed formalism, add Eq. (9.73) for all single-particle orbitals and demand that the chemical potential be such that the total number of occupants in all orbitals is equal to N:

$$\sum_i n_i(\mu, T) = N. \tag{9.74}$$

Upon carrying out the sum, we find that

$$\exp(\mu/k_BT)Z_1 = N, \quad \text{or} \quad \exp(\mu/k_BT) = \frac{N}{Z_1}. \tag{9.75}$$

This is the same as what we had for classical ideal gases; see Eqs. (4.56) and (4.67). So bosons and fermions do behave alike in the classical regime and follow the Boltzmann statistics. The occupation numbers in the classical regime may be obtained by substituting Eq. (9.75) into Eq. (9.73),

$$n_i = N\frac{\exp(-\mathcal{E}_i/k_BT)}{Z_1}, \tag{9.76}$$

where the fraction gives the probability of finding any particle in orbital i. Since the particle can be any one of the N particles, n_i is the occupation number. This is called the Boltzmann distribution function. Remember that being distributed here are not probabilities but particle numbers.

To summarize, we have three distribution functions:

$$n_i = \frac{1}{e^{(\mathcal{E}_i-\mu)/k_BT}+1}, \quad \text{Fermi–Dirac Distribution} \tag{9.77}$$

$$n_i = \frac{1}{e^{(\mathcal{E}_i-\mu)/k_BT}-1}, \quad \text{Bose-Einstein Distribution} \tag{9.78}$$

$$n_i = \exp(\mu/k_BT)\exp(-\mathcal{E}_i/k_BT), \quad \text{Boltzmann Distribution} \tag{9.79}$$

This is the average occupation numbers. If we have a way of peeking into the orbitals, we would count: (i) 0 or 1 in the case of F-D; (ii) any number in the case of B-E; (iii) 0 most of the times but rarely 1 and even more rarely greater than 1 in the case of Boltzmann. In other words, there are far more orbitals than there are particles to occupy them, and the average occupation number is therefore $n_i \ll 1$.

We started out our journey with ideal gases and learned thermal physics mainly using them. We played with them when they were so playful or very disorderly at high temperatures. We also admired them how orderly they become at low temperatures. We have to leave them now.

Chapter 10

Thermodynamics

10.1 Introduction

Let me start the chapter with a little bit of history. The subject of thermal physics started as thermodynamics primarily during the period when heat engines were built. It is remarkable how much was learned purely on a phenomenological basis. The most triumphant example was probably the discovery of Carnot that there exists an upper limit to the efficiency of heat engines. Carnot showed that not all absorbed heat can be converted into work even in his imaginary ideal heat engine. No doubt that this was important to those who were working on heat engines, but it had far more important implications than that. It appears to have inspired Clausius to introduce the notion of 'entropy' and ultimately introduce the second law. Lord Kelvin, also inspired by Carnot, came out with the notion of absolute temperature. Joule, Clapeyron and Thompson were the other luminaries during this period around the mid 1800s. Then there emerged in the late 1800s two giants Boltzmann and Gibbs who laid the foundation of thermodynamics, which is also called statistical mechanics, statistical physics, or statistical thermodynamics. Until the mid 1960s, thermodynamics was first taught to undergraduates, reserving statistical mechanics for graduates. Then an educational revolution took place combining the two right from the beginning and the title thermal physics was coined.

Thermodynamics is based on two very simple principles, namely the first and second laws, but it is amazing how much one can do solely based on these two principles. The Carnot engine is a good example illustrating this point. To practice thermodynamics efficiently, one has to know the inter-relationships between thermodynamic coefficients. This will constitute a good portion of this chapter. The free energies or the thermodynamic potentials serve as a bridge between thermodynamics and statistical physics. Their mathematical shapes as a function of their independent variables determine the stability of thermal systems, which will be our last topic.

10.2   Carnot Heat Engine I

Heat engine consists of a working medium, two heat reservoirs, one at a high temperature T_h and one at a low temperature T_l, and a work source. The work source is purely mechanical and does not carry any entropy. Heat energy is drawn from the hot reservoir to the working medium. The working medium then converts it into work to the work source, as depicted in Fig. 10.1.

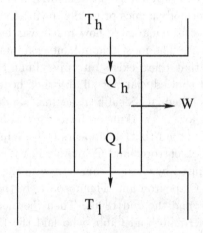

Fig. 10.1   The heat engine.

We would of course like to build a heat engine which can deliver the maximum possible work output for a given heat input from the hot reservoir. The Carnot engine is a hypothetical engine which achieves this objective. The maximum work theorem says that in order to achieve that limit, all processes have to be reversible. This reversibility requirement immediately demands that the heat transfer between the working medium and the heat reservoirs must occur while the two are at the same temperature. Recall that when two objects at two different temperatures are brought into contact, heat flows from the hot object to the cold object, and that is an irreversible process because the receiving party increases its entropy more than the donor decreases its entropy resulting in a net positive change in entropy. In order to carry out a heat exchange process with zero net entropy change, the two parties have to be at the same temperature. Then they can change their entropies in the same amount but in opposite directions for a net zero entropy change.

Carnot engine carries out the following four processes. (i) Bring the working medium (WM) at temperature T_h and put it in a direct contact with the hot reservoir so that it may draw heat Q_h from the reservoir isothermally and reversibly. (ii) Take WM away from the hot reservoir and cool it adiabatically and reversibly down to T_l. 'Adiabatically' means 'without changing the entropy'. This requires that the WM can receive or perform work but cannot receive or give away heat. (iii) Put WM in a direct contact with the lower temperature reservoir so that it will dump out Q_l of heat to the reservoir isothermally and reversibly. (iv) Remove the working medium from the low temperature reservoir and raise its temperature back to T_h adiabatically and isothermally. Once all these four processes are completed, the working medium is back to its initial state so that the cycle can be repeated as many times as desired.

How should we apply the second law? Since the working medium returns to its initial state after each cycle, its final entropy after one cycle has to be the same as its initial entropy. Therefore when we address the second law, we only need to consider the two reservoirs. The net change of entropy after one cycle is

$$\Delta S_{(i)} + \Delta S_{(iii)} \geq 0 \qquad \text{or} \qquad \Delta S_{(iii)} \geq -\Delta S_{(i)} \qquad (10.1)$$

where the first term is the increase of entropy of the hot reservoir during process (i) and the second term that of the cold reservoir during process (iii). The equality applies if the heat transfers are reversible while the inequality applies if the heat transfers are irreversible. We shall keep track of this inequality more like a cautious pedestrian than an error-prone high flying wizard.

The heat input, work output, and heat output may all be written in terms of the above entropy changes:

$$Q_h = -T_h \Delta S_{(i)}$$
$$Q_l = -T_l \Delta S_{(iii)}$$
$$W = Q_h - Q_l = -T_h \Delta S_{(i)} - T_l \Delta S_{(iii)} \qquad (10.2)$$

where the minus sign in the first equation is needed because we strictly follow the sign convention: both Q and W always mean 'received' quantities. Q_h represents the heat received by the working medium, but since it is equal to the heat given away by the hot reservoir we have to say that it is equal to the minus of the heat 'received' by the hot reservoir.

Substituting the second equation of Eq. (10.1) into the equation for W, we have

$$W \leq -T_h \Delta S_{(i)} + T_l \Delta S_{(i)} \qquad (10.3)$$

It then follows that the efficiency of the engine is given by

$$\eta = \frac{W}{Q_h} \leq \frac{-T_h \Delta S_{(i)} + T_l \Delta S_{(i)}}{-T_h \Delta S_{(i)}} \qquad (10.4)$$

or

$$\eta \leq \frac{T_h - T_l}{T_h} \qquad (10.5)$$

Thus the efficiency cannot surpass the limit $1 - T_l/T_h$ called the Carnot efficiency.

10.3   Carnot Heat Engine II

In the above we have derived the Carnot efficiency purely from thermodynamic principles without checking whether or not the proposed processes can actually be carried out. In this section, we will use an ideal gas as a working medium and calculate the work output to check the Carnot efficiency equation.

How may we carry out the processes (i) and (iii)? The medium and the reservoir are brought into contact while they are at the same temperature. If they do nothing, they will be just looking at each other, which cannot cause a heat transfer; the working medium will have to do some thing to induce a heat transfer. We know by experience that if we confine air in a piston and compress it fast, then the temperature will rise. If we let the air expand, the temperature will fall. So what we need to do for (i) is to let it expand very slowly. Then, as the temperature falls, heat is immediately drawn from the reservoir to offset the falling temperature. Thus the desired process may be realized by an isothermal reversible expansion. The process of (iii) is just the reverse of it, namely, an isothermal reversible compression.

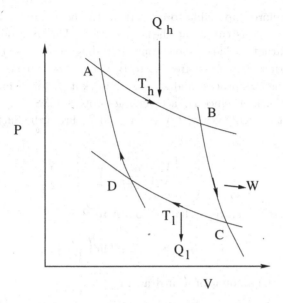

Fig. 10.2 The basic equation depicted here represents a stable system. The fact that point A is above point B reflects the inequality in Eq. (10.65). Notice that all the tangent lines are above the curve, which does not allow a common tangent line anywhere.

How about the processes of (ii) and (iv)? For (ii), it is an adiabatic reversible expansion. Insulate the working medium so that no heat can come in or leave and let it expand slowly. Imagine again that the working medium is a gas confined in a piston. Insulate the piston and let the gas push out the piston very slowly. The gas is now performing work against you at the expense of its internal energy, which therefore cause a drop in temperature. Process (iv) is the reverse of (ii). In both (ii) and (iv), entropy does not change because the working medium merely receives or performs work without receiving or giving away any heat energy.

The total entropy of the four parties returns to its initial value after the four processes because every process is a reversible process. After the four processes, the working medium returns to its initial thermodynamic state to repeat the cycle. Then, so should the net entropy change of the reservoirs. The entropy lost by the hot reservoir in process (i) should be equal to the entropy gained by the cold reservoir after process (iii). That is to say that all of the entropy that came into the working medium with Q_h must leave the medium during process (iv). Thus it is unavoidable that a part of the initial heat input is 'wastefully' dumped out in process (iv), and it is therefore impossible to convert all the heat input into work 100 percent! Now let us carry out these processes. The four processes that make up a Carnot cycle are shown on a P-V diagram in Fig. 10.2.

(i) Isothermal Expansion from A to B. Let us find out how much heat is drawn during this process and how much work it performs for the expansion. For the former, since for a reversible process $T\Delta S = Q$, we need to know how entropy changes with V for a fixed T. From the Sackur-Tetrode equation,

$$S = k_B N \left\{ \ln \left[\frac{V}{N} \left(\frac{mk_B T}{2\pi\hbar^2} \right)^{3/2} \right] + \frac{5}{2} \right\}, \tag{10.6}$$

we can see that the entropy change from A to B is

$$\Delta S = S_B - S_A = k_B N \ln \left(\frac{V_B}{V_A} \right), \tag{10.7}$$

and therefore the amount of heat drawn is

$$Q_h = k_B N T_h \ln \left(\frac{V_B}{V_A} \right). \tag{10.8}$$

To calculate the work performed, use the ideal gas equation of state $PV = Nk_B T$ to write

$$W_{AB} = - \int_{V_A}^{V_B} P dV = -Nk_B T_h \int_{V_A}^{V_B} \frac{dV}{V} = -Nk_B T_h \ln \frac{V_B}{V_A}. \tag{10.9}$$

(ii) Adiabatic Expansion from B to C. Since no heat may enter or leave, the expansion is an isentropic process. When the volume reaches V_C, the temperature falls to T_l. In between, the Sackur-Tetrode equation again says that the increasing volume and the decreasing temperature must do so following

$$VT^{3/2} = \text{constant along the path}, \tag{10.10}$$

which gives

$$V_B T_h^{3/2} = VT^{3/2} \quad \text{or} \quad T = V_B^{2/3} T_h V^{-2/3}. \tag{10.11}$$

Using the ideal gas equation of state $PV = Nk_B T$, we can then tell that the pressure falls with the increasing volume following

$$P = Nk_B T_h V_B^{2/3} V^{-5/3}, \tag{10.12}$$

whence the precipitously falling pattern depicted in the figure. The work performed is

$$W_{BC} = -\int_{V_B}^{V_C} p\,dV = -Nk_BT_hV_B^{2/3}\int_{V_B}^{V_C} V^{-5/3}dV = \frac{3}{2}Nk_B(T_l - T_h),$$

$$(10.13)$$

which says $W_{BC} = \Delta U = \frac{3}{2}Nk_B(T_l - T_h)$.

(iii) Isothermal Compression from C to D. Repeating the same argument for (i), we find

$$\Delta S = S_D - S_C = k_BN\ln\left(\frac{V_D}{V_C}\right) \qquad (10.14)$$

which gives

$$Q_l = k_BNT_l\ln\left(\frac{V_D}{V_C}\right) \qquad (10.15)$$

and

$$W_{CD} = -Nk_BT_l\ln\frac{V_D}{V_C}. \qquad (10.16)$$

Before we proceed for the next process, we need some juggling to do. The process from B to C is an isentropic process and therefore $V_BT_h^{3/2} = V_cT_l^{3/2}$. The process from D to A is also an isentropic process, and therefore $V_DT_l^{3/2} = V_AT_h^{3/2}$. Combining these two, we arrive at $V_D/V_C = V_A/V_B$. Thus we may write, for Q_l,

$$Q_l = k_BNT_l\ln\left(\frac{V_A}{V_B}\right). \qquad (10.17)$$

(iv) Adiabatic Compression from D to A. Following the same argument as in (ii), we find

$$W_{DA} = \frac{3}{2}Nk_B(T_h - T_l). \qquad (10.18)$$

Combining all the above results,

$$W = W_{AB} + W_{BC} + W_{CD} + W_{DA} = -Nk_B(T_h - T_l)\ln\frac{V_B}{V_A} \qquad (10.19)$$

and therefore

$$\eta = \frac{-W}{Q_h} = \frac{T_h - T_l}{T_l}, \qquad (10.20)$$

where $-W$ because all those works we calculated are works done to the working medium; the work done by the medium is $-W$. The efficiency is in agreement with Carnot.

We may get the same result by taking a shortcut:

$$\eta = \frac{W}{Q_h} = \frac{Q_h - Q_l}{Q_h} = \frac{T_h - T_l}{T_l}, \tag{10.21}$$

where we used Eq. (10.17) to obtain the same result. Remember that this efficiency is based on the assumption that we can carry out all the four processes reversibly. Moreover, we assumed that the working medium did not have to expend any energy to overcome any kind of friction or that no additional heat energy was added to the working medium due to internal friction. That is why the engine is an ideal engine.

Finally, we have studied an isothermal and reversible, and thus very slow, expansion and compression. In passing we should also mention at least a word on a fast irreversible expansion. Suppose that a cylinder is divided into two parts by a partition with a gas in one side and the other side in vacuum. If we make a hole in the partition allowing the gas to fill up the other side, that is an irreversible expansion. No work has been done by the gas nor to it and therefore the total internal energy remains the same after the sudden expansion. For ideal gases, therefore, the temperature remains the same but volume has increased. According to the Sackur-Tetrode equation, the only change is in entropy:

$$\Delta S = N k_B \ln \left(\frac{V_f}{V_i} \right) \tag{10.22}$$

where V_f and V_i are the final and initial volumes, respectively. Our ideal gas model works so well and can represent real gases quite well in all other cases, but unfortunately not in this case. The real gases behave differently; the temperature does change, which is called the Joule effect.

10.4 Relationships Between Thermodynamic Derivatives

To practice thermodynamics successfully, it is vital to know all the relationships between thermodynamic derivatives. And there are many. Here are several trickeries which anyone can practice to collect them. To save writings, I will limit to systems of fixed number of particles and thus μ and N will be omitted, but you may easily include them to collect more.

• Trick I. This is simple and straightforward. Start from the first law,

$$dU = TdS - PdV \qquad (10.23)$$

where

$$T = \left(\frac{\partial U}{\partial S}\right)_V, \quad \text{and} \quad -P = \left(\frac{\partial U}{\partial V}\right)_S. \qquad (10.24)$$

Taking cross derivatives, it follows that

$$\left(\frac{\partial T}{\partial V}\right)_S = \frac{\partial^2 U}{\partial V \partial S} = \frac{\partial^2 U}{\partial S \partial V} = -\left(\frac{\partial P}{\partial S}\right)_V \qquad (10.25)$$

whence

$$\left(\frac{\partial T}{\partial V}\right)_S = -\left(\frac{\partial P}{\partial S}\right)_V. \qquad (10.26)$$

Or write the first law in the form of

$$dS = \frac{1}{T}dU + \frac{P}{T}dV \qquad (10.27)$$

and then take cross derivatives to obtain

$$\frac{\partial}{\partial V}\left(\frac{1}{T}\right)_U = \frac{\partial}{\partial U}\left(\frac{P}{T}\right)_V. \qquad (10.28)$$

We may do the same with the Helmholtz free energy, Gibbs free energy, and the heat function to obtain

$$\left(\frac{\partial P}{\partial T}\right)_V = \left(\frac{\partial S}{\partial V}\right)_T, \qquad (10.29)$$

$$\left(\frac{\partial V}{\partial T}\right)_P = -\left(\frac{\partial S}{\partial P}\right)_T, \qquad (10.30)$$

$$\left(\frac{\partial T}{\partial P}\right)_S = -\left(\frac{\partial V}{\partial S}\right)_P. \qquad (10.31)$$

These are the most basic relationships called Maxwell's relationships. We only wrote down five of them, but if you include the μdN term in the first law, you can collect more.

• Trick II. The three variables P, T and V must satisfy an equation of state $\mathcal{G}(P, T, V) = 0$. So any one may be regarded as a function of the other two: $P = P(T, V)$, $T = T(P, V)$ and $V = V(T, P)$. We may then write

$$dP = \left(\frac{\partial P}{\partial T}\right)_V dT + \left(\frac{\partial P}{\partial v}\right)_T dV, \tag{10.32}$$

$$dT = \left(\frac{\partial T}{\partial P}\right)_V dP + \left(\frac{\partial T}{\partial v}\right)_P dV, \tag{10.33}$$

$$dV = \left(\frac{\partial V}{\partial T}\right)_P dT + \left(\frac{\partial V}{\partial P}\right)_T dP. \tag{10.34}$$

Now let me act drunk. I am going to substitute one of the above three into one of the other two, say, Eq. (10.32) into Eq. (10.34) to obtain

$$dV = \left[\left(\frac{\partial V}{\partial T}\right)_P + \left(\frac{\partial V}{\partial P}\right)_T \left(\frac{\partial P}{\partial T}\right)_V\right] dT + \left(\frac{\partial V}{\partial P}\right)_T \left(\frac{\partial P}{\partial V}\right)_T dV. \tag{10.35}$$

You may say that the best I can get out of this is $dV = dV$. Watch! the dT term must go, whence

$$\left(\frac{\partial V}{\partial T}\right)_P = -\left(\frac{\partial V}{\partial P}\right)_T \left(\frac{\partial P}{\partial T}\right)_V \tag{10.36}$$

or to symmetrize it,

$$\left(\frac{\partial T}{\partial V}\right)_P \left(\frac{\partial V}{\partial P}\right)_T \left(\frac{\partial P}{\partial T}\right)_V = -1. \tag{10.37}$$

Notice the cyclic pattern which makes it easy to remember.

We may play the same game but with one of the three variables replaced by a new variable, say, entropy. Then entropy is regarded as a function of P and V, or $S = S(P, V)$. Inverting it, we also have $P = P(S, V)$ and $V = V(S, P)$. Repeating the above with these three, we obtain

$$\left(\frac{\partial P}{\partial V}\right)_S \left(\frac{\partial V}{\partial S}\right)_P \left(\frac{\partial S}{\partial P}\right)_V = -1. \tag{10.38}$$

Continue with U instead of S to obtain,

$$\left(\frac{\partial P}{\partial V}\right)_U \left(\frac{\partial V}{\partial U}\right)_P \left(\frac{\partial U}{\partial P}\right)_V = -1. \tag{10.39}$$

• Trick III. Entropy and internal energy may be regarded as a function of any two variables among V, T and P, the reason for not taking all three being that they must satisfy an equation of state. So consider the entropy. We may write:

$$dS = \left(\frac{\partial S}{\partial V}\right)_T dV + \left(\frac{\partial S}{\partial T}\right)_V dT, \tag{10.40}$$

$$dS = \left(\frac{\partial S}{\partial T}\right)_P dT + \left(\frac{\partial S}{\partial P}\right)_T dP, \tag{10.41}$$

$$dS = \left(\frac{\partial S}{\partial V}\right)_P dV + \left(\frac{\partial S}{\partial P}\right)_V dP. \tag{10.42}$$

Obviously, the entropy is regarded as a function of V and T in the first equation, and as a function of T and P in the second equation, and as a function of V and P in the third equation. We can switch from one to another by substituting any one of its two independent differentials with those of the other two. For example, we can switch from Eq. (10.40) to Eq. (10.41) by getting rid of dV in favor of dT and dP using Eq. (10.34). The result is

$$dS = \left[\left(\frac{\partial S}{\partial T}\right)_P + \left(\frac{\partial S}{\partial V}\right)_T \left(\frac{\partial V}{\partial T}\right)_P\right] dT + \left(\frac{\partial S}{\partial V}\right)_T \left(\frac{\partial V}{\partial P}\right)_T dP. \tag{10.43}$$

Comparing the result with Eq. (10.41), we find

$$\left(\frac{\partial S}{\partial T}\right)_P = \left(\frac{\partial S}{\partial T}\right)_V + \left(\frac{\partial S}{\partial V}\right)_T \left(\frac{\partial V}{\partial T}\right)_P. \tag{10.44}$$

That is how we hunt for relationships between all those partial derivatives in Eqs. (10.40) through (10.42).

Now let us pursue the same trick but so as to involve the internal energy U.

$$dU = \left(\frac{\partial U}{\partial T}\right)_V dT + \left(\frac{\partial U}{\partial V}\right)_T dV. \tag{10.45}$$

Consider the entropy

$$dS = \frac{1}{T} dU + \frac{P}{T} dV. \tag{10.46}$$

Substituting Eq. (10.45) into Eq. (10.46) we have

$$dS = \frac{1}{T} \left(\frac{\partial U}{\partial T}\right)_V dT + \left[\frac{P}{T} + \frac{1}{T} \left(\frac{\partial U}{\partial V}\right)_T\right] dV. \tag{10.47}$$

Taking Cross derivatives, we have

$$\frac{\partial}{\partial V}\left[\frac{1}{T}\left(\frac{\partial U}{\partial T}\right)_V\right]_T = \frac{\partial}{\partial V}\left[\frac{P}{T} + \frac{1}{T}\left(\frac{\partial U}{\partial V}\right)_T\right]_V. \qquad (10.48)$$

Carrying out the differentiations and canceling out the two second order cross derivatives,

$$T\left(\frac{\partial P}{\partial T}\right)_V = P + \left(\frac{\partial U}{\partial V}\right)_T \qquad (10.49)$$

which is another useful equation and in fact we will use it shortly. In the above, we chose (T, V) as independent variables. One may make other choices to find other similar relationships.

• Trick IV. If you practice the above a few times, it becomes clear that there is a shortcut. Choose any three variables and write the differential of any one in terms of those of the other two. For example,

$$dU = TdS - PdV. \qquad (10.50)$$

Divide the equation by, say, dV and impose any condition that seems desirable, say, that all three differentials must be such that the changes will keep the temperature unchanged. Then we have

$$\left(\frac{\partial U}{\partial V}\right)_T = T\left(\frac{\partial S}{\partial V}\right)_T - P \qquad (10.51)$$

which is Eq. (10.49).

As another example, take

$$dS = \left(\frac{\partial S}{\partial V}\right)_T dV + \left(\frac{\partial S}{\partial T}\right)_V dT. \qquad (10.52)$$

Divide the equation by dT imposing the condition that the three differentials must be such that the pressure will remain unchanged. This gives

$$\left(\frac{\partial S}{\partial T}\right)_P = \left(\frac{\partial S}{\partial V}\right)_T\left(\frac{\partial V}{\partial T}\right)_P + \left(\frac{\partial S}{\partial T}\right)_V \qquad (10.53)$$

which is Eq. (10.44).

• Trick V. Do not forget to use the chain relationships such as

$$\left(\frac{\partial S}{\partial P}\right)_T = \left(\frac{\partial S}{\partial V}\right)_T\left(\frac{\partial V}{\partial P}\right)_T. \qquad (10.54)$$

If you will try more with these trickeries, you can collect quite a few more yourself. They should be put on a list so that you can use them when you need. You may have an issue which calls for a particular derivative. That derivative as it stands may be very difficult to measure, but it may be transformed into other derivatives which may be much easier to measure or may be physically more revealing, or both. So, it should be handy to have them all on a list. If you tour through some condensed matter physics laboratories, you can see on a wall in front of someones desk a list of equations involving many partial derivatives. There is a good chance that it is their list.

10.5 Important Thermodynamic Derivatives and Their Mutual Relationships

There are several important derivatives which serve useful purposes. The first one is the coefficient of thermal expansion

$$\alpha = \frac{1}{V}\left(\frac{\partial V}{\partial T}\right)_P \tag{10.55}$$

which shows how the material expands when the temperature is raised under a fixed pressure.

The next are two heat capacities (specific heats) under two different conditions. They are defined by

$$C_V = \left(\frac{\partial U}{\partial T}\right)_V = T\left(\frac{\partial S}{\partial T}\right)_V, \tag{10.56}$$

$$C_V = \left(\frac{\partial U}{\partial T}\right)_V = T\left(\frac{\partial S}{\partial T}\right)_V. \tag{10.57}$$

To be introduced next are two compressibilities under two different conditions:

$$\kappa_S = -\frac{1}{V}\left(\frac{\partial V}{\partial P}\right)_S, \tag{10.58}$$

$$\kappa_T = -\frac{1}{V}\left(\frac{\partial V}{\partial P}\right)_T, \tag{10.59}$$

where κ_S is called the adiabatic compressibility and κ_T the isothermal compressibility.

It follows that

$$C_P - C_V = T \left(\frac{\partial S}{\partial V}\right)_T \left(\frac{\partial V}{\partial T}\right)_P ,$$

$$= T \left(\frac{\partial P}{\partial T}\right)_V \left(\frac{\partial V}{\partial T}\right)_P ,$$

$$= T \frac{-1}{(\partial T/\partial V)_P (\partial V/\partial P)_T} \left(\frac{\partial V}{\partial T}\right)_P . \tag{10.60}$$

Equation (10.44) was used for the first line; Eq. (10.29) for the second line; Eq. (10.37) for the third line. Referring to the definitions of α and κ_T, we finally find that

$$C_P - C_V = \frac{TV\alpha^2}{\kappa_T} . \tag{10.61}$$

One can express how the internal energy U changes with T using $dU = C_V(T)dT$. If one wishes to find out how the entropy changes with T, remember $dS = [C_V(T)/T]dT$. Representing the fluctuation of energy is C_V, not C_P. So, C_V is busy telling us about the inner workings of materials. On the other hand, C_P is easier to measure between the two, and most of the times measured specific heat is C_P. Eq. (10.61) is saying something, but I have no enlightening explanation about it.

I will leave it for your exercise to show that

$$\frac{\kappa_T}{\kappa_S} = \frac{C_P}{C_V} . \tag{10.62}$$

10.6 Thermodynamic Stability

Thermal systems suffer constant fluctuations, but they are never driven far away. We may even attempt to disrupt their equilibrium state, but they always find a way to a new equilibrium. As an example, let us consider the way energy is distributed in a system. It never happens that one part of the system gobbles up all the energy from the rest. If we add an excessive amount of energy to one part, then the added energy is immediately distributed to everywhere for a new equilibrium. The stability must be be built in the fundamental function $S(U, V, N)$ and all the energy-like functions. We wish to find out the characteristics of these functions which ensure the stability.

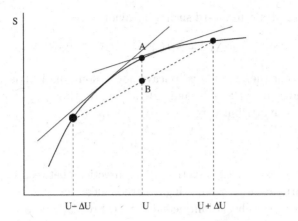

Fig. 10.3   The basic equation depicted here represents a stable system. The fact that point A is above point B reflects the inequality in Eq. (10.65). Notice that all the tangent lines are above the curve, which does not allow a common tangent line anywhere.

Divide the system into two equal parts. The two parts are now in equilibrium. Let the entropy of each part be $S(U, V)$ — we are again considering only systems of fixed number of particles. Should one part steal energy ΔU from the other part, then the function $S(U, V)$ must be shaped so as not to award such a behavior, i.e. the total entropy must then be less than the previous total $2S(U, V)$,

$$S(U + \Delta U, V) + S(U - \Delta U, V) \leq 2S(U, V). \tag{10.63}$$

This says that $S(U, V)$ is larger than the average of $S(U + \Delta U, V)$ and $S(U - \Delta, V)$ as depicted in Fig. 10.3, i.e. the function $S(U, V)$ must be of a concave-down shape everywhere along the U axis. Draw a tangential line anywhere on the $U(S, V)$ curve. The line is always below the curve no matter where the tangential line is drawn.

Expand the two terms on the left-hand side to second order. The first order terms are canceled out as do the two $S(U, V, N)$'s leaving only the second order terms. The final result is

$$S_{UU} \equiv \left(\frac{\partial^2 S}{\partial U^2} \right)_V \leq 0. \tag{10.64}$$

We have just explored the function $S(U, V)$ along the axis of U. Let us explore it along the direction of volume V. Can one part steal the others territory to swell while the other part just quietly shrinks? The function must be shaped not to award such a behavior, i.e.

$$S(U, V + \Delta V) + S(U, V - \Delta U, V) \leq 2S(U, V) \qquad (10.65)$$

which says that the function is concave down along the V axis as well. By Taylor expansion of the left-hand side as before, this geometrical characteristics may be attributed to

$$S_{VV} \leq 0. \qquad (10.66)$$

Now let us explore the function along directions between the two axes. The two parts are now attempting a pathological behavior both in U and V simultaneously. The stability condition still reads

$$S(U + \Delta U, V + \Delta V) + S(U - \Delta U, V - \Delta V) \leq 2S(U, V), \qquad (10.67)$$

which says that the entropy curve is concave-down. Now let us view $S(U, V)$ as a plane in three dimensions, U and V axes on the horizontal plane and the S axis along the vertical direction, the three findings above may be summarized by saying that the entropy plane $S(U, V)$ is concave everywhere. Draw a tangential plane anywhere. The plane always lies above the entropy plane.

Expand again the two terms to second order (see Eq. (2.3) in Chapter 2). The zeroth and first order terms are all canceled out as before and the final result is

$$\phi \equiv S_{UU} \Delta U^2 + 2S_{UV} \Delta U \Delta V + S_{VV} \Delta V^2 \leq 0. \qquad (10.68)$$

This may be written in the form of

$$\phi S_{UU} = (S_{UU} \Delta U + S_{UV} \Delta V)^2 + (S_{UU} S_{VV} - S_{UV}^2) \Delta V^2 \geq 0, \qquad (10.69)$$

which has to be positive because $S_{UU} \leq 0$. The condition that we seek is

$$S_{UU} S_{VV} - S_{UV}^2 \geq 0. \qquad (10.70)$$

Apparently this is what it takes to ensure that $S(U, V, N)$ is concave-down in every intermediate direction.

Let us reformulate the stability arguments using $U(S, V)$. Since U follows a minimum principle, stability requires

$$U(S + \Delta S, V + \Delta V) + U(S - \Delta S, V - \Delta V) \geq 2U(S, V), \qquad (10.71)$$

which gives three conditions:

$$\frac{\partial^2 U}{\partial S^2} = \frac{\partial T}{\partial S} \geq 0, \qquad \frac{\partial^2 U}{\partial V^2} = -\frac{\partial P}{\partial V} \geq 0, \qquad (10.72)$$

and

$$U_{SS}U_{VV} - U_{SV}^2 \geq 0. \qquad (10.73)$$

The function $U(S, V, N)$ must be convex in every direction. Thus the shapes of the two surfaces described by $S(U, V, N)$ and $U(S, V, N)$ are opposite.

Next let us find out the shapes of all other energy-like functions. We cannot continue the same argument here. For example, for $F(T, V)$, because T is an intensive variable, we cannot divide it up between the two subsystems as if we had to conserve it. But, since $F(T, V)$ is generated from $U(S, V, N)$ by a Legendre transformation, there must be a way to relate a daughter function's second derivative to its mother function's. To this end, consider a function $A(x, y)$ or $dA = u\, dx + v\, dy$. Transform it to $B(u, y) = A(x, y) - ux$. Take note that

$$u = \frac{\partial A}{\partial x}, \qquad x = -\frac{\partial B}{\partial u}, \qquad (10.74)$$

from which it follows that

$$\frac{\partial^2 B}{\partial^2 u} = \frac{\partial}{\partial u}(-x) = -\frac{1}{\partial u/\partial x} = -\frac{1}{(\partial^2 A/\partial x^2)}. \qquad (10.75)$$

It states that if the old function (A) is convex with respect to the old variable (x), the new function (B) is concave with respect to the new variable (u), and vice versa.

It then follows:

$$\left(\frac{\partial^2 F}{\partial T^2}\right)_V \leq 0, \qquad \left(\frac{\partial^2 F}{\partial V^2}\right)_T \geq 0, \qquad (10.76)$$

$$\left(\frac{\partial^2 H}{\partial S^2}\right)_P \geq 0, \qquad \left(\frac{\partial^2 H}{\partial P^2}\right)_S \leq 0, \qquad (10.77)$$

$$\left(\frac{\partial^2 G}{\partial T^2}\right)_P \leq 0, \qquad \left(\frac{\partial^2 G}{\partial P^2}\right)_T \leq 0. \qquad (10.78)$$

The energy and all the energy-like functions are therefore concave functions of their extensive variables and convex functions of their intensive variables.

The stability conditions are all given in such a straightforward form that they can easily be reduced to simple basic thermodynamic derivatives. Returning to Eq. (10.64), we find

$$S_{UU} = \left(\frac{\partial^2 S}{\partial U^2}\right)_{V,N} = \frac{\partial}{\partial U}\left(\frac{1}{T}\right)_{V,N} = -\frac{1}{T^2}\frac{1}{C_V} \leq 0, \qquad (10.79)$$

which gives

$$C_V \geq 0. \qquad (10.80)$$

This simple condition guarantees that the system will remain stable energy-wise. Thermal fluctuations could cause some inhomogeneous energy distribution, but as long as the specific heat is positive, the system will by itself return to a more normal distribution. It is interesting to ask: what would it be like if $C_V \leq 0$? Upon receiving energy, the receiver will lower its temperature, meaning more appetite for energy. As this hungry bum eats more and more, its appetite only increases. The negative specific heat also means that upon losing energy, the substance will want to lose even more! Then each of the two parts can do what it wants at the courtesy of the other. Such a system, if one ever existed, is doomed to vanish.

Similarly, the convexity of the Helmholtz potential with respect to volume V gives

$$\left(\frac{\partial^2 F}{\partial V^2}\right)_T = -\left(\frac{\partial P}{\partial V}\right)_T = \frac{1}{V\kappa_T} \geq 0, \qquad (10.81)$$

$$\kappa_T > 0. \qquad (10.82)$$

Again just imagine what it would be like if the isothermal compressibility were negative. Suppose that a fluctuation increased the volume of one part of the system at the expense of the remainder. The pressure then increases in the receiver while it decreases in the donor. The receiver wants more and more volume while the donor wants less and less. If such a system ever existed anywhere, it was doomed to vanish (or the shrinking part became a black hole and the expanding part a part of the expanding universe? Just a joke! It is intended to help you remember what a negative isothermal compressibility means.)

Referring to Eq. (10.61), the positive isothermal compressibility also means that

$$C_P \geq C_V \geq 0, \qquad (10.83)$$

which in turn means, from Eq. (10.62), that

$$\kappa_T \geq \kappa_S \geq 0. \qquad (10.84)$$

To conclude the chapter, I wish to tell you what I have just read from this week's Time magazine. This is a conversation between two Apollo lunar astronauts. One was overwhelmed to see the lunar surface through the window of the orbiting capsule, and said to the other "Where do you suppose a planet like this comes from? Do you suppose it broke away from the earth like a lot of people say?" The other answers "I ain't no cosmologist. I don't care nothing about that." The latter astronaut felt clearly that they were there at that moment for a very urgent task and it was no time to become emotional. Yes, the two astronauts successfully did their job and brought the astronauts who walked on the moon back to the earth. I am wondering if I may say that it was in such a highly pragmatic spirit that all those nineteenth century thermodynamicists did their work. Just reflect on what we are able to do thanks to their efforts. But we never hear a single fancy title we became accustomed to from more modern day sciences. Not even one. They were so focused on their urgent job that they did not have time to engage in fancy decoration.

Exercise. *Derive the stability condition including the particle number N and thus μ.*

Chapter 11

Energy Versus Free Energy

11.1 Introduction

To pursue thermal physics from a basic point of view, one only has to calculate the partition function which is a sum of Boltzmann factors which involve only energy, not free energy. Once the partition function is calculated, one can calculate free energy and all other thermodynamic quantities from it. In this approach, the key player is energy and the free energy minimum principle has no role to play, which is why the extremum principle of free energy is not even mentioned in many textbooks on statistical mechanics. Unfortunately, however, the partition function is in general very difficult to calculate. Under these circumstances, a wise alternate approach is to make a model for free energy and proceed from there, guided by the minimum free energy principle. Indeed, if many of our heroes did not take this approach, we would not be where we are today, and the title "thermal physics" would have never been coined. The difference between energy and free energy is that free energy takes into account the multiplicity of macrostates. The purpose of this chapter is to highlight how the multiplicity factor replaces energy with free energy.

In Sec. 11.2, we will construct macrostates in terms of energy and look for the probability of the system visiting each macrostate. When all the Boltzmann factors in each macrostate are added to take into account the multiplicity factor, the result looks like a Boltzmann factor but with the microstate energy replaced by a free energy. Landau makes a model of this free energy in his celebrated mean field theory.

In Sec. 11.3, we will start from the work-energy relationship that we have learned from mechanics and add to it the multiplicity factor. The multiplicity factor once again replaces energy with free energy and turns the work-energy relationship into one of thermodynamic work and free energy. This is the same issue that we pursued from the point of view of thermodynamics in Chapter 7: Do the free energies have any physical significance on their own or are they just a tool to be used for the minimum principle? The same issue is discussed here but from the point of view of statistical mechanics.

11.2 Boltzmann Factor for Macrostates and Macrostate Free Energy

We introduced the notion of free energy as a gauge by which systems in contact with a thermal reservoir can decide how much energy to claim to help maximize the total entropy of the combined system. To go to a déeper level of thermal physics, since these systems follow the modified postulate, it is desirable to see the notion of free energy from the standpoint of the modified postulate. According to the modified postulate, the probability of finding the system in microstate j is proportional to the Boltzmann factor, i.e.

$$\text{probability} \sim \exp(-E_j/k_B T). \tag{11.1}$$

The Boltzmann factor is a not-yet-normalized probability, but let us simply call it the "probability" without the qualifier. Their sum over all microstates, namely, the normalization factor

$$Z = \sum_j \exp(-E_j/k_B T) \tag{11.2}$$

is the partition function.

We now wish to construct a Boltzmann factor appropriate for macrostates, namely the probability of finding the system in a specified macrostate. Macrostates may be defined in many different ways. Let us first define it by specifying energy. In the sum in Eq. (11.2), there may be multiple terms corresponding to the same value of energy to account for possible degeneracies. For example, suppose that the first three microstates are degenerate, i.e. $E_1 = E_2 = E_3 \equiv U_1$. We lump the three microstates together to define macrostate 1. We do likewise to define macrostates 2, 3, 4, etc. We may then write

$$Z = \sum_j \Gamma(U_j) \exp(-U_j/k_B T), \tag{11.3}$$

where $\Gamma(U_j)$ is the degeneracy (multiplicity) corresponding to U_j. Absorbing Γ into the exponential,

$$Z = \sum_j Z_j = \sum_j \exp(-F_j/k_B T), \qquad (11.4)$$

where

$$F_j = U_j - Tk_B \ln \Gamma(U_j) = U_j - TS_j. \qquad (11.5)$$

The probability of finding the system in macrostate j, i.e. the probability of finding the system with energy U_j is given by

$$\text{Probability} \sim Z_j = \exp(-F_j/k_B T), \qquad (11.6)$$

which differs from the Boltzmann factor of microstates in that E_j has been replaced with F_j.

S_j, Z_j, and F_j are inter-related attributes of macrostate j. Since j may or may not be the label of the largest macrostate, S_j and F_j may or may not represent their respective equilibrium properties characteristic of the largest macrostate. When we wish to make this distinction, we may call Z_j the partial partition function because it is only one term in Eq. (11.4), S_j the macrostaste entropy and F_j the macrostate free energy. In the literature, such distinctions are not made. The implication is that the difference is obvious, and it is. For example, suppose that we are given free energy $F(M, T)$ for an Ising model. This cannot be the equilibrium free energy because if it were, M should be a function of T and therefore it has no business to be sitting there next to T. When free energy is given in this manner with a seemingly inappropriate variable, take it as a sign that what you are given is a macrostate free energy; in the above example M is there to specify the macrostate and therefore is an 'independent' variable. It is independent because we may wish to select the largest macrostate, the smallest macrostate or anything in between. All the free energies that carry the name Landau or Landau and Ginzburg in Sec. 8.6 and throughout Chapter 9 are all macrostate free energies.

From Eq. (11.6), it is apparent that **more probable macrostates are associated with a lower free energy, and less probable macrostates with a higher one.**

For isolated thermal systems, what matters for each macrostate is its size Γ_j. As we stated in Sec. 4.4, the role played by Γ for isolated systems is played by Z for systems in contact with a thermal reservoir. Here, the analogy is clearer. We may regard Z_j as representing the "size" of macrostate j.

A system in contact with a thermal reservoir visits all macrostates of different energies. How is it then that the most probable macrostate alone can represent the equilibrium state? Suppose that $j = 64$ is the most probable macrostate. The corresponding free energy F_{64} is the lowest that the free energy can be, i.e. the equilibrium free energy. Recall that F is an extensive quantity; it increases with N like $N\hat{f}$, where $\hat{f}$ is the free energy per molecule. Because N is so large, it is safe to approximate the partition function with Z_{64}, i.e.

$$Z \approx Z_{64} = \exp(-N\hat{f}_s/k_BT). \qquad (11.7)$$

For example, $\hat{f}_5$ may only be slightly larger than $\hat{f}_{64}$, but the factor N makes Z_5 look like an ant and Z_{65} a fully grown redwood tree. What if $\hat{f}_9$ is so close to $\hat{f}_{64}$ that F_9 is actually comparable to F_{64}? It could be, but do not worry about it. Suppose that there are not one but as many as N such macrostates. We should then instead write $Z \approx NZ_{64}$. But take the logarithm and see how much it adds to the free energy. The additional factor that we bring in to account for such macrostates is $\ln N$ which is negligible compared to what the largest state brings by itself, namely, $-N\hat{f}_s/k_BT$.

In the above analysis, we defined macrostates in terms of energy. We may define them in terms of any variable, say, A. The partial partition function corresponding to $A = a$, i.e. the probability of finding $A = a$ is

$$Z_a = \sum_i \Delta(A_i - a)\exp(-E_i/k_BT), \qquad (11.8)$$

where A_i is what A would be when the system is in microstate i, and $\Delta(x)$ is unity when $x = 0$ and zero when x is not zero. The corresponding free energy is given by

$$F_a = -k_BT\ln\sum_i \Delta(A_i - a)\exp(-E_i/k_BT). \qquad (11.9)$$

We shall return to this equation in Sec. 8.6 to learn Landau's mean field theory. In terms of F_a, Eq. (11.8) may be rewritten as

$$Z_a = \exp[-F_a/k_BT]. \qquad (11.10)$$

The relationship between the probability Z_a and the corresponding free energy F_a is worth repeating. **More probable macrostates are associated with lower free energy while less probable macrostates are associated with higher free energy.**

In view of the parallel between Eqs. (11.1) and (11.10), the equipartition theorem applies to free energy just as well as it does to energy. To give an example, suppose that we define macrostates with a space-dependent function, say, $A(\vec{r})$ or its Fourier transform $A(\vec{q})$. F should then be a functional of $A(\vec{q})$, i.e. a function of a function. Furthermore, suppose that the corresponding free energy is given by

$$F\{A(\vec{q})\} = \sum_{\vec{q}} \frac{1}{2} K_{\vec{q}} |A(\vec{q})|^2, \qquad (11.11)$$

where $|A(\vec{q})|^2$ can be anything from zero to infinity. The equipartition theorem gives us

$$<|A(\vec{q})|^2> = \frac{k_B T}{K_{\vec{q}}}, \qquad (11.12)$$

a result which appears quite often in advanced literature.

11.3 Thermodynamic Work and Thermodynamic Potential

We have just discussed how the multiplicity or degeneracy turns energy into free energy in the Boltzmann factor. From the same point of view of statistical mechanics, we now wish to find out if the free energy satisfies the same type of work-energy relationship that prevails in mechanical systems. It was shown in Chapter 7 that any change in free energies may be related to the work delivered to a work source. If this relationship is to be the same as in mechanical systems, all changes have to be reversible. This condition makes the issue murky. If the temperature or pressure remains fixed, there is only one equilibrium state and there are no reversible changes to speak of. In addition, the work that a thermal system does to or by the volume reservoir was not included in the work. So, the work source was assumed to be exchanging energy with the system via a different mode of work, a type of work different from the expansion work PdV. To be free from these two restrictions, we shall let the pressure change slowly so that the system will be in equilibrium for each value of pressure, and therefore we shall have plenty of reversible changes in free energy to speak of. The work source shall exchange energy with the system via the mode of PdV type only, i.e. the work that the system can receive from the the volume reservoir or deliver to it.

When we perform work on a thermal system, say a gas, we cannot guarantee that our work will be completely consumed to increase their energy (by altering its microstate energies). Thermal systems are driven to increase their entropy or to decrease their free energy. If our action is contrary to this prevailing tendency of nature, our work may have to be consumed at least in part to overpower that tendency. To check this possibility on a more rigorous basis than we argued in Sec. 3.13, let us assume for a moment that the gas is stuck in one microstate i of energy E_i. In order to change the volume by ΔV, the amount of work to be done is

$$\Delta W = \frac{\partial E_i(V)}{\partial V} \Delta V. \tag{11.13}$$

If our assumption that the gas is stuck in one microstate were true, the gas pressure $\partial E_i/\partial V$ would be a mechanical property. We know that the gas does not stay stuck in one microstate. This is the difference between mechanical systems and thermal systems. For mechanical systems, multiplicity plays no role because there is no pertinent degeneracy to speak of. By contrast, degeneracy or the multiplicity of macrostates is in fact the essential element of thermal systems. Let us take into account this crucial element and recognize that the gas visits all of its microstates according to the Boltzmann factor. Taking the average of Eq. (11.13) over all microstates, we arrive at

$$\Delta W = \frac{1}{Z} \sum_i \frac{\partial E_i(V)}{\partial V} \exp[-\beta E_i(V)] \Delta V, \tag{11.14}$$

Using the trick of Eq. (2.71), we can write this entirely in terms of Z, namely,

$$\Delta W = -\frac{1}{\beta} \frac{1}{Z} \frac{\partial Z}{\partial V} \Delta V = -\frac{1}{\beta} \frac{\partial \ln Z}{\partial V} \Delta V. \tag{11.15}$$

Upon substituting $Z = \exp(-\beta F)$, we arrive at

$$\Delta W = \frac{\partial F}{\partial V} \Delta V. \tag{11.16}$$

In comparison with Eq. (11.13), it is notable the microstate energy has been replaced with free energy. Denoting the initial volume with V_1 and the final volume with V_2, we may write

$$\Delta W = F(V_2) - F(V_1).\tag{11.17}$$

The work is not consumed just to increase its energy. It is consumed to increase its free energy, and therefore the work may be called thermodynamic work and $-\partial F/\partial V = P$, the force against which we push the piston, may be called a thermodynamic force, and F a thermodynamic potential. In fact, since the temperature remains fixed (assuming that we pushed the piston very slowly), there is actually no change in energy and the change in free energy is entirely in entropy. To check this, use Eqs. (4.63) and (4.65) to obtain

$$\Delta W = -k_B T N [\ln V_2 - \ln V_1] = -k_B T (S_2 - S_1),\tag{11.18}$$

which confirms what we suspected. It takes work to increase free energy or to decrease entropy. If we wish to act against the "desire" of nature, we have to sweat. This is what we claimed in Sec. 3.13. The restoring force is entirely driven by entropy.

We may make Eq. (11.16) look just like an ordinary mechanical potential energy. Referring to Fig. 11.1, since $V = Ax$, the work that the gas does to the piston is given by

$$\Delta W_{\text{gas}\to\text{piston}} = -\frac{\partial F}{\partial x}\Delta x,\tag{11.19}$$

where the restoring force is

$$-\frac{\partial F}{\partial x} = \frac{\partial F}{\partial V}\frac{\partial V}{\partial x} = -k_B T N \frac{1}{x}.\tag{11.20}$$

The restoring force of the gas has a potential energy to associate with it, namely,

$$U_{\text{eff}}(x) = -k_B T N \ln x.\tag{11.21}$$

Written in this form, the free energy should be regarded as a thermodynamic potential.

Let us compare this to the way we describe a mechanical system, say, a rock of mass m. The earth pulls the rock with a force of mg. We describe this force as $-\partial mgz/\partial z$ where mgz is the gravitational potential energy. Personalizing the rock as we do for thermal systems, the rock really really wants to decrease its mgz. It takes work to move a rock from the bottom of a hill to the top against this nature's tendency. Once at the top, the rock is unstable because it really wants to come down. We have all seen rocks sliding down hills but none spontaneously sliding up. Similarly, try to increase the thermodynamic potential of a gas. It costs work as we have seen a moment ago. Once placed at a high thermodynamic potential, it is unstable because it really wants to decrease its thermodynamic potential.

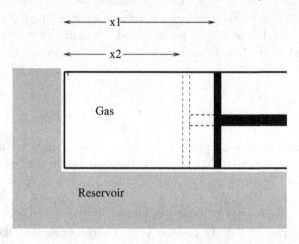

Fig. 11.1   A gas confined to a cylinder by a piston.

Let us generalize Eq. (11.16) where V is just one of several variables that determine the microstate energy E for molecular systems. Call such a parameter of other systems X. The amount of work that we have to do to change X by ΔX is given by

$$\Delta W = \frac{\partial F(X)}{\partial X} \Delta X, \qquad (11.22)$$

where $-\partial F(X)/\partial X$ may be regarded as the corresponding thermodynamic force.

11.4   Is Some Energy Really Free?

The answer to this question is clearly no, but it does appear as if we are getting some thing out of nothing for free. The fact that ΔF has no energy component (at a fixed temperature) is bewildering. if one tries to focus on where the energy resides.

To see this, push the piston from $x = x_1$ to $x = x_2$, as shown in Fig. 11.1. Do this very slowly to ensure that the process is isothermal, that is, to ensure that the gas remains in equilibrium with the reservoir at all times. By pushing the piston, we will have performed work in the amount of

$$\Delta U = U_{\text{eff}}(x_2) - U_{\text{eff}}(x_1) \qquad (11.23)$$

to the gas. But the gas has no more energy at the end than it had at the beginning because the total energy U does not depend on volume. What happened to the work? It has been transferred to the reservoir. By the way, the process does not have to be isothermal. If we perform the action quickly, the gas will have more energy at the end than at the beginning with increased temperature, but soon the energy will be transferred to the reservoir and the gas temperature will return to the reservoir temperature T. It is then interesting to ask: can the disposed energy be recovered as useful work for us? The answer is yes. If we allow the gas to push the piston back to x_1, it will perform exactly the same amount of work performed on it earlier, namely, ΔU, but again oddly enough, the gas has no less energy at the end of the expansion than it had at the beginning. At whose expense did the gas do the work if not its own? The disposed energy has been brought back from the reservoir. Here again the process does not have to be isothermal. If the gas is allowed to expand quickly, it will have less energy at the end than at the beginning of the expansion with a lower temperature, but soon the gas will replenish its depleted energy from the reservoir and the temperature will return back to T.

The relationship between the reservoir and the gas appears to be like that between a bank (reservoir) and a customer (system) who enjoys complete trust from his bank and never tries to take advantage of it. When the customer gets some extra money that he does not need, he deposits it in the bank, but the bank makes no record of the deposit. When the customer needs the money to pay a bill, he can go ahead and pay it out of his pocket because he knows that the bank will reimburse it without asking for any record. To a third party who does not understand the unusual business practice between the two parties, it would appear as if money were coming for *free* out of nowhere. But nothing in life is free. Neither the bank nor the customer gets anything for free.

Another good example is the rubber bands which we studied in Chapter 4 with an ideal rod model. Remaining in the same model, Eq. (3.129) is entirely a thermodynamic force. The corresponding potential energy is

$$U_{\text{eff}}(L) = k_B T \frac{L^2}{2Ns^2} \tag{11.24}$$

where N is the number of rods, s is the length of the unit rods, and L is the length of the rubber band. When the rubber band is stretched slowly, the work we do turns into a potential energy, but where does it reside? The energy is actually in the air. Again if we stretch it slowly, the air is not any warmer than before, and yet we are assured that it is there and ready to return and perform work for us when we want it.

Chapter 12

Phases and Phase Transitions

12.1 Introduction

When the temperature is high, the constituent particles act independently and freely, and therefore they can be treated as a collection of ideal gas molecules, ideal spins or ideal rods, etc. Alas, the gas molecules actually interact with one another as do the magnetic spins, and the interactions can no longer be ignored when the temperature is not sufficiently high. Those molecules come close to each other and turn into a different phase, namely liquid. We will not study interacting molecules *per se*, but we will find a way to take a glimpse into the thermal world of interacting molecules and observe how the free energy is altered by the molecular interactions. We will then observe how the altered free energy alters molecular behavior. Here, molecules are effectively cornered so that the only way to minimize the free energy is to change their basic structure. Yes, they decide that they are better off by coming closer to each other and to condense into a liquid. They are just like school children running around on their play ground. When the bell rings, they too know that they are better off by stop running and come to their classrooms. What a turnaround! That is also the trend for magnetic spins. They too interact with each other, and when the temperature is low, they act together and orient in the same direction.

Such an orderly behavior at low temperatures, as opposed to the random and disorderly behavior at high temperatures, is our theme from now on. The same maximum entropy or the minimum free energy principle apply, but under the changed condition, the orderly behavior is the best way to maximize their entropy or to minimize their free energy.

12.2   Liquid State and Solid State

When we studied the gas phase with the ideal gas model, the spatial part of the multiplicity had nothing but the incredibly simple V^N factor. The real atoms and molecules do interact with each other, and if we wish to take the interactions into account, we no longer have that kind of simplicity. Using the idea of Eq. (3.14) in Chapter 3, the partition function of classical interacting particles is given by

$$Z = \frac{1}{N!h^{3N}} \int \cdots \int \exp\left(-\sum \frac{p_i^2}{2mk_BT}\right) \exp\left(-\frac{U_{int}\{r\}}{k_BT}\right) d\{p\}\, d\{r\}$$

(12.1)

where

$$U_{int}\{r\} = \sum_{i<j} \phi(r_{ij})$$

(12.2)

represents the interaction energy between particles; $\{r\} = (\vec{r}_1, \vec{r}_2 \cdots \vec{r}_N)$ denotes the particle's positional configuration; $d\{r\}$ and $d\{p\}$ represent the spatial part and the momentum part of the differential volume element of the phase space, respectively, i.e. $d\{p\} = d\vec{p}_1 d\vec{p}_2 \cdots d\vec{p}_N$ and $d\{r\} = d\vec{r}_1 d\vec{r}_2 \cdots d\vec{r}_N$; $d\vec{r}_i$ is a differential volume element taken around $\vec{r}_i$. The differential phase space volume has been divided by h^{3N} according to the rule given by Eq. (3.14) and $N!$ is there to correct the over counting error.

In spite of all the difficulties that we will have to deal with due to the particle interactions, we should be glad that the momentum part is still completely factorized from the spatial part, and therefore, the arguments that we did in Chapter 4 for the equipartition theorem still holds but for the kinetic energy, not for the total energy. Thus we may still say that

$$<\text{Kinetic Energy}> = N\frac{3}{2}k_BT.$$

(12.3)

Let us do what we can for the partition function. The momentum part of the integrations may be carried out in the same way as before to give

$$Z = \frac{1}{N!}\left(\frac{1}{\Lambda^3}\right)^N Q,$$

(12.4)

where Q is the remaining spatial part,

$$Q = \int \cdots \int \exp\left(-\frac{U_{int}\{N\}}{k_BT}\right) d\vec{r}_1 d\vec{r}_2 \cdots d\vec{r}_N,$$

(12.5)

which is called the configurational integral.

Examine Q. The contribution of a given particle is different depending on where all the other particles are. Put it another way, each particle knows where other particles are, and therefore, the particles now take up their positions in a correlated fashion or in an orderly fashion; they no longer say that 'I shall be wherever I want regardless of others'. The consequent positional order may be captured by a very important correlation function called the radial distribution function. Let me do some preparation for its introduction.

Take a good look at Q. It starts with the potential energy part of the Boltzmann factor $\exp(U_{int}\{r\})$ which may be regarded as the not-yet-normalized probability distribution for the spatial configuration $\{r\}$. Then we are integrating it over all configurations. Now let us ask: what is the probability of finding the first particle in $d\vec{r}_1$ and the second particle in $d\vec{r}_2$, etc.? The answer is

$$P^{(N)}(\vec{r}_1, \vec{r}_2, \cdots \vec{r}_N)d\vec{r}_1 d\vec{r}_2 \cdots d\vec{r}_N = \frac{1}{Q}\exp(-U_{int}\{N\}/k_B T)d\vec{r}_1 d\vec{r}_2 \cdots d\vec{r}_N.$$

(12.6)

Now let us say that we only care about the first two particles and not the rest. The probability of finding the first particle in $d\vec{r}_1$ and the second particle in $d\vec{r}_2$ regardless of the rest is

$$P^{(2)}(\vec{r}_1, \vec{r}_2)d\vec{r}_1 d\vec{r}_2 = \frac{1}{Q}d\vec{r}_1 d\vec{r}_2 \int \cdots \int \exp(-U_{int}\{N\}/k_B T)d\vec{r}_3 d\vec{r}_4 \cdots d\vec{r}_N.$$

(12.7)

We shall now demand less and say that what we called the first particle can actually be any particle and likewise for the second particle. The probability of finding any particle in $d\vec{r}_1$ and any particle in $d\vec{r}_2$ is given by

$$\rho^{(2)}(\vec{r}_1, \vec{r}_2) = N(N-1)P^{(2)}d\vec{r}_1 d\vec{r}_2 \equiv n^2 g(\vec{r}_1, \vec{r}_2)d\vec{r}_1 d\vec{r}_2 \qquad (12.8)$$

where we have N because the first particle can be any one of the N particles and the second one any of the remaining $(N-1)$ particles; $n = N/V$ is the average particle number density, and $g(\vec{r}_1, \vec{r}_2)$ is called the pair correlation function and, is given by, from Eqs. (12.8) and (12.7),

$$g(\vec{r}_1, \vec{r}_2) = \frac{V^2}{N^2}N(N-1)\frac{1}{Q}\int \cdots \int \exp(-U_{int}/k_B T\{N\})d\vec{r}_3 d\vec{r}_4 \cdots d\vec{r}_N$$

(12.9)

which makes it clear that it should be normalized as

$$\frac{1}{V^2}\int g(\vec{r}_1, \vec{r}_2)d\vec{r}_1 d\vec{r}_2 = 1 + O\left(\frac{1}{N}\right).$$

(12.10)

For a homogeneous fluid, the pair correlation function should depend only on the relative distance $r_{12} = |\vec{r}_1 - \vec{r}_2|$. If we write $\vec{r}_2 = \vec{r}_1 + \vec{r}_{12}$, the pair correlation function should depend on r_{12} only. The integration over $\vec{r}_1$ is then trivially carried out to give V. The remaining integral gives

$$\frac{1}{V} \int g(r_{12}) d\vec{r}_{12} = 1 + O\left(\frac{1}{N}\right). \tag{12.11}$$

This allows us to say something very useful about the correlation function, but let us arrive at it more cautiously. To that end, ask: what is the probability to find a specific particle, say, particle 1 to be found in $d\vec{r}_1$ *and* any other particle in $d\vec{r}_2$? The answer is:

$$(N-1)P^{(2)}(\vec{r}_1, \vec{r}_2)d\vec{r}_1 d\vec{r}_2 = (N-1)\frac{n^2}{N(N-1)}g(r_{12})d\vec{r}_1 d\vec{r}_{12}$$

$$= \frac{n^2}{N}g(r_{12})d\vec{r}_1 d\vec{r}_{12} \tag{12.12}$$

where Eq. (12.8) has been used. This is a joint probability but the two events are not independent of each other, and therefore we cannot expect the correlation function to be a product of two parts. In fact the correlation function has no dependence on $\vec{r}_1$. So, it is wise not to ask anything about the first particle. To that end, we now ask: if the first particle is known to be at $\vec{r}_1$, what is the probability of finding any particle in $d\vec{r}_{12}$? That probability must be

$$\text{constant} \times g(r_{12})d\vec{r}_{12}. \tag{12.13}$$

Integrate this out over all $d\vec{r}_{12}$

$$\text{constant} \times \int g(r_{12})d\vec{r}_{12} = N - 1 \tag{12.14}$$

because the integral should encounter all the other $N - 1$ particles. But the integral is, from Eq. (12.11), equal to V. Thus, the constant is approximately n, and the sought-after probability is $ng(r_{12})d\vec{r}_{12}$, or putting the first particle at the coordinate origin,

$$ng(r)d\vec{r}. \tag{12.15}$$

Written in this form, $g(r)$ is called the radial distribution function. It is a correlation function. Given that a particle is located at the origin, it shows how the rest of the particles recognize its presence. Thus, the local density changes with the distance following

$$n(r) = g(r)n, \qquad (12.16)$$

where n is again the global average number density.

Suppose that we go to somebody's house for a party and find a friend standing at a certain spot. What do we do? We do not go to him and knock him off to claim his spot. We go near to that friend but not too close. The next guest then comes and joins the group similarly, and the group grows larger and larger. The first arrived guest chose that particular spot for no compelling reason, but the rest of the all the guests position themselves in correlation to that spot. But for the person who arrives fifth or tenth, there is no way to tell who arrived there first. This is depicted in Fig. 12.1. Indeed, if the gas phase is compared to a group of wild birds who fly around randomly irrespective of the rest, the highly civilized partying guests are a good representation of the liquid phase.

The positional order that we anticipate from the interactions should be manifest in $g(r)$. A typical liquid state radial distribution is as depicted in Fig. 12.1. First see (a). It shows that the molecules are ordered enough around the shaded first molecule to speak of first neighbor, second neighbor, and somewhat poorly third neighbor, after which there is no order. The spherical particles seem to be trying to come together close to each other enough to pack up the space, but that is only partially successful; some neighboring spheres touch each other while others are away. As a result, the particles around the central one has some short-range positional order, and the resultant radial distribution function is as depicted in (b). If the spherical particles packed the space more perfectly to make a triangular crystal, the radial distribution would make spikes at $r = n\sigma$ for all n. What about the radial distribution in the gas phase? For an ideal gas, it will be shown shortly that $g(r) = 1$; there is no correlation. So, the liquid state is between the well-ordered solid phase and badly disordered gas phase, which is why it is more difficult to study than the other two. We will return to this point after learning more about different phases and why matters choose to be in different phases under different conditions.

To better understand the radial correlation, we may rewrite Eq. (12.14) as

$$n \int g(r)d\vec{r} \approx N - 1 \qquad (12.17)$$

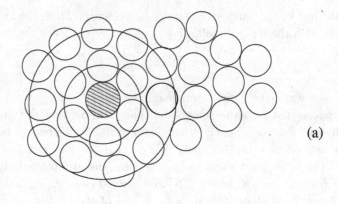

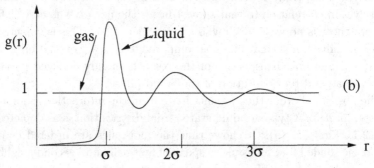

Fig. 12.1   A radial distribution.

or

$$n \int [g(r) - 1] \, d\vec{r} + 1 \approx 0. \tag{12.18}$$

The reader wondered why I even kept on using the approximation sign for neglecting 1 from $N-1$, which we did routinely before but without using the approximate sign. As is shown by the fact that $ng(r)$ is the local density, the above equation addresses the question of how the local particle numbers are fluctuating. We stayed in the above in the N-fixed canonical ensemble formalism, but in order to account for the particle number fluctuations correctly, we should have used the grand Canonical formalism. If we do, it turns out that

$$n \int [g(r) - 1] \, d\vec{r} + 1 = n k_B T K_T \ll 1. \tag{12.19}$$

The fact that the isothermal compressibility is involved here makes a good sense, but its derivation is too long and involved for us.

Finally, let us find out what $g(r)$ is for ideal gases. The configurational integrations are carried out trivially to give $Q = V^N$. The integrations in Eq. (12.9) are similarly carried out to give

$$g(r) = \frac{V^2}{N^2} N(N-1) V^{N-2} \frac{1}{V^N} = 1. \qquad (12.20)$$

If the temperature is lowered further, most liquids turn into a solid with a crystalline structure which possesses the most ubiquitous long range order. Pick any lattice site and walk in any direction. Thanks to the repeating pattern, you can predict when you will encounter a lattice site. The long range order extends as far away as the sample size allows.

The radial distribution function may be measured in laboratory by X-ray scattering. Since it probes the structure of matter in liquid or solid phase, one may wonder why it is not called the structure function. That title is reserved for its Fourier transform,

$$S(k) = 1 + n \int \exp(i\vec{k} \cdot \vec{r}) \, (g(r) - 1) \, d\vec{r}, \qquad (12.21)$$

which is called the structure factor.

The experiment works like this. Send an X-ray beam or a beam of neutrons into a sample. The sample scatters the beam into various directions. Let the incoming beam's wave vector be q_0. For x-ray, $q_0 = 2\pi/\lambda$ where λ is the wavelength. For neutron beam, $q_0 = p/\hbar$ where p is the momentum of each neutron. The intensity of the scattered beam is different depending on the angle of scattering and thus on the change in the wave vector. If the angle of scattering is θ, the change in wave vector is

$$\Delta q = 2q_0 \sin(\theta/2) \qquad (12.22)$$

and the intensity of the scattered beam into this scattering angle is

$$I(\Delta q) = I_{id} S(\Delta q) \qquad (12.23)$$

where I_{id} is what the scattered intensity would be if the sample were an ideal gas, i.e. $g(r) = 1$. Once the structure factor is measured in this way, an inverse Fourier transformation will bring to light the radial distribution function $g(r)$.

12.3 Phase Diagrams

Figure 12.2 shows the phase diagram of a typical substance consisting of a single component. The substance can exist in three different phases. In the low temperature and high pressure region, the substance is a solid (S). In the low pressure and high temperature region, it is a gas (G). In the intermediate region of p and T, it is a liquid (L). The sharp lines dividing different phases are called coexistence curves. When the system is placed exactly on these curves, the neighboring phases coexist. The coexistence curve of L and S is called the melting curve. The coexistence curve of G and L is called the vapor pressure curve. The coexistence curve between solid and gas is called the sublimation curve.

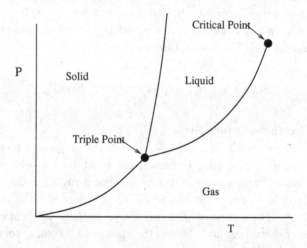

Fig. 12.2   A schematic phase diagram of a normal substance.

The fact that different phases are divided by a sharp line is remarkable because it means that even a small amount of change in temperature or pressure across the coexistence lines can drive the substance from one phase into another. Since we know that the minimum free energy principle is the lone guiding principle for the molecules, this means that the free energy landscape changes completely upon crossing the lines, forcing the molecules to change the way they aggregate, namely, their density. Such an abrupt transition is called a first-order phase transition; it occurs when we cross a coexistence line. When we cross a coexistence line from one phase to another, because molecules aggregate differently in the two phases requiring different amount of energy, we cannot just cross the line; we have to provide or remove the energy difference as needed. The energy difference is called the latent heat.

All three coexistence curves meet at the triple point which means that the three phases can coexist there. The vapor pressure curve ends at a point called the critical point, (P_C, T_C). As one moves along the vapor pressure curve toward the critical point, the distinction between the liquid phase on one side and the gas phase on the other gradually decreases and finally disappears at the critical point. The phase transition that occurs exactly at the critical point but by changing T through T_C is called a second-order phase transition and all phenomena taking place near the critical point are called critical phenomena. Unlike in first-order transitions, the transition is continuous and does not require any latent heat.

The two types of phase transitions differ in a more important way. Second-order phase transitions are "stormy" in the sense that all the response functions diverge at $T = T_C$! This means large fluctuations in extensive quantities. The fluctuations grow larger and larger as T approaches T_C from either direction as if to warn us that a transition is forthcoming, and finally diverge at $T = T_C$. Ironically, the abrupt first-order phase transitions have no such warning signs.

We have a fairly good intuitive idea what molecules and atoms are doing in each phase. In the gas phase, they are moving around randomly and individually. In the solid phase, they are highly organized in the way they arrange themselves; they form various lattice structures. In the liquid phase, molecules are doing something in between, which is why the liquid state is more difficult to study than the other two states.

In some substances, molecules and atoms have some additional characteristic properties which lead to another type of phases. Two good examples are magnetic materials and liquid crystals. In magnetic materials, the atoms have magnetic dipole moments which give rise to magnetism. Depending on how the dipole moments are oriented with respect to each other, their magnetic properties are different and hence have different magnetic phases. In the case of liquid crystals, the molecules have various distinctive geometrical shapes. In some liquid crystals they look like a plate and in some others they look like a rod. Not surprisingly, the positional entropy is quite different depending on how they are oriented with respect to each other. The different orientational orders lead to different phases in which the molecules exhibit different electro-optical properties.

12.4   Phase Rule

We can learn a great deal about the gross structure of phase diagrams, including some crucial information, by merely stating that the 'unlivability' conditions posed by coexisting phases are all the same. To be most general, suppose that n different species (components) of particles are coexisting in r different phases. If different phases coexist, it means that those different phases pose the same 'unlivability' to particles of each species. So representing the species as subscript and the phases as superscript,

$$\mu_1^I = \mu_1^{II} = \cdots = \mu_1^r$$
$$\mu_2^I = \mu_2^{II} = \cdots = \mu_2^r$$
$$\vdots$$
$$\mu_n^I = \mu_n^{II} = \cdots = \mu_n^r. \qquad (12.24)$$

How many independent variables do we need to specify each chemical potential? From the extensivity of the Gibbs free energy, we learned it earlier: $\mu_1^I = \mu_1^I(P, T, c_2, c_3, \cdots, c_n)$ where $c_2 = N_2/(N_1 + N_2 + N_3 + \cdots + N_n)$, etc. So each chemical potential needs, in addition to P and T, $(n-1)$ fractions. So the answer is: $2 + r(n-1)$. To determine these unknowns, how many equations do we have? The answer is: $(r-1)n$. The number of equations should be equal to or less than the number of unknowns, which gives

$$. \quad (r-1)n \le 2 + r(n-1), \qquad \text{or} \qquad r \le n+2. \qquad (12.25)$$

This is called Gibb's phase rule. For $n = 1$, the maximum number of phases that can coexist is 3 in agreement with Fig. 12.2. In this case, since the number of unknowns is equal to the number of the demanding equations, P and T are determined to specific values. In other words, the coexistence may occur only at the triple point. Should we go back to Eq. (12.24) and demand the coexistence of any two phases as we will do in the next section, the number of equation is less than that of the unknowns and therefore the equation may be satisfied by an infinite number of combinations of P and T, whence the coexistence lines.

12.5   Coexistence Curves

The fact that molecules can coexist in two different phases, say 1 and 2, means that a molecule in phase 1 has no compelling reason to migrate to join those in phase 2, and vice versa. The "unlivability" must be the same in both phases:

$$\mu_1(P,T) = \mu_2(P,T) \,. \tag{12.26}$$

Under this condition, the two phases are in a state of diffusive equilibrium; there are as many molecules migrating from phase 1 to phase 2 as there are migrating from phase 2 to phase 1. Now consider another adjacent point on the same coexistence line at $(P+dP, T+dT)$. The diffusive equilibrium exists there as well, whence

$$\mu_1(P + dP, T + dT) = \mu_2(P + dP, T + dT) \,. \tag{12.27}$$

To prepare for the next step, since $\mu = G/N$, write Eqs. (12.26) and (12.27) in the form

$$\left(\frac{\partial \mu}{\partial T}\right)_P = -\frac{S}{N} = -s \,, \tag{12.28}$$

$$\left(\frac{\partial \mu}{\partial P}\right)_T = -\frac{V}{N} = -v \,. \tag{12.29}$$

Now expand Eq. (12.27) to first order in dP and dT. Remembering Eq. (12.26), the zeroth-order terms on the left and right side cancel out. Separate the remaining terms involving dP from those involving dT to obtain

$$\frac{dP}{dT} = \frac{s_1 - s_2}{v_1 - v_2} \,, \tag{12.30}$$

which is called the Clausius–Clapeyron equation.

The slope dP/dT of a coexistence line holds a rather precious piece of information. It tells us how entropy and volume change upon crossing the coexistence lines. Since the change in entropy is divided by the change in volume, the information is not as complete as it would be if the two came separately. But considering how little it costs to map the coexistence lines, what we get from Eq. (12.30) is a bargain. At a session of an American Physical Society meeting sometime ago, a speaker opened up his talk with the question "You have seen the Clausius–Clapeyron equation already six times. Would you like to see it once more?" That demonstrates how busy the equation was; it was during the period when the low temperature physics of helium was extremely intense. The equation is still busy.

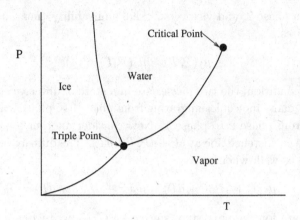

Fig. 12.3   A schematic phase diagram of water.

For most normal substances, the slope of the melting curve is positive. The liquid phase requires more volume and more entropy than the solid phase. Water is not normal. The slope of the melting curve of water is negative, as shown in Fig. 12.3. The liquid phase (water) requires more entropy than the solid phase (ice) as in all normal substances, but the volume requirement is the other way around; the liquid phase (water) requires less volume than the solid phase (ice), which makes the slope negative.

Figure 12.4 shows the phase diagram of helium-3. The coexistence line between the solid phase and the liquid phase has a negative slope at very low temperatures. In this case, the liquid phase requires more volume than the solid phase does as in normal substances, but the liquid phase has less entropy than the solid phase. In other words, the liquid is more ordered than the solid, and therefore it takes heat to change the liquid to a solid! The molecules do move around in the liquid phase, but not individually as they do in the normal liquid phase. They behave collectively in much the same way as the members of a flawless marching band do on a football field to execute a routine which allows little or no room for error. If you wish to break them up, give them a dose of highly disruptive heat. What do they do then? The atoms crystallize and all the energy is used to vibrate, which turns out to give them more entropy.

Exercise. *Calculate the solid-gas coexistence curve using the Einstein model for a solid phase and the ideal gas model for a gas phase. For the solid phase, since it takes up only a small amount of volume, ignore the PV term in $G = N\mu = F + PV$.*

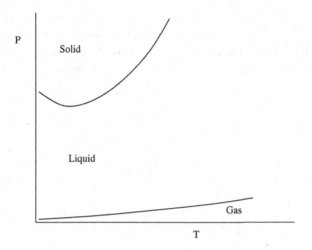

Fig. 12.4   A schematic phase diagram of Helium-3.

12.6   Interface

Two different phases can coexist, but their coexistence comes at a price. In order for two phases to coexist, there has to be a region where the two phases meet. The key question is whether or not it is possible to make a sharp dividing surface so that all the extensive thermodynamic variables may be written as the sum of two bulk variables representing the two coexisting phases like

$$N = N_1 + N_2, \quad V = V_1 + V_2, \quad U = U_1 + U_2, \quad \text{etc.} \tag{12.31}$$

This can easily be done for N and V by exercising the spirit of give and take between the two phases, but not for U and S. For U and S, there is a leftover which cannot be attributed to the two bulk phases. One way to proceed under these circumstances is to envision a sharp dividing surface, the interface, and then attribute the leftover energy and entropy to the interface. We may then write

$$U = U_1 + U_2 + U^{\sigma}, \tag{12.32}$$

and

$$S = S_1 + S_2 + S^{\sigma}, \tag{12.33}$$

where σ refers to the interface.

The surface energy is a function of the surface entropy S^σ and the surface area σ. The interface has neither molecules nor volume because N and V have been completely allocated to the two bulk phases, leaving nothing for the interface. The first law for the interface is therefore

$$dU^\sigma = TdS^\sigma + \gamma d\sigma, \tag{12.34}$$

where

$$\gamma = \left(\frac{\partial U^{(\sigma)}}{\partial \sigma} \right)_{S^{(\sigma)}} \tag{12.35}$$

is called the surface tension. The surface tension is to the surface what the pressure is to the bulk phases.

The surface area σ, the surface energy $U^{(\sigma)}$, and the surface entropy $S^{(\sigma)}$ are all extensive quantities, and therefore the first law gives

$$U^\sigma = TS^\sigma + \gamma\sigma, \tag{12.36}$$

or equivalently

$$F^\sigma = U^\sigma - TS^\sigma = \gamma\sigma, \tag{12.37}$$

which says that the surface tension γ is equal to the surface free energy per area. What do molecules do to minimize the surface free energy? The molecules can do nothing to change the surface tension, but they can change the surface area. The molecules in the two coexisting phases will rearrange themselves so as to minimize the interface area σ to minimize the surface free energy. Consequently the interface acts like a stretched rubber sheet. This is why raindrops never come in a cubic or any other sharp-angled shape. If they did, it would be hazardous to go out on a rainy day without protection!

A study of the interface would not be complete without the following experiment. Fill a bowl with water and sprinkle black pepper onto the surface. The pepper grains float, thanks to the surface tension of the water. Now touch the surface with a soap bar near the middle. The detergent in the soap effectively punctures the "rubber sheet" of the interface. The broken sheet is then pulled towards the edge of the bowl, carrying the pepper grains with it. This happens very quickly — as soon as the soap bar touches the surface. Some insects living on water surfaces know how to take advantage of this phenomenon. When they are in dangerously close proximity to a predator, they shoot a drop of detergent-rich liquid to the surface in the direction of the predator. The punctured sheet quickly carries them away from the predator.

12.7   The Van der Waals Theory

If I have to be stranded in an island far away and I can only take one section of this book with me, I will choose this section. We have sung quite a few verses of the free energy song, but none can match the one we will sing in this section. It embodies everything that I wish to emphasize in this book. It attempts to find out the effects of particle interactions, but in the spirit of "Let me try what I can" on the back side of an idle envelope.

For atoms and most molecules, the interaction potential may be represented with the Lennard–Jones potential,

$$\phi(r) = 4\epsilon \left\{ \left(\frac{\sigma}{r}\right)^{12} - \left(\frac{\sigma}{r}\right)^{6} \right\}, \qquad (12.38)$$

which is sketched in Fig. 12.5. Two important ingredients to be captured are: (a) the strong short-range repulsion, and (b) the weak long-range attraction. Let us see how (a) and (b) affect the free energy.

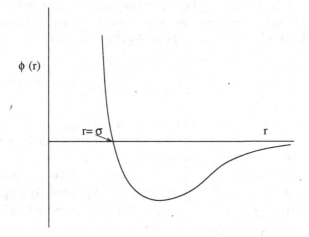

Fig. 12.5   The Lennard–Jones potential.

Consider (a). For $r < \sigma$, the interaction is strongly repulsive. No molecule allows other molecules to come within this range. When we have N such molecules, the effective volume of space in which the molecules can move around freely like an ideal gas is not V, but $V - Nb$ where b is approximately $4\pi\sigma^3/3$.

Turning to (b), assume that there is a molecule at the coordinate origin. Its interaction with all other molecules in the vicinity gives

$$\delta U_1 = \int_0^\infty dr 4\pi r^2 n(r)\phi(r) , \qquad (12.39)$$

where $n(r)$ is the number density of molecules at $\vec{r}$ given that a molecule is at the origin. The local density $n(r)$ is essentially zero for $r < \sigma$, and is expected to rise and fall periodically with increasing r, as we saw in Fig. 12.1 and Eq. (12.16). If we ignore this local fluctuation and approximate it with its global average value, $n = N/V$, then the interaction energy may be approximated as

$$\delta U_1 \approx n \int_\sigma^\infty dr 4\pi r^2 \phi(r)$$

$$\equiv -2na , \qquad (12.40)$$

where $-2a$ represents the definite integral. Since every one of the N molecules may be chosen as the coordinate origin, every molecule causes all other molecules to acquire the same amount of energy. Counting each pair twice in this way and then later dividing the result by 2, the total interaction potential energy is

$$U_{int} \approx N\delta U/2 = -N^2 a/V . \qquad (12.41)$$

This is the approximate total amount of interaction energy to be associated with all the configurations $(\vec{r}_1, \vec{r}_2, \cdots \vec{r}_N)$ in which no pair of particles are separated from each other by a distance less than σ. Let us refer to these configurations as allowed configurations. The rest of the configurations are forbidden by the Boltzmann factor as they call for infinitely large energy.

Armed with these two pieces of information, let us go back to Eqs. (12.4) and (12.5) and see if we may calculate the configurational integral Q approximately. The integral has been brought below for your convenience.

$$Q = \int \cdots \int \exp\left(-\frac{U_{int}\{r\}}{k_B T}\right) d\vec{r}_1 d\vec{r}_2 \cdots d\vec{r}_N .$$

Among all the configurations $\{r\}$ that the integrations cover, making any contribution to Q are those that we called in the above allowed configurations. Among them, $(-U_{int}\{r\})$ would be different from configuration to configuration, but let us ignore the fluctuation and approximate it using Eq. (12.41). The integrations are then performed trivially to give

$$
Q \approx \exp\left(-\frac{N^2 a}{V k_B T}\right) \int \cdots \int d\vec{r}_1 d\vec{r}_2 \cdots d\vec{r}_N
$$

$$
= \exp\left(-\frac{N^2 a}{V k_B T}\right)(V - Nb)^N \tag{12.42}
$$

where all the integrations have been limited to the effectively reduced volume $(V - Nb)$. Substitute it into Eq. (12.4) to obtain

$$
Z = \frac{1}{N!}\left(\frac{1}{\Lambda^3}\right)^N Q
$$

$$
= \frac{1}{N!}\left(\frac{1}{\Lambda^3}\right)^N (V - Nb)^N \exp(-\Delta U / k_B T). \tag{12.43}
$$

The corresponding free energy is

$$
F_{\text{vdw}} = -k_B T \ln Z
$$

$$
= -k_B T \ln\left\{\frac{(V - Nb)^N}{N! \Lambda^{3N}}\right\} - \frac{a N^2}{V}
$$

$$
= -k_B T N \ln\left\{(m k_B T / 2\pi\hbar^2)^{3/2}(V - Nb)\right\}
$$

$$
+ k_B T N \ln N - k_B T N - N^2 a / V. \tag{12.44}
$$

Comparing with the ideal gas free energy, the volume V has been replaced with $V - Nb$ and there is an additional term $-N^2 a/V$. The reduced volume of space available for the particles has reduced the spatial part of entropy, which can be taken into account by merely replacing V with $V - Nb$. Since $F = U - TS$, the reduced energy due to the attractive interactions has reduced the free energy by the same amount.

Let us calculate the pressure:

$$
P_{\text{vdw}} = -\left(\frac{\partial F}{\partial V}\right)_T
$$

$$
= \frac{N k_B T}{V - Nb} - \frac{N^2 a}{V^2}. \tag{12.45}
$$

The molecular interactions have two opposite effects on the pressure: the first term shows that the reduced effective volume increases the pressure, as

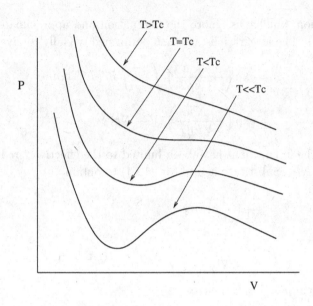

Fig. 12.6   The van der Waals pressure.

anticipated, while the second term shows that the reduced energy decreases the pressure. The reduced energy is due to the attractive interactions. Since all molecules are pulling each other, they now try to form a more bound state which effectively reduces their "appetite" for volume, namely, their pressure.

The van der Waal's pressure equation has a very important message. The message is like this: "There is a big problem. I cannot handle it, but if you carefully examine what I do forced by this problem, you will know what that problem is and what you need to do". Plot P_{vdw} as a function of V holding T fixed at various values. The results are as depicted in Fig. 12.6. At sufficiently high temperatures, the pressure increases with decreasing volume monotonically as it should. As the temperature falls, however, the system exhibits the opposite behavior starting from a certain temperature T_C to be called the critical temperature; in a certain region of V the pressure curve makes a dip whereby the pressure decreases with decreasing volume. How could the pressure decrease when you squeeze it? It suggests that the system is suffering an instability. It is like when you play tug of war with your opponent. For a while the rope is mechanically stable because, when you pull harder, so does your opponent. But at some point your increased pull is not met with an increased opposition but with

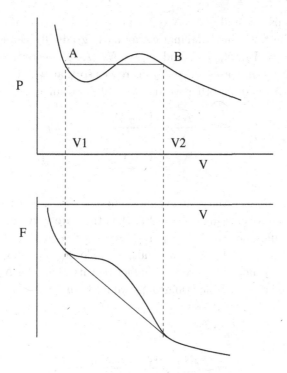

Fig. 12.7   The van der Waals pressure and free energy curves.

a decreased one. Either your opponent lost his footing or he just decided to yield. Either way, it is a mechanical instability. A similar thermodynamic instability is developing in the thermal system.

Examine the corresponding free energy curves in Fig. 12.7. They should also exhibit a similar sign of instability, and they do very loudly. Unlike its ideal gas counterpart, there is a region in which F makes a concave protrusion, namely, $(\partial^2 F/\partial V^2) < 0$, in the middle of a convex curve. This cannot happen in any real thermal system. Oddly enough, the most significant about the van der Waals theory is its breakdown.

How do real systems then behave in the presence of this instability? Due to this protrusion, it is possible to draw a common tangent line so that it just touches the bottom of the left dip at $V = V_1$ and another dip to the right at $V = V_2$; let F_1 denote the free energy at $V = V_1$ and F_2 the free energy at $V = V_2$.

The fact that the free energy curve allows a common tangent line has far reaching consequences. Suppose that the N molecules divide themselves into two groups, N_1 and $N_2 = N - N_1$, and that the size of the two groups changes in the following systematic way as V gradually changes from V_2 to V_1. At $V = V_2$, $N_1 = 0$ and $N_2 = N$. As V decreases toward V_1, N_1 progressively increases while N_2 decreases so that when V reaches V_1, $N_1 = N$ and $N_2 = 0$. In other words, N_1 and N_2 change with V via

$$\frac{N_2}{N} = \frac{V_1 - V}{V_1 - V_2},$$

$$\frac{N_1}{N} = \frac{V - V_2}{V_1 - V_2}. \tag{12.46}$$

Furthermore, suppose that the N_1 molecules behave like a part of an N-molecule system occupying volume V_1 (thus they are very compact) while the N_2 molecules behave like a part of another N-molecule system but occupying volume V_2 (thus they are much more loosely held together); since $V_2 > V_1$, the N_2 molecules are at a higher density than the N_1 molecules. The total free energy of the two groups would then be

$$F = F_2 + \frac{F_1 - F_2}{V_1 - V_2}(V - V_2)$$

$$= \frac{F_2(V_1 - V)}{V_1 - V_2} + \frac{F_1(V - V_2)}{V_1 - V_2}, \tag{12.47}$$

which we recognize as the common tangent line.

Two important observations should be made. First, since the common tangent line lies *below* the van der Waals free energy curve, real molecules follow it rather than the van der Waals free energy curve to save free energy. In this way, they refuse to be in a single homogeneous phase in the region between V_1 and V_2 and prefer to be in two coexisting phases, one in a gas phase and one in a liquid phase. Being faithful to the minimum free energy principle, some of the molecules are changing the way they assemble together. It is important to recognize that in the transient region of volume, $V_1 < V < V_2$, the free energy is a mixture of F_2 and F_1. The molecules divide themselves into two groups, one in a gas phase with density $n_2 = N/V_2$ and one in a liquid phase with density $n_1 = N/V_1$. The van der Waals theory is not equipped to handle this situation. At $V = V_2$, it is all gas, but as V is further reduced, a small amount of liquid appears while the fraction of the gas phase decreases, and finally at $V = V_1$, it is all liquid. At no point in the transient region do we have a group claiming a density between n_1 and n_2. There is a gap in the density. The transition here from the gas phase to the liquid phase is a first order transition. Second, since the common tangent line maintains a fixed slope between V_1 and V_2, the pressure remains constant between V_1 and V_2. In other words, the line connecting points A and B in Fig. 12.7 is straight and horizontal, and thus the two coexisting phases are in equilibrium against each other. The fact that the volume may be changed with no change in pressure means that the isothermal compressibility is infinite, which may be taken as a signature of the first order phase transition.

Is it possible to find points A and B directly from the van der Waals pressure curve without drawing the free energy curve? The answer is yes, thanks to the following curious geometrical property of the line connecting points A and B. Call the pressure corresponding to the horizontal line P_{12}. Since $P = -(\partial F / \partial V)_{T,N}$, the area under this horizontal line is $p_{12}(V_2 - V_1) = -(F_2 - F_1)$. But $-(F_2 - F_1)$ is equal to the area under the van der Waals pressure curve, namely, $\int_{V_2}^{V_1} P dV$. This means that the area bound by the horizontal line and the van der Waals pressure curve above it is equal to the area bound by the left portion of the same horizontal line and the van der Waals pressure curve below it. The answer is simple: draw the horizontal line so that it encloses two equal areas as explained above. This is called the Maxwell construction.

The locus of the two ends of the horizontal line in the top figure of Fig. 12.7 gives the boundary of the two-phase region. Figure 12.8 depicts how the Maxwell construction may be extended to construct the two-phase

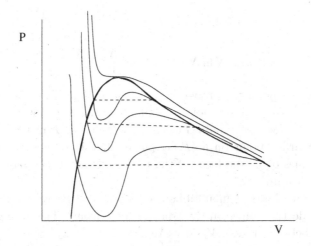

Fig. 12.8 The graph depicts how the two-phase boundary may be constructed using the Maxwell construction method.

boundary. Inside this boundary represented by a thick line, a single homogeneous phase is not allowed. Instead, a gas phase and liquid phase
coexist. The temperature of the isotherm that just touches the top of
the phase boundary is the critical temperature T_C which we will calculate
shortly. Let me repeat again what has been already said above. At $T > T_C$,
the N molecules can exist in a single phase in any volume V, namely, with
any density $n = N/V$. At $T < T_C$, on the other hand, the story is quite
different. They can exist in a homogeneous phase either in volume $V < V_1$
with any density greater than $n_1 = N/V_1$, or in volume $V > V_2$ with any
density less than $n_2 = N/V_2$. There is a gap in the density allowed for a
single homogeneous phase. Should we confine the N molecules at $T < T_C$
in volume V which lies between V_1 and V_2, the molecules then split into
two groups to form two coexisting phases, one in a liquid phase and one
in a gas phase. The density gap emerges at $T = T_C$ and widens as the
temperature is further reduced. The two boundary lines, one on the gas
side and one on the liquid side, are simply called the coexistence lines. The
exterior of the "dome" bound by the two coexistence lines is called the
one-phase region and the interior the two-phase region. If we start from a
point slightly above the critical point, the apex, and lower the temperature
so that the system will pass through the apex straight down, the ensuing
phase transition is a second order phase transition.

Let us examine the van der Waals theory in terms of the Gibbs free
energy G. Since $G = F + PV$, Eqs. (12.44) and (12.45) give

$$G = -k_B T N \ln \left\{ (m k_B T / 2\pi\hbar^2)^{3/2} (V - Nb) \right\}$$

$$+ k_B T N \ln N - k_B T N + \frac{N k_B T V}{V - Nb} - \frac{2N^2}{V}. \qquad (12.48)$$

Since V is not one of the proper variables of G, it needs to be eliminated
in favor of P. We have no choice but the following brutal way of doing it.
Choose a value for V, compute G using Eq. (12.48) and P using Eq. (12.45),
and then mark a point on a (P, G) plane. If this is repeated for many
different values of V, the locus of the points on the (P, G) plane turn out
to be as shown in Fig. 12.9.

Because of the triangular ribbon, G is not a single-valued function of
P. Where do the points on the triangle come from? They are generated
when V is between V_1 and V_2. If we replace G in this range of V with $G =$
Eq. (12.47) $+ PV$ while retaining the same G as given above by Eq. (12.48)
in other ranges of V, then we get the same curve without the triangle.

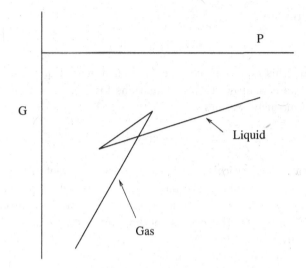

Fig. 12.9   Gibbs free energy as a function of p.

While V runs from V_1 to V_2, P and G do not change; the results repeatedly go to the vertex to which the triangle is attached. The free energy curve then consists of two line segments. Where the two lines meet, the slope changes abruptly. Here V and thus the concentration change abruptly, which signifies that there is a first-order phase transition.

As we move along the gas-liquid coexistence line toward the critical point in the P-T phase diagram, the first-order phase transition should become less and less abrupt. The critical point should mark the end point at which the abruptness finally disappears. Bearing this in mind, let us return to the van der Waals pressure equation, Eq. (12.45), and rewrite it in the form of

$$V^3 - \left(Nb + \frac{Nk_BT}{P}\right)V^2 + \frac{N^2aV}{P} - \frac{abN^3}{P} = 0\,. \qquad (12.49)$$

Since this equation relates P, T, and V, it is an equation of state to replace the ideal gas equation of state for interacting molecules. Given P and T, the ideal gas equation of state give us a unique corresponding value of V. The van der Waals equation of state does not. For temperatures below a certain temperature, the van der Waals equation of state gives us three different values for V as Fig. 12.7 shows. The three solutions should merge into one at $T = T_C$. In other words, at the critical point, Eq. (12.49) should therefore turn into

$$(V - V_C)^3 = 0. \tag{12.50}$$

If we expand this equation and then match it with Eq. (12.49) for each order of V, we end up with three equations for V_C, P_C and T_C. Solving them simultaneously, we find

$$P_C = \frac{a}{27b^2}, \quad V_C = 3bN^2, \quad T_C = \frac{81}{27b}. \tag{12.51}$$

We have located the critical point in terms of the parameters a and b. Since these parameters come from the inter-particle interaction potential of the chosen species of molecules, different species of molecules and atoms possess different critical points. However, something interesting emerges if we scale P, V, and T by writing $P = \hat{P}P_C$, $V = \hat{V}V_C$, and $T = \hat{T}T_C$. Then the van der Waals pressure equation becomes

$$\left(\hat{P} + \frac{3}{\hat{V}^2}\right)\left(\hat{V} - \frac{1}{3}\right) = \frac{8\hat{T}}{3}, \tag{12.52}$$

where all the parameters that characterize different species, namely a and b, have disappeared. In other words, when P, V and T are viewed in such scales, all species of molecules look alike. This is called the law of corresponding states. It suggests that everything looks the same if we see them in the proper scales and offers a glimpse of a much broader scaling phenomenon which occurs in second-order phase transitions. We will discuss this in more detail in Sec. 8.7.

Let me then summarize the limitation of the theory. First, return to Eq. (12.39) and recall what we did for $n(\vec{r})$. The local density fluctuates, but we replaced it with its global average value $n = N/V$, ignoring the fluctuation. Second, when we integrate over all particle coordinates approximating the integrand with the global average, we effectively treated the particles as if everybody is interacting with everybody else. The Lennard–Jones interaction is not an infinite ranged interaction and there is no configuration in which everybody is in the range of interact with everybody else. It would be so only if the spatial dimensionality of the system is infinite. Those are the two limitations. Such theories are called a mean field theory. In mean field theories, each particle receives the same mean field from all the others regardless of where they are located in reference to its own position. As imperfect as they may be, however, all mean field theories offer something useful as we have seen here and as we will see more later on. At the least, they can tell us that the real thermal system is doomed to have a phase transition. Considering how little we put in, that is a bargain.

12.8 Dynamics of First Order Phase Transitions

Shown in Fig. 12.10 is a phase diagram in a T-V plane. If the system enters the dome straight down through the apex where $V = V_C$, the transition is of second order; the phase diagram makes it clear that the density cannot change abruptly. If the entrance is anywhere else where $V \neq V_C$, the transition is of first order. For example, if the entrance is through the two-phase boundary line on the gas side, it is obvious that the emerging liquid phase requires an abrupt change in density. Here all changes were assumed to be gradual and slow so that the system is always in some equilibrium state. What if the entrance is abrupt? Prepare a system in equilibrium at a temperature above T_C and above the dome. Then suddenly quench the temperature to displace it into the dome. What happens? That is the subject of this section.

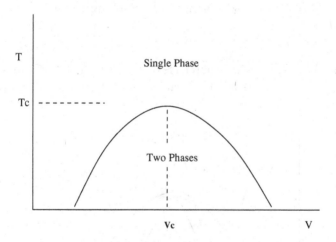

Fig. 12.10   Schematic T-V phase diagram.

Since the system is in the two-phase region, a phase separation has to occur, but the actual dynamics of the phase separation is different depending on exactly what part of the two-phase region the molecules are thrown into. This seemingly "playful" question captivated scientists across a broad spectrum for nearly half a century and is still an active area of research.

Since the system was initially in a gaseous equilibrium state, it will attempt to behave like the van der Waals pressure curve but at a reduced temperature. Figure 12.7 explains what happens. The pressure curve first goes down below the horizontal line, goes up above it, and then finally keeps going down. There is a valley and a peak. Call the volume corresponding to the bottom of the valley $\tilde{V}_1$ and that corresponding to the top of the peak $\tilde{V}_2$. At these two points,

$$\frac{\partial P}{\partial V} = 0 \,, \tag{12.53}$$

or equivalently,

$$\frac{\partial^2 F}{\partial V^2} = 0 \,, \tag{12.54}$$

where F is the van der Waals free energy. For $\tilde{V}_1 < V < \tilde{V}_2$, notice that

$$\frac{\partial P}{\partial V} > 0 \,, \tag{12.55}$$

which suggests a negative isothermal compressibility

$$K_T < 0 \,. \tag{12.56}$$

which means, as we saw in Chapter 11, that the system has lost its stability.

The locus of $(\tilde{V}_1, T)$ and $(\tilde{V}_2, T)$ at various temperatures form two lines which divide the two-phase region into two parts as shown in Fig. 12.11;

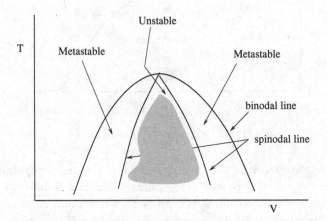

Fig. 12.11　The binodal and spinodal lines in the two-phase region.

these lines are called spinodal lines and are characterized by Eqs. (12.53) and (12.54). When molecules are displaced into the central region bound by the two spinodal lines, they become **unstable** against the phase separation. When molecules are displaced into the outer region bound by the two spinodal lines and the two coexistence lines, they become **metastable** against the phase separation. Let us find out why.

12.8.1 *Unstable Region*

When molecules are displaced into the unstable region, they become unstable against phase separation. To see why, examine Eq. (12.55). It says that **pressure increases with increasing volume and decreases with decreasing volume**! This means that the more volume it gets its "appetite" for volume goes up, which is pathological and puts the molecules in a dangerously explosive condition. The same equation also means that if a system loses its volume, its appetite for volume also goes down, which means that it wants to lose more volume. What will they do then? The molecules divide themselves into two groups. One group expands to turn into a further rarefied gas while the other compresses itself to turn into a more condensed liquid phase. One group wants more volume while the other wants less volume, and therefore there is no shortage or excess of volume. Each party can do what it wants to do at the courtesy of the other. This explains the instability. The initial prompt phase separation is not, however, a run-away process. Soon the molecules are no longer homogeneously mixed as they were at the beginning and therefore the pathological feature $\partial P/\partial V > 0$ no longer applies. The mechanism that drives the ensuing phase separation is called the spinodal decomposition. One phase appears everywhere like tangled seaweeds in the back ground of the other phase. The pattern coarsens as time progresses.

12.8.2 *Metastable Region*

Consider what happens when molecules are quenched into the metastable region, say, the one on the right hand side facing the gas phase in Fig. 12.11. The initial homogeneous mixture is to be transformed into two phases, a large gas phase and a small liquid phase. By doing so, they would save free energy by the amount of ΔF which is the difference between the van der Waals free energy and the common tangent line in Fig. 12.7. In order to do so, N_1 molecules should go into the liquid phase with density $n_1 = N_1/V_1$

while the rest should go into the vapor phase with density $n_2 = N_2/V_2$. Because the liquid phase constitutes only a small fraction, the liquid phase has to occur as droplets in the midst of a vapor. Thus immediately after the quench, many droplets are nucleated. To see what then happens to them, consider a droplet of radius R assuming that its density is $n_1 = N_1/V_1$. The free energy saved by the droplet is $(\Delta F/N)n_1 4\pi R^3/3$. Meanwhile, a liquid droplet in the midst of a vapor has an interface which costs free energy; the interface cost is proportional to the surface area of the droplet. Thus the net free energy cost is given by

$$\delta F = \gamma 4\pi R^2 - (\Delta F/N)n_1 4\pi R^3/4 \,, \qquad (12.57)$$

where the first term represents the interface cost and the second the savings.

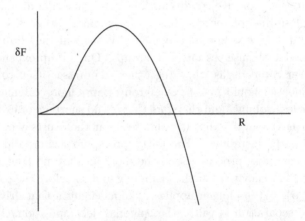

Fig. 12.12   The free energy cost for a droplet of radius R.

Figure 12.12 shows how δF changes with R. For a small R, the surface term dominates and $\delta F > 0$. For a sufficiently large radius, the second term dominates and it pays to phase separate because $\delta F < 0$, but a barrier has to be overcome to get there, which explains the metastability. Most of the droplets fail to overcome the barrier and evaporate back into the vapor, but rare thermal fluctuations carry some of the droplets over the barrier for further growth. This process is called nucleation and growth. So only at a small number of spots phase separated domains grow bigger and bigger until they merge for a complete phase separation.

12.9 Binary Systems: Mixing and Demixing

We studied the entropy of mixing in Chapter 3. We now wish to study the first order phase transition from a mixed phase to a demixed phase of binary systems. The ultimate purpose is to manipulate the demixing process to produce materials of desirable properties, which is why this is an important subject for material science. After an introduction here, we will return to the subject in Chapter 14 where we will develop a tool appropriate to study the actual dynamics of the demixing process.

We know that water and oil do not mix at normal temperatures but they do at very high temperatures. We wish to find out how the free energy minimum principle leads to such behavior. Mixing and demixing phenomena occur in all phases of matter, but for atoms in a solid phase, it is relatively simple to construct a simple mean field theory. Since the atoms are in a solid phase, the volume will be assumed to remain constant and therefore will not be taken as a variable.

Consider N_A atoms of species A and $N_B = N - N_A$ atoms of species B. Let $N_B/N = x$. For the sake of simplicity, assume that both species of atoms share the same lattice structure and that each atom supports p bonds irrespective of species. Let u_{AA}, u_{BB}, and u_{AB} represent the bond energy between $A - A$ pairs, $B - B$ pairs, and $A - B$ pairs, respectively. In the actual mixtures in a laboratory, the mixing would not be uniform throughout the sample. But in the spirit of mean field theory, let us ignore the local fluctuations. Then on average, an A-atom supports $p(1-x)$ bonds of $A - A$ type and px bonds of $A - B$ type. The average energy of an A atom is

$$u_A = p(1 - x)u_{AA} + p\,x\,u_{AB}\,, \tag{12.58}$$

and the average energy of a B atom is

$$u_B = p(1 - x)u_{AB} + p\,x\,u_{BB}\,. \tag{12.59}$$

The average energy per atom is given by

$$u = \frac{1}{2}\left\{(1 - x)u_A + x\,u_B\right\}\,, \tag{12.60}$$

where the factor $1/2$ corrects the fact that each bond has been doubly counted. The total energy is then

$$U = Nu = \frac{Np}{2}\left\{(1 - x)^2 u_{AA} + 2x(1 - x)u_{AB} + x^2 u_{BB}\right\}\,. \tag{12.61}$$

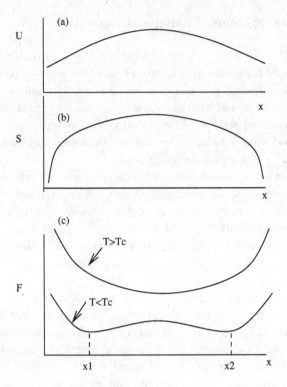

Fig. 12.13   (a) Energy, (b) entropy, and (c) free energy.

Assuming that $U_{AB} > U_{BB} = U_{AA}$, Fig. 12.13(a) depicts how U changes with x. It is significant that U has an upward bulge.

Let us now calculate the entropy. Assume that the entropy comes entirely from mixing. The total number of ways of distributing the two species of atoms over N lattice sites is given by

$$\Gamma = \frac{N!}{(N - N_B)! N_B!} . \qquad (12.62)$$

Assuming that the three numbers are large, we may use the Stirling formula to obtain

$$S = k_B \ln \Gamma$$

$$= k_B (N - N_B + N_B) \ln N - (N - N_B) \ln(N - N_B) - N_B \ln N_B$$

$$= k_B (N - N_B) \ln \left(\frac{N - N_B}{N} \right) - N_B \ln \left(\frac{N_B}{N} \right)$$

$$= -N k_B \left((1 - x) \ln(1 - x) + x \ln x \right) . \qquad (12.63)$$

This is a concave-downward function as shown in Fig. 12.13(b). Notice that the slope is infinite at both ends, and therefore the entropy of mixing is going to be the dominant factor near $x = 0$ and $x = 1$.

Combining Eqs. (12.61) and (12.63), the free energy $F = U - TS$ is given by

$$F = \frac{Np}{2} \left[(1 - x^2)u_{AA} + 2x(1 - x)u_{AB} + x^2 u_{BB} \right]$$
$$+ Nk_B T \left[(1 - x)\ln(1 - x) + x\ln x \right]. \tag{12.64}$$

Next let us calculate the chemical potential

$$\mu_B = \left(\frac{\partial F}{\partial N_B} \right)_{N_A}$$

$$= \frac{F}{N} + \frac{p}{2} \left[2xu_{AA} + 2(1 - 2x)_{AB} + 2x2u_{BB} \right] (1 - x)$$

$$+ k_B T \left[-\ln(1 - x) + \ln x \right] (1 - x). \tag{12.65}$$

$$\mu_A = \left(\frac{\partial F}{\partial N_A} \right)_{N_B}$$

$$= \frac{F}{N} + \frac{p}{2} \left[2xu_{AA} + 2(1 - 2x)_{AB} + 2x2u_{BB} \right] (-x)$$

$$+ k_B T \left[-\ln(1 - x) + \ln x \right] (-x). \tag{12.66}$$

The free energy takes the shape shown in the bottom of Fig. 12.14. Starting from a temperature to be called T_C, there is an upward bulge in the mid range of x which suggests the same instability that we saw in the van der Waals theory. Draw a common tangent line which touches the F curve at x_1 and x_2. The two species of atoms do not mix with any mixing ratio in the range $x_1 < x < x_2$, which creates a miscibility gap. If x is in this range, they instead split into two different mixtures, one mixed at the ratio of x_1 and the other mixed at the ratio of x_2. The miscibility gap emerges at T_C and widens as the temperature is further lowered.

In literature the word 'segregation' is often used to describe the phase separation, but I do not like it. It is true that the free energy cost is high for atoms of either species to be mixed with the other group at any mixing ratio between x_1 and x_2, and therefore they divide themselves into two groups to avoid paying the cost. If this is called a segregation, notice that the free energy cost is just as expensive if they wish to segregate themselves near completely below the mixing ratio x_1 or above x_2. Where does the latter come from? It is the precipitously falling entropy curve near $x = 0$ and $x = 1$. The best one can say then is that, when particles do not have enough of other type of particles to dislike, they dislike themselves just as much. I do not mind personalizing particles for some harmless fun, but certainly not at the expense of their integrity.

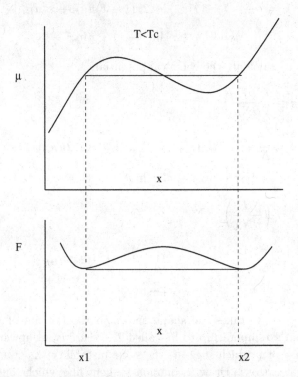

Fig. 12.14 The chemical potential and the free energy of binary atomic systems. Here x represents the ratio of one species to the total number of atoms.

The primary thermodynamic variables are x, T and $\mu = (\partial F/\partial x)_T$. The variable x plays the same role that V plays for liquid-gas systems while μ plays the role that pressure P plays for liquid-gas systems. The chemical potential for species B is shown in the top figure of Fig. 12.14. It looks like the van der Waals pressure curve flipped over. It has a peak at $x = \tilde{x}_1$ followed by a valley at $x = \tilde{x}_2$. At these two points of x,

$$\frac{\partial^2 F}{\partial x^2} = 0, \quad \text{or} \quad \frac{\partial \mu}{\partial x} = 0. \tag{12.67}$$

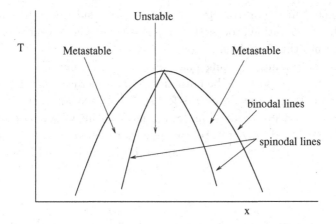

Fig. 12.15   The phase diagram in T-x plane.

The locus of these two points for various temperatures define the spinodal lines which divide the two-phase region of the phase diagram into two parts, as shown in Fig. 12.15. Examine the chemical potential in the region of x between $\tilde{x}_1$ and $\tilde{x}_2$. In this region,

$$\frac{\partial \mu}{\partial x} < 0 \qquad (12.68)$$

which says that the unlivability for B atoms decreases with increasing population of B atoms, or that the livability increases for B atoms with an increasing fraction of B atoms. The chemical potential of A atoms as a function of the same x looks like μ_B flipped over. So the livability of A atoms increases with decreasing fraction of B atoms. So the situation is the same as for gas-liquid systems. The atoms split into two phases, one rich in B and one rich in A; each group can grow thanks to the courtesy of the other. This is called the spinodal decomposition. This explains the instability in the interior of the two-phase region.

Examine the phase diagram shown in Fig. 12.15. If the mixture is thrown into the outer regions, the mixture is metastable against the phase separation. The reason for the metastability is the same as before. Droplets rich in one species have to be formed in a sea of the phase rich in the other majority species, but the interface cost poses a free energy barrier which the droplets have to overcome for further growth. As a result, most of the small droplets evaporate away, but only a small number of them can grow further. Again this process is called the nucleation and growth. The phase diagram is as depicted in Fig. 12.15. In Sec. 9.3, we will find a way to study the actual phase separation process that happens when a homogeneous mixture is thrown into this forbidden region.

When the mixing-demixing phase diagram is as shown in Fig. 12.15, the phase diagram is said to be normal with an upper critical temperature, which means that demixing occurs for $T < T_C$. Not all binary mixtures have this type of phase diagram. Some have an inverted phase diagram with a lower critical temperature, meaning that demixing occurs when $T > T_C$. Some have a closed phase diagram with both upper and lower critical temperatures.

12.10　More on Binary Mixtures: Their Phase Equilibria

There are binary mixtures which are miscible when they are in a single phase. But when they are allowed to be in more than one phase, there emerges a mixing gap and a variety of interesting phenomena including which can be significant for technical reasons. Figure 12.16 is a typical phase diagram of such a mixture.

Understand what this says about the pure-A ($x = 0$) system and the pure-B system ($x = 1$). Moving vertically up along the temperate axis with x fixed at $x = 0$ or $x = 1$, it is clear that the pure-A solid melts at T_A while the pure-B solid melts at T_B. At a very high temperature, the free energy for liquid, f_L is always lower than f_S and the mixture exists in the liquid phase. By the same reason when the temperature is very low, the liquid free energy lies above the solid free energy, and therefore, the mixture exists in the solid phase. At intermediate temperatures, the two free energy curves cross each other at a certain mixing ratio, which makes it possible to draw a double tangent line to create a forbidden mixing. In the bottom frame for the phase diagram, the shown free energy crossing is responsible for the two points, one on the liquid boundary curve called liquidus line and one on the solid boundary called the solidus curve.

It should be noticed that (i) at any temperature between T_A and T_B, the two-coexisting liquid phase and solid phase have different mixing ratios, (ii) the melting starts at T_A for A particles but not all A particles are melted away at that temperature. As the temperature further rises, the solid component B needs A friends for coexistence. For the same reason, the liquid component dominated by A also needs some B friends, which means that some of the B solids also started melting starting from T_A. It

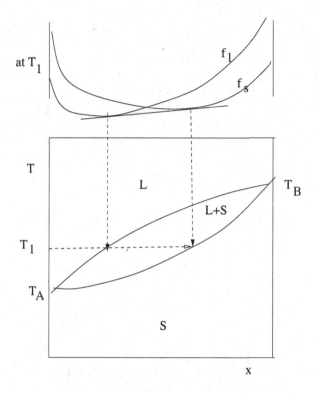

Fig. 12.16   A radial distribution.

is interesting that not all A particle melt at T_A. Until the temperature reaches T_B, a good number of A particles are remaining in the solid phase mixed with the B particles. Similarly, a good number of B particles actually melt into liquid along with A particles before the temperature reaches T_B. This is all entropy effect. When A particles melt at temperatures above T_A, the entropy cost would be severe if they were to melt into a pure-A liquid, which is why they bring some B friends to avoid the entropy cost. The favor goes in both directions. The small number of A particles remaining in the solid phase even at temperatures above T_A are there because the B particles need a small number of A friends to avoid the same entropy cost.

In the above we assumed that A and B atoms share the same lattice structure in their solid phase. That is not always true. For example, gold and silicon can be mixed to make an alloy. Pure gold favors a face centered cubic lattice while silicon favors a diamond structure. A small number of gold atoms may be substituted in otherwise pure silicon lattice, and vice

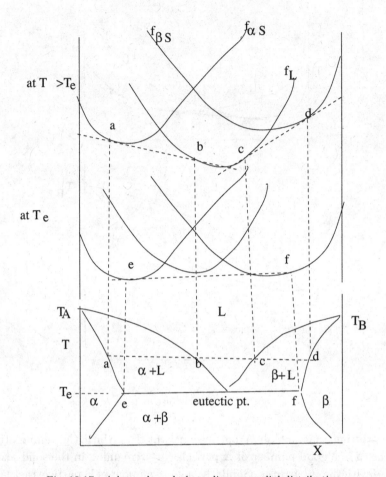

Fig. 12.17   A horn-shaped phase diagram radial distribution.

versa, but as the amount of substitution increases, it creates an unfavor-
able strain in the lattice. The phase diagram is therefore as depicted in
Fig. 12.17. Here α denotes the lattice structure favored by A atoms and
β that favored by B atoms. The homogeneous liquid phase is now bound
by two liquidus lines, which together with two solidus lines form what look
like two horns. Where the two liquidus lines merge is called the eutectic
point, and the straight line passing through the eutectic point is called eutectic line. So, we have two tilted triangles or horns, the left one is for two
coexisting phases of α and liquid, the right one is for two existing phases
of *beta* and liquid. Below the eutectic line, the two solid phases and α and

β coexist. So the eutectic point is a triple point; actually each of the three coexisting phases represents two coexisting phases but the liquid phases merge into one and therefore five different phases coexist

This phase behavior may be understood easily if we assume that the various free energies may be modeled as as shown in the top panel: $f_{S\alpha}$ and $f_{S\beta}$ represent the free energies of the two solid phases while f_L represent the homogeneous fluid mixture. Let T_e represent the eutectic temperature. For $T_1 > T_e$, there are two possibilities: (i) the fluid free energy curve is below either one of the two solid free energy curves for all x. (ii) the fluid free energy curve protrude below the two solid free energy curves in the central part of x, but crosses over the solid free energies curves near $x = 0$ or near $x = 1$, as shown in the panel. This allows two double tangent lines which explains the horn-shaped two liquid us lines and two solidus lines bounding from the homogeneous liquid part. The second panel explains T very near T_e. As the temperature further falls, soon the liquid free energy curve remains sufficiently above the solid free energy curves enough to allow a double tangential line touching the two solid free energy curve, which explains the bottom part of the phase diagram panel where the two co-existing solid phases are surrounded by two pure solid phases near $x = 0$ and $x = 1$.

Chapter 13

Second-Order Phase Transitions

13.1 Introduction

When a thermal system is cooled or heated through its critical point, a second order phase transition occurs. This transition is different in two important ways from those first order transitions that we discussed in the previous chapter. (a) The transition occurs without any abrupt change. (b) But all response functions, the specific heat, isothermal compressibility and susceptibility, become larger and larger as the temperature approaches T_C and diverge at T_C. The first item (a) is understandable since the critical point marks the end of the coexistence line of the two phases that we are talking about, and therefore as the critical point is approached, the difference between the two phases across the coexistence line disappears. Item (b), on the other hand, is a different story. It is not so simple to understand it. Take, for example, a liquid-gas system. The average energy changes smoothly through T_C, but just imagine that you are actually monitoring the energy at each instant. As T approaches T_C, it begins to fluctuate up and down more wildly. The fluctuation diverges at T_C or exhibits a singular behavior of some sort. Suppose that the system is in contact with a pressure reservoir and imagine that you are monitoring the volume at each instant. The volume is fluctuating up and down wildly and the fluctuation diverges at T_C. In spite of all these diverging fluctuations, the order parameter $n_L - n_G$ smoothly changes; it is non-zero below T_C but gradually decreases reaching zero at T_C. So, the order parameter and the response functions are the main actors to watch. What do they do at the critical point? As far as they are concerned, the 'cool' thing to do is either to exhibit some sort of singular behavior or just simply to vanish to zero, and any thing normal in between is not 'cool' enough. How crazy, but do you sense the fun that is about to ensue? We bet that there is still a regularity even in this crazy behavior. Welcome to the world of critical point phenomena!

Let us try if we can make some sense out of this madness, focusing on the role of the temperature in the Helmholtz free energy equation $F/T = U_s/T - S_s$. We argued in Chapter 5 that F/T is the net entropy cost that the heat reservoir must pay to allow the system to have energy U_s. Think of the first term U_s/T as the investment cost or rather the fear that the reservoir feels when it decides to invest a small portion of its energy into the system and the second term S_s as the return of the investment. Low temperature brings fear and discourages investment, which results in a state of low energy and low entropy, and thus an order. High temperature removes the fear and encourages investment, which results in a state of high energy and large entropy, and thus a disorder. The low temperature states of low energy are like a bear stock market while the high temperature states of high energy are like a bull market. I hardly know anything about the way stock markets work, but when they change from one type to the other, they seem very volatile. This is presumably because investors are nervous to find out when to sell or buy and react more to any sign of business activity than they do at other times.

The way the symmetry changes is also different depending on whether the transition is of the first kind or the second kind. Consider crossing the same coexistence line from one side to the other. One symmetry prevails on one side of the line and another on the other side, and both on the coexistence line. So, upon crossing the line, the symmetry changes abruptly from one to the other. In the second kind, on the other hand, the symmetry changes gradually. For liquid-gas systems, for example, if the system is cooled through the critical point, the symmetry that prevails in the high temperature gas phase remains but a new reduced symmetry of the liquid phase gradually builds up.

There is no formula telling us what the order parameter should be for a given transition. There is only a guide line as to what it should do. It should capture the onset of the order that prevails in the more ordered phase. For magnets, the order is in the direction of the spins, and therefore the obvious choice is the magnetization. It is zero in the disordered symmetric phase and non-zero in the ordered non-symmetric phase. For liquid-gas systems, the order is in the degree of how much the particles are gathered. Accordingly, the order parameter should involve the density. It makes no difference whether we fix N and describe the degree of otherness with V

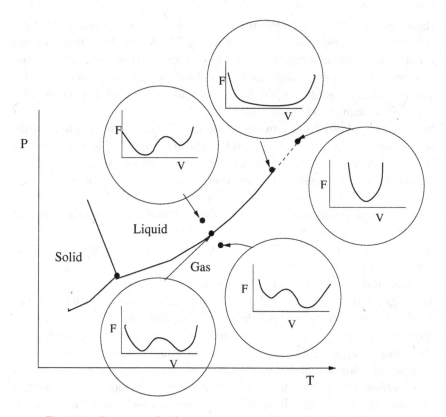

Fig. 13.1   Free energy landscape at various points in the phase diagram.

as we did earlier in Chapter 11, or fix V and represent the changing order with changing N. This correlated fluctuation is not a result of an external force.

It would be helpful to let a free energy say what we have said about the phase behavior. In particular what feature should it have in order to support the diverging fluctuations? See the phase diagram in Fig. 13.1. Since it is on a (P, T) plane, it stipulates a liquid-gas system subject to a temperature and pressure reservoir. Fluctuating are, however, not P and T but rather energy and volume. Thus the appropriate free energy suitable for our purpose of explaining the diverging fluctuation in volume is the macrostate Helmholtz free energy $F(V, T)$ where the particle number N has been omitted because it stays fixed and has no role to play here. Consider points A. B, D which lie away from the critical point but near or on the gas-liquid coexistence line. The macrostate free energy corresponding to these points are shown in the inserts. At point A which lies on the gas side, the free energy has two local minima. The lower of the two favoring a much larger volume represents the largest macrostate and thus it is the equilibrium free energy while the higher of the two favoring a much smaller volume (thus favoring a liquid phase) is a free energy of a much smaller and thus a highly metastable state. The latter is like someone trying to float in the middle of a stormy ocean on a small float. Let the temperature and pressure at point A be T_A and P_A, respectively. When the system is subject to P_A and T_A, the volume fluctuates around the average given by the lower minimum, but since there is no diverging fluctuation, the minimum must be reasonably sharp. Now move to point B on the liquid side. The corresponding free energy also has two local minima, but this time the lower of the two favors a much smaller volume. In the mean time, at point D which lies exactly on the line, the two competing local minima are at the same height, hence the two phases coexist.

Now move toward the critical point staying on the coexistence line. We would guess that the two local minima will come closer to each other and the hump between the two will become lower and lower. Now we have to guess really hard. By the time we land on the critical point C, would the the hump disappear only when the two local minima come close enough to make one normal minimum (like the bottom of a wine glass), or would it disappear fast enough to let the two make a wide nearly flat basin (like the bottom of a bowl)? If the former were the case, there should be no diverging fluctuation. The answer has to be the latter as amply exaggerated in the figure. Extend the coexistence line further out to point E. The minimum should become reasonably sharp again. To summarize, if we move from point E toward points C and D, the minimum 'bifurcates' into two with a wide flat minimum at the critical point C. The flatness of the minimum as opposed to the sharp minima elsewhere is responsible for the large fluctuations. The flat basin frustrates the system which is endlessly moving around on the basin in search for a sharp minimum which does not exist, whence the diverging fluctuation. We will see in a later section how Landau actually constructs such a form of macrostate free energy.

We will discuss in Secs. 8.2 and 8.3 how we learn the critical behavior in laboratory. Section 8.3 is a survey for liquid-gas molecular systems and Sec. 8.4 for magnetic systems. Such studies for many different systems lead to the notion of universality in Sec. 8.5 and the scaling laws in Sec. 8.6. The static scaling hypothesis in Sec. 8.8 is an effort to explain the scaling laws. A good portion of this chapter is actually beyond the scope of this book. It should be read as a story. You will often find equations following a statement like 'it turns out' or 'it has been found', etc. They are never meant to say anything like 'one can show easily'; just let the ensuing equations help tell the story.

13.2 Liquid-Gas Molecular Systems

Return to the phase diagram in Fig. 12.10 for liquid-gas systems. Take a point above the peak of the dome, where $T > T_C$. The molecules are in a homogeneous single gas phase. Now enter the two-phase region straight down through the peak by lowering the temperature. At $T = T_C$, i.e. on the peak of the dome, the molecules are still in one homogeneous phase. As the temperature is lowered below T_C, the molecules gradually turn into a mixture of liquid and gas and the difference between their densities, n_l and n_g, becomes larger and larger. The onset of the accompanying order is described with

$$\tilde{\rho} = \rho_L - \rho_G,\tag{13.1}$$

which serves as the order parameter; it is zero above T_C and non-zero below T_C. While the order parameter changes smoothly, the specific heat C_V and compressibility K_T diverge or exhibit a singular behavior of some sort. Assuming that the system is in contact with a heat and pressure reservoir, this means that the energy and volume will violently fluctuate. We wish to quantify this behavior with critical exponents. A set of Greek letters are reserved for the exponents: β, α, γ, ν, δ, and η.

Suppose that we have found a certain critical quantity changing with temperature near $T = T_C$ like $f(t)$ where $t = |T - T_C|/T_C$, the relative deviation of T from T_C. We do not give the full exposition $f(t)$. Instead we write

$$f(t) \sim t^{-\lambda}.\tag{13.2}$$

Given $f(t)$, how do we determine the exponent λ? The answer is:

$$\lambda = \lim_{t \to 0} \frac{\ln f(t)}{\ln t}.\tag{13.3}$$

This calls for a careful attention for two reasons. First, the exponent relies only on the leading order. For example, suppose that

$$f(t) = t^{-\lambda}(1 + At^y + \cdots), \qquad y > 0.\tag{13.4}$$

According to the definition, the higher order terms do not make any contribute to λ. Second, what if $f(t)$ exhibits a singularity which is, however, too weak to be described by a power law? A good example is the logarithmic singularity, namely, $f(t) \sim \ln t$. Substitute it into Eq. (13.3) and apply the L'Hôpital's rule, which gives $\lambda = 0$. Similarly, if $f(t)$ shows a finite discontinuity or a cusp-like singularity, Eq. (13.3) again gives $\lambda = 0$. Long experience has shown that writing the temperature dependence in the format of Eq. (13.2) captures the essence of the critical properties without requiring to carry around any unnecessary baggage. Except in some special cases, we do not care about the proportionality constant, and thus we do not use the equality sign; instead we use the symbol $\sim$ which should be read as "goes like".

Experiments have shown that

$$\tilde{\rho} \sim t^{\beta}, \qquad (13.5)$$

where β is called the order parameter exponent. For most liquid-gas systems, $\beta \approx 1/2.9$, regardless of the intermolecular interactions. Take this as the first sign that there is indeed a regularity in the seemingly wild critical behavior.

Experiments have also shown that the specific heat shows a singular behavior which may be characterized with exponent α

$$C_V = \left(\frac{\partial <U>}{\partial T} \right)_V \sim t^{-\alpha}. \qquad (13.6)$$

The specific heat exponent α can, however, be zero, and the zero exponent does not necessarily mean that C_V is independent of temperature, as explained a moment ago.

The compressibility diverges with exponent γ

$$K_T = -\frac{1}{V} \left(\frac{\partial V}{\partial P} \right)_{N,T} \sim t^{-\gamma}. \qquad (13.7)$$

The diverging compressibility means that V can change with no change in P. Thus if we allow V to vary while holding P and N fixed, the density $\rho = N/V$ should exhibit large fluctuations. What does this indicate? If the local density fluctuations are random, then there should be as many localities where the density is above average as there are localities where it is below average. Therefore, there cannot be a large global or overall density fluctuation. The fact that there are large global density fluctuations means that the local density fluctuations are not random but correlated. Indeed, the local density fluctuations are correlated over a length scale which increases as T approaches T_C and finally diverges at $T = T_C$. To describe the correlated density fluctuations, define the local density deviation by

$$\delta\tilde{\rho}(\vec{r}) = \tilde{\rho}(\vec{r}) - <\rho>, \qquad (13.8)$$

where $<\rho>$ is the average density over all different localities. The correlations may then be described by the average

$$G(r) = <\delta\tilde{\rho}(\vec{0})\delta\tilde{\rho}(\vec{r})>. \qquad (13.9)$$

The correlation function questions whether or not distant localities tend to act in unison for fluctuations in the same direction. If they do, there are bound to be many patches of different densities, and the patches should be sized comparably to the length scale over which they act together. This assertion can be checked experimentally by observing how the sample scatters light. When T is near T_C, according to a theory (called the Ornstein-Zernike theory), the correlation function exhibits the pattern given by

$$G(r) \sim \frac{\exp(-r/\xi)}{r}, \qquad (13.10)$$

which says that $G(r)$ is non-zero approximately up to $r = \xi$ and then drops off quite rapidly as r increases further. In other words, the correlation extends up to the length scale of ξ. Therefore, ξ is called the correlation length. It is the most important player in critical phenomena. Experimental results indicate that

$$\xi \sim t^{-\nu}, \qquad (13.11)$$

where ν is called the correlation length exponent. The correlation length grows as T approaches T_C and diverges at T_C. This means that at exactly T_C, the patches are of all sizes up to infinity. This is why the compressibility diverges at T_C. This is just a crude verbal version of what will be proven shortly in the name of the static susceptibility sum rule.

Since the molecules spontaneously form patches of liquids (in which molecules are more compressed than in the gas phase) by themselves without any external force, not much change in pressure is required to force a large change in density.

This correlated fluctuation is not a result of an external force. Groups of molecules act together spontaneously. They are like good friends. When

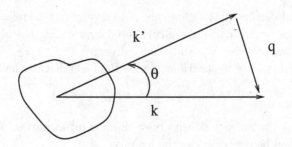

Fig. 13.2   Light scattering.

something happens to make one of them sad, it tends to also make the others sad, and when something good happens to make one of them happy, it tends to also make the others happy; their moods tend to fluctuate in a correlated fashion. How do we study this correlation in the laboratory? A monochromatic light beam of wavelength λ is sent into a sample and the intensity of the scattered light is measured at various angles with respect to the incoming light as depicted in Fig. 13.2. The wave vector (momentum divided by $\hbar$) of the incoming photons $\vec{k}$, the wave vector of the scattered photons $\vec{k}'$, and the wave vector transferred to the sample from the light $\vec{q}$ form a triangle. Here, the scattering is elastic and therefore $k = k' = 2\pi/\lambda$. It is clear from the triangle that the transferred wave vector q is given by

$$q = 2k\sin(\theta/2) = \frac{4\pi}{\lambda}\sin(\theta/2)\,. \tag{13.12}$$

The intensity of the scattered light at the angle corresponding to the wave vector q is found to follow

$$\text{Intensity} \sim S(q) = \frac{k_B T n K_T}{1 + q^2 \xi(T)^2}\,, \tag{13.13}$$

where $S(q)$ is the Fourier transform of the correlation function $G(r, T)$ and may be written as

$$S(q) = <|\delta n(\vec{q})|^2>\,, \tag{13.14}$$

where $\delta n(\vec{q})$ is the Fourier transform of the local density fluctuations $\delta n(\vec{r})$. If the sample has a prominent repeating structure due to a particular size of patches, $S(q)$ should exhibit a peak at the corresponding q, hence it is the same structure factor that we met earlier.

By measuring the light intensity with the detector placed at an angle corresponding to q, we are probing the density fluctuation (df) in the length

Fig. 13.3 An incomplete and random diffraction grating. Someone must have been drunk.

scale $\lambda_{df} = 2\pi/q$. If $S(q)$ is large for a particular value of q, it means that the density fluctuation has a prominent pattern which tends to repeat itself with the wavelength given by $2\pi/q$. As it turns out, there is not just one such pattern but rather many of them. Therefore, near $T = T_C$ the sample may be thought of as a "random" diffraction grating, as depicted in Fig. 13.3. Unlike a "uniform" diffraction grating where lines are drawn with the same spacing and in the same direction everywhere, many different groups of lines are drawn here and there with different spacings and in different directions. Such diffraction gratings would scatter white light in all directions. That is exactly what samples do near $T = T_C$. Away from the critical point, samples look transparent. But as the critical point is approached, samples look milky because all of the components of white light are scattered. This phenomenon is called critical opalescence. Watching this is great fun. Go to any physical chemistry laboratory and ask for a demonstration. A binary liquid mixture with an inverted phase diagram appears to be the easiest way to see it. If you are as lucky as I was, you will find someone to do this for you. My host poured a binary liquid mixture into a test tube and then put the tube on a Bunsen burner to raise the temperature. Initially the mixture looked transparent like a wine, but as the temperature approached the critical temperature, it turned into a white milk.

The fact that the patches come in many different length scales is the root of the difficulty of studying the critical point phenomena. Here in America, they say "Be careful what to wish for because you may get it." Boy! Is that not true? We asked for a regularity. What did we get? Far more than we asked for! It is like asking for a cup of water and getting drowned. Well, almost. No other physical phenomena involve such a vast range of different length scales. Examine the atomic physics of hydrogen. You will notice that in all of the energy eigenfunctions the variable r always appears in the form of r/a_0 where a_0 is the Bohr radius, and therefore a_0 is the unique characteristic length scale. Examine the electric field in the presence of, say, a conductor of radius R. Again, r appears in the electric field in the form of r/R, meaning that R is the lone characteristic length scale. Can you imagine how difficult it would be if there were an infinite number of Bohr radii? How would you then construct a Bohr theory? Critical opalescence was observed in the late 19th century, but a satisfactory explanation was not found for nearly a century. That ultimate successful explanation was provided by the renormalization group theory. The fact that there are infinitely many different length scales ranging from zero to infinity provides a unique symmetry, and the renormalization theory became successful because it was designed so as to capture that symmetry, as we will find in a later chapter where we will take a glimpse into this advanced topic.

13.3   Spin Systems

Now consider uniaxial ferromagnets using Ising models. Above the transition temperature, the spins are oriented randomly with no net magnetization, and therefore they are in a paramagnetic phase. As the temperature is lowered below the critical temperature (called the Curie temperature), the spins spontaneously begin to align and go into a ferromagnetic phase. The net magnetization begins to become nonzero from $T = T_C$; upon passing through T_C, two symmetric phases emerge, one of M and the other of $-M$, as shown in Fig. 13.4. This is the magnetic analogue of Fig. 12.10. With a magnetic field, the two coexisting phases can be turned into one. Figure 13.5 is a schematic H–T phase diagram. This is the magnetic analogue of Fig. 12.2. The thick line represents the coexistence of up ($+M$) and down ($-M$) phases, and therefore is a line of first-order phase transitions. Now put the system on the coexistence line and move towards the

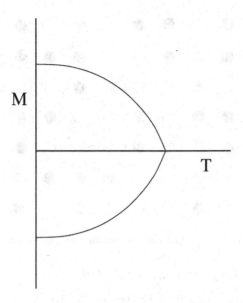

Fig. 13.4   The phase diagram of ferromagnetic in an M–T plane.

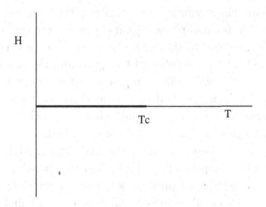

Fig. 13.5   The phase diagram of ferromagnets in a T–H plane.

right while staying on the line. The difference between the up phase and the down phase becomes less and less and finally disappears at the critical temperature T_C; the end of the line is a point of second-order phase transition. Since M is zero above T_C and non-zero below T_C, it serves as the order parameter.

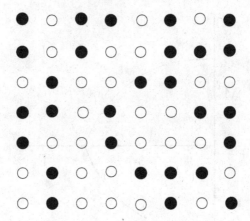

Fig. 13.6   Free energy landscape at various points in the phase diagram.

13.3.1   *Broken Symmetry, Order Parameter*

Even in the zero magnetic field, the two coexisting phases are actually only metastable. Given enough time, they will ultimately turn into one of the two phases, breaking the up-down symmetry. Let us talk about this symmetry. The appropriate symmetry operation is the spin reversal, i.e. s_i to $-s_i$ for all spins. In the one-phase region above T_C, the spin configurations are symmetric (invariant) to the symmetry operation. In what sense? Suppose that we just took a snapshot of the spins and the result is as shown in Fig. 13.6, where the white circles represent up spins and the dark circles down spins. Go to each spin and turn around to reverse its direction, i.e. white circles into black circles, and black circles into white circles. Let me refer to this configuration the inverted configuration. In a naive view, the inverted configuration looks different from its original counterpart, but it could have been the snapshot if we had taken the snapshot at some other time. Monitor the spin configuration for a sufficiently long time. It is a guaranty that the two configurations will appear with equal frequency. In that sense, we say that the original spin configuration is invariant to the symmetry operation. Now turning to $T < T_C$, consider a spin configuration far below T_C in which all spins are up (or down). Now apply the symmetry operation to obtain a configuration in which all spins are down (or up). Unlike at $T > T_C$ above, we cannot claim that this inverted configuration could have been the snapshot. No matter how long we may wait, they will not spontaneously turn around. Nothing in the spin interaction or in the microstate structure favors one spin direction over the other, but once they choose one for all spins over the other, they do not change the decision. You can see why. It is not a matter of one spin flipping. It is a matter of all the spins conspiring to flip, which is very difficult to expect. The symmetry in the paramagnetic phase is broken in the ferromagnetic phase in this way.

One can argue similarly for liquid-gas systems or liquid-solid systems, but it needs more complex arguments, which is why we omit it.[12]

If the spin interaction is symmetric to spin reversal, how do they then decide to be up or down. A truly minuscule impurity in the sample which favors one direction over the other or a truly insignificant up-down asymmetry which may have existed while the system was in the homogeneous phase is now enough to break the symmetry and allows all the spins to turn in the favored direction. This symmetry breaking process is very slow. We can accelerate it with an external magnetic field. Apply an infinitesimal magnetic field in one direction. Given enough time, all spins align in the direction of the field. Now remove the field. The spins **remain aligned** in the same direction even though the field has been removed. In this way, we can put the system on the $+M$ curve or on the $-M$ curve. In order to save the interface cost, the spins prefer to break the up-down symmetry.

Since the order means that all spins are in the same direction, the magnetization M is the obvious choice for the order parameter. The transition to the ordered state at T_C is smooth in the sense that M changes continuously with the temperature. However, the continuous change at T_C is accompanied by diverging specific heat and diverging susceptibility, as in the liquid-gas system. Experiments have shown that the order parameter M vanishes, or begins to grow, like

$$M \sim t^\beta,$$ (13.15)

and the response functions diverge like

$$C_H = \left(\frac{\partial U}{\partial T} \right)_V \sim t^{-\alpha}, \qquad H = 0$$ (13.16)

and

$$\chi_T = \frac{1}{N} \left(\frac{\partial M}{\partial H} \right)_T \sim t^{-\gamma}.$$ (13.17)

Notice that the susceptibility exponent shares the same symbol with that of the compressibility of liquid-gas systems. This is because they represent

---

[12] See the book by James Thethna.

the fluctuation of their respective order parameters, and therefore they represent essentially the same critical property.

Playing a big role is the correlation function

$$G(r) = <(s_i - <s>)(s_j - <s>)>, \tag{13.18}$$

where r represents the distance between site i and site j. It exhibits the same pattern of correlation as in Eq. (13.10). In particular, the spins tend to act together in unison over the correlation length which changes with temperature like

$$\xi(r, t) \sim t^{-\nu}. \tag{13.19}$$

Exactly at $T = T_C$, the spatial correlation decays much more slowly,

$$G(r, t = 0) \sim \frac{1}{r^{d-2+\eta}}, \tag{13.20}$$

where d is the dimensionality of the space and η is yet another exponent.

Let us introduce one more exponent. At $T = T_C$, the magnetization follows the field like

$$M \sim H^{1/\delta}. \tag{13.21}$$

Critical exponents do not depend on the minute atomic details, but in general different systems have different critical exponents. So, α, β, γ, etc., are in general different for different systems, but they all satisfy certain relationships such as:

$$\alpha + 2\beta + \gamma = 2, \tag{13.22}$$

$$\gamma = \beta(\delta - 1), \tag{13.23}$$

$$\gamma = (2 - \eta)\nu, \tag{13.24}$$

$$\nu d = 2 - \alpha. \tag{13.25}$$

Such relationships among the critical exponents are called the scaling laws. We will return to these relationships in Sec. 8.7. In a way, aren't the scaling laws similar to the third law of Kepler? Different planets maintain different average distances (R) from the sun and take different amounts of time (T) to revolve around the sun, but for all planets, T^2/R^3 is the same. Just like Kepler's laws provided clues to the laws of motions and the law of gravitation, these scaling laws imply something which ultimately helped solve the puzzle of critical phenomena.

13.4 Universality

Many different systems exhibit second-order phase transitions. Experiments have shown that these systems form several universality classes. Those belonging to the same class share the same set of critical exponents. This is remarkable because the member systems belonging to the same class can be quite different from atomic and molecular points of view. For example, two-dimensional planar ferromagnets and liquid helium-4 belong to the same universality class. The former is a solid (in which the dipole moments are on a three-dimensional lattice but their direction vectors form a plane) while the latter is a liquid. The former is in its classical regime while the latter is in its quantum mechanical regime. But near the critical point the atomic and molecular characteristics do not seem to play a role in the critical behavior of these systems, and what determines the critical behavior are the spatial dimensionality of the system and the number of components of the order parameter, or the dimensionality of the order parameter space. For both planar ferromagnets and liquid He-4, $d = 3$. The order parameter for planar ferromagnets has two components because the direction vectors of the magnetic dipole moments are restricted to lie on a plane and therefore the order parameter is a two-dimensional vector of two components. The order parameter for liquid He-4 is given by the condensate wave function which has two components, the real and imaginary parts.

The liquid-gas molecular systems and the three-dimensional uniaxial ferromagnets (the Ising model) provide another example. They belong to the same class because their order parameters have only one component. And $d = 3$ for both.

13.5 Susceptibility and Spatial Correlations

The root cause of critical phenomena is the correlation between constituents. This is true in all second order phase transitions, but it is most clear, visually, in the magnetic phase transition as it is from a paramagnetic phase where all magnetic dipole moments are directed randomly without any order to a ferromagnetic phase where they are all aligned to the same direction. This occurs spontaneously without an external field, i.e. without being forced by an external force. Such a manifest order cannot not possibly occur unless there is a correlation among the spins. Thus correlation is everything. The divergence of susceptibility may be attributed to a long-range correlation.

We shall study a system of interacting spins placed on a lattice. Its microstate energy, or Hamiltonian, is given by

$$E\{s\} = -J \sum_{(i,j)\in nn} s_i s_j - H \sum s_i \qquad (13.26)$$

where J is the coupling constant, $(i,j) \in nn$ is meant to say that i and j are nearest neighbors of each other, and H is the external magnetic field.

Let me rewrite the correlation function here.

$$G(\vec{r}_i - \vec{r}_j) = <(s_i - <s_i>)(s_j - <s_j>)> = <s_i s_j> - <s_i><s_j> . \qquad (13.27)$$

Its purpose is to find out whether or not two spins, one at $\vec{r}_i$ and one at $\vec{r}_j$, fluctuate independently or have any tendency to fluctuate in the same direction as if they wanted to act in unison. If s_i and s_j are not coupled, directly or indirectly, the two should behave independently and $<s_i s_j>$ should factorize into $<s_i><s_j>$, and therefore there should be no correlation. In order to have any correlation, the two have to interact with each other directly or indirectly through one or more third parties. When the correlation extends to a long range even though the spin interaction extends only to nearest neighbors, the correlation is primarily through third parties. When the correlation extends to a long range, the spins tend to act together and become more responsive to an external field.

To see why, let us start from the partition function

$$Z = \sum_{\{s\}} \exp\left[\beta J \sum_{\{s\}}\right]. \qquad (13.28)$$

The average spin would be

$$\left\langle \sum s_i \right\rangle = \frac{1}{Z} \sum_{\{s\}} \left(\sum s_i\right) \exp\left[\beta J \sum_{\{s\}} s_i s_j + \beta H \sum s_i\right]$$

$$= \frac{1}{\beta Z} \frac{\partial Z}{\partial H} \qquad (13.29)$$

and likewise,

$$\left\langle \sum s_i s_j \right\rangle = \frac{1}{Z} \sum_{\{s\}} \left(\sum s_i s_j \right) \exp \left[\beta J \sum_{\{s\}} s_i s_j + \beta H \sum s_i \right]$$

$$= \frac{1}{\beta^2 Z} \frac{\partial^2 Z}{\partial H^2}. \tag{13.30}$$

Remembering that $M = m \sum <s_i>$, it follows that

$$\chi_T = \frac{1}{N} \frac{\partial M}{\partial H} = \frac{m}{N} \frac{\partial}{\partial H} \left(\frac{1}{\beta Z} \frac{\partial Z}{\partial H} \right)$$

$$= \frac{m}{N} \frac{1}{\beta} \left[\frac{1}{Z} \frac{\partial^2 Z}{\partial H^2} - \frac{1}{Z^2} \left(\frac{\partial Z}{\partial H} \right)^2 \right]$$

$$= \frac{m}{N} \frac{1}{\beta} \left[\beta^2 \sum_{i,j} <s_i s_j> - \beta^2 \left(\sum <s_i> \right)^2 \right]$$

$$= \frac{m}{N} \frac{1}{k_B T} \sum_{i,j} G(\vec{r}_i - \vec{r}_j)$$

$$= \frac{m}{k_B T} \sum_i G(\vec{r}_i)$$

$$= \frac{m}{k_B T} \frac{N}{V} \int G(\vec{r}) d^d r, \tag{13.31}$$

where it has been assumed that our spin system is isotropic and $G(\vec{r}_i, \vec{r}_j) = G(|\vec{r}_i - \vec{r}_j|)$, i.e. the spin correlation between two spins depends only on the distance between them. Thus for the next line to the last we took for spin j the spin located at the coordinate origin and multiplied it by N. This is because no matter which spin we may choose as j, its correlation with the rest is the same. For the last line, note that the density of the spin sites is N/V; if our system is on a plane the density should be N/A.

With this important relationship between the susceptibility and the correlation function in hand, it is now clear that if the susceptibility is to diverge, the only way it can happen is a macroscopically long ranged correlation between spins. This is called the static sum rule. We argued verbally for the same idea for liquid-gas systems earlier.

I am wondering if shepherds find something similar to this on a sheep farm. If all the sheeps act independently and randomly, it would be difficult to handle many of them. But if they tend to act together, would they not be more responsive? If they try to move in a wrong direction, all he has to do is to force just a few to change its direction. The correlation among them would do the rest of the job for the shepherd. Maybe?

13.6   Interacting Ising Spins on a Chain

When the correlation extends over a sufficiently long range, the response functions diverge. It means that all spins are 'dancing' together even though their interactions extend only to nearest neighbors. Ising spins exhibit such a spectacle in two dimensions, and seeing it would be like watching a rabbit coming out of someones hat. Unfortunately, that is beyond our means. In the mean time, we will work with a much simpler model, one-dimensional Ising model. The rabbit is much smaller, but if we can pull it out of our hat, we can certainly claim that we know something about the correlation, which is a respectable goal for us. So, bring some paper and pencil and get ready.

The mathematics is simple. Suppose that η can only be $+1$ or -1. Under that condition and only under that condition, we may write

$$e^{\eta A} = \cosh A + \eta \sinh A = \cosh A (1 + \eta \tanh A). \tag{13.32}$$

Notice that on the left-hand side η is high up on the second floor in the exponent seat but the identity brings it down to the floor on the right-hand side. That is the virtue of this identity. In our applications, η is a product of Ising spins, and therefore no matter how many spins are involved in the product and no matter what each spin may be, the product is either $+1$ or -1. So, when we sum over all possible microstates by carrying out the summations,

$$\sum_{\{s\}} = \sum_{s=-1}^{+1} \sum_{s_2=-1}^{+1} \cdots \sum_{s_N=-1}^{+1}, \tag{13.33}$$

the result for each spin product will be:

$$\sum_{s_i=-1}^{+1} (s_i)^k = \begin{cases} 2 & \text{if k is even} \\ 0 & \text{if k is odd} \end{cases}. \tag{13.34}$$

Let us consider N Ising spins on a chain. Depending on how we choose to handle the two ends of the chain, we have two different Hamiltonians;

we may leave the two ends open or fold and close the chain to make a ring. The corresponding Hamiltonians are given by

$$H = -J(s_1s_2 + s_2s_3 + s_4s_5 + \cdots + s_{N-1}s_N), \quad \text{Open} \tag{13.35}$$

$$H = -J(s_1s_2 + s_2s_3 + s_4s_5 + \cdots + s_{N-1}s_N + s_Ns_1), \quad \text{Closed} \tag{13.36}$$

As an exercise, let us calculate the partition function Z_3 of a closed chain with $N = 3$,

$$Z_3 = \sum_{s_1=-1}^{1} \sum_{s_2=-1}^{1} \sum_{s_3=-1}^{1} e^{J(s_1s_2+s_2s_3+s_3s_1)/k_BT} . \tag{13.37}$$

There are three interaction terms in the exponent. Applying Eq. (13.32) for each,

$$Z_3 = \sum_{s_1=-1}^{1} \sum_{s_2=-1}^{1} \sum_{s_3=-1}^{1} \cosh^3(J/k_BT)(1 + vs_1s_2)(1 + vs_2s_3)(1 + vs_3s_1) \tag{13.38}$$

where $v = \tanh(J/k_BT)$. We have three products to expand. We will be picking from each parenthesis either a unity or a v term involving two spins. Only two terms survive the spin summations. In the first term only unity is picked throughout while in the second term only the v is taken throughout. The net result is

$$Z_3 = \cosh^3(J/k_BT)(1 + v^3)2^3 . \tag{13.39}$$

If the three spin chain is open, it is obvious that $Z_3 = \cosh^3(J/k_BT)2^3$.

Extending the same arguments to N spins, the partition function is given by

$$Z_N = \cosh^{N-1}(J/k_BT)2^N \qquad \text{Open Chain,} \tag{13.40}$$

$$Z_N = \cosh^{N-1}(J/k_BT)(1 + v^N)2^N \quad \text{Closed Chain.} \tag{13.41}$$

Next let us calculate the magnetization:

$$M = \mu_B \left\langle \sum_i s_i \right\rangle$$

$$= \mu_B \frac{1}{Z_N} \sum_{\{s\}} \left(\sum_i s_i \right) e^{J(s_1s_2+s_2s_3+\cdots+s_{N-1}s_N)/k_BT} . \tag{13.42}$$

Notice that there is one s_i already on the floor. When those on the second floor come down, it will be multiplied by unity or by $v \times$ *even number of spins*. Either way, it cannot survive the spin summation. Hence, $M = 0$ at any finite temperature. This reflects the symmetry between the ordered state with all spins up and that with all spins down. At finite temperatures, the spins will visit the state of 'all up' as often as that of 'all down' and therefore the equilibrium average magnetization is zero for all finite temperatures. We will return to this at the end.

Now let us calculate the correlation function for an open chain,

$$G(i, i+r) = \frac{1}{Z_N} \sum_{\{s\}} (s_i - <s>) (s_{i+r} - <s>) e^{J(s_1 s_2 + s_2 s_3 + \cdots + s_{N-1} s_N)/k_B T}$$

$$= \frac{1}{Z_N} \sum_{\{s\}} s_i s_{i+r} e^{J(s_1 s_2 + s_2 s_3 + \cdots + s_{N-1} s_N)/k_B T} \tag{13.43}$$

where $<s>$ is the average spin and is zero in the paramagnetic phase. The two spin variables are already down on the 'floor' waiting to be multiplied by those to come down. Proceeding as before, there is only one term that survives the spin summations. For this term, only 1 is picked until the ith spin is reached, from there only the v terms are picked until the $(i + r)$th spin, and then from there back to unity til the last spin is reached. Therefore the correlation between any two spins at distance r (in units of of the lattice spacing) is

$$G(i, i+r) = <s_i s_{i+r}>$$

$$= \frac{1}{2^N \cosh^{N-1}(J/k_B T)} \cosh^{N-1}(J/k_B T) v^r 2^N$$

$$= v^r. \tag{13.44}$$

At finite temperatures, since $v = \tanh(J/k_B T) < 1$, the correlation vanishes for large r, and therefore there is no long range correlation. As T approaches zero, however, v approaches unity, and therefore we do find a long range correlation but only at $T = 0$, not at a finite temperature as in real magnets in two or three dimensions. The correlation may be expressed more conveniently:

$$G(i, i+r) = [\tanh(J/k_B T)]^r = e^{r \ln \tanh(J/k_B T)} = e^{-r/\xi}, \tag{13.45}$$

where

$$\xi = 1/\ln \coth(J/k_B T). \tag{13.46}$$

is the length scale over which spin fluctuations are correlated, and hence is called the correlation length. By the way, the distance r here is the distance along the chain, not the shortest Pythagorean distance off the chain. This is so because the correlation is mediated along the chain. Such distances are often called 'cow' distance or 'taxi' distance. When cows want to go from one side of a pond to the opposite side, they walk around the pond, and so the distance between the two opposite ends should be measured along the path they take, not along the Pythagorean shortest path over the pond — they do not have an airplane yet. And likewise, taxi drivers charge us according to the path they take.

Let us find out how the correlation length depends on the temperature. Since

$$\coth(J/k_BT) = \frac{e^{J/k_BT} + e^{-J/k_BT}}{e^{J/k_BT} - e^{-J/k_BT}}$$

$$= 1 + 2e^{-2J/k_BT}(1 + e^{-2J/k_BT} + e^{-4J/k_BT} + \cdots), \quad (13.47)$$

it follows that

$$\xi = \frac{1}{2}e^{2J/k_BT}. \quad (13.48)$$

So the correlation length diverges at $T = 0$ like $e^{1/T}$ and not like $|T-T_C|^{-\nu}$. The severe singularity is called the essential singularity.

Next, let us calculate the susceptibility following Eq. (13.31)

$$\chi_T = \frac{1}{k_BT}\int_0^\infty \exp(-r/\xi)dr = \xi = \frac{1}{2}\frac{1}{k_BT}\exp(2J/k_BT) \quad (13.49)$$

which exhibits again the essential singularity of the type $e^{1/T}$. So the susceptibility does diverge as T approaches zero but much more severely than it does in real magnets in two or three dimensions at finite temperatures.

To conclude, the Ising model in one-dimension can be solved analytically, but at a finite temperature the correlations between spins are not strong enough to support a phase transition. Instead, the phase transition occurs at $T = 0$ with an essential singularity.

The Ising model in two dimensions requires a more intense analysis. Draw a two dimensional square lattice and put a spin on each lattice site. Each interacts with its nearest four neighbors. Starting from any spin site, keep on drawing v lines and ultimately return to the initial site to make a connected closed loop. Only such graphs survive the spin summations.

The length of the loop and its topology (shape) determine how much each type of graph contributes and how many such graphs there are. So, one has to be engaged in a lengthy mathematical issues but it is doable. The end result does show a finite temperature phase transition with power-law type of singularities as explained earlier. The solution is called the Onsager solution, but Onsager did not follow the simple method we have followed.[13] Extending this analysis to three dimensions has proven illusive for all these years and frustrated so many. It is waiting for a new idea.

The history of Ising model itself is interesting. Ernst Lenz suggested the model to his student Ising. Ising solved the model in one-dimension for his thesis but did not find a finite temperature phase transition, and they never returned to the model as far as I know. The model, however, inspired so many others over the years. In fact, it would be hard to find an issue of any major physics journal section devoted to condensed matter physics that does not contain one paper carrying the title 'Ising Model'. The trend will continue because Ising model is adaptable to so many different systems beyond magnets. Ising model embodies the principle that a model should capture the essence of physics in the simplest possible way without requiring to carry around any unnecessary baggages.

13.7　Mean-Field Theory: Magnetism

If one cannot calculate the partition function and correlation function exactly for any model, the next best thing one can do is a mean field theory. Let us start with an Ising model. The Hamiltonian, or the microstate energy, depends on the spin configuration in the following way:

$$E\{s\} = -Hm \sum_i s_i - J \sum_{<ij>} s_i s_j \,, \tag{13.50}$$

where H is the external magnetic field, $<ij>$ means that the two spins are nearest neighbors. The coupling constant J is positive, meaning that it costs energy for neighboring spins to oppose each other.

Suppose that our Ising spins are on a three dimensional simple cubic lattice where each has 6 neighbors to interact with, and imagine that we can actually see the spins in a slow motion movie. Each spin would be flipping up and down at each instant influenced by its neighboring spins. Let us examine this in more detail. Suppose that all of its 6 neighboring spins are

---

[13]Two good references are the book by Stanley and that by Landau and Lifshitz.

up. The energy cost would be $-6J$ if the spin chooses to be up following its neighbors, and $6J$ if it chooses to be down against its neighbors. This particular neighbor environment therefore strongly influences the spin to be up. On the other hand, if half of them are up while the other half are down, it does not cost anything for the spin to flip. In this way, depending on the specific spin configuration of its neighbors, the environment could strongly influence the spin to be down, slightly influence it to be down, or have no influence at all, etc. It is this fluctuating environment from site to site that makes it so difficult to solve the model in two and three dimensions. We may describe this situation more clearly by rewriting Eq. (13.50) in the following way.

$$E\{s\} = -\sum_i s_i \left\{ mH + J \sum_j s_j \right\} = -\sum_i s_i m\hat{H}_i , \qquad (13.51)$$

where

$$m\hat{H}_i = mH + J \sum_j s_j . \qquad (13.52)$$

may be regarded as the effective local field consisting of the external applied field and the sum of its nearby spins. The latter cannot be regarded as a field, but written in this form it acts as if it were. Because of the latter, the effective field fluctuates from site to site. One way to handle this difficulty is to ignore the fluctuation and replace the local field with its global average. Thus we approximate the effective field with

$$m\hat{H}_i = mH + Jz<s> , \qquad (13.53)$$

where z is the number of nearest neighbors (called the coordination number) which is 6 for the simple cubic lattice. Then the spins may be treated as ideal spins subject to a constant field. We may rely on Eq. (4.41), i.e. $M = Nm \tanh(mH/k_BT)$. Since $M = Nm<s>$, we have

$$<s> = \tanh \left(\frac{mH}{k_BT} + \frac{<s>}{a} \right) \qquad (13.54)$$

where

$$a = k_BT/Jz . \qquad (13.55)$$

We have ended up with a transcendental self-consistency equation. Since the phase transition takes place in the absence of an external field, set for

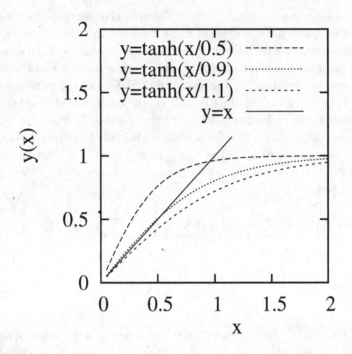

Fig. 13.7   Graphic plots of both sides of Eq. (7.4.4) for various values of a and with H set to zero.

a moment $H = 0$. The equation to solve is in the form of $x = \tanh(x/a)$. Let us first solve this graphically with a as a parameter. To that end, consider two functions: $y(x) = x$ representing the left-hand side, and $y(x) = \tanh(x/a)$ representing the right-hand side; we may find graphically where the two functions meet. Plotting the two functions with $a = 0.5, 0.9, 1.1$, the results are as shown in Fig. 13.7. It is clear that a self-consistent non-zero solution exists only for

$$a = k_B T / Jz < 1. \tag{13.56}$$

The solution as a function of a, i.e. as a function of T, is as shown in Fig. 13.8. To be complete, $x = 0$ is the solution for $a > 1$. The transition temperature is given by $a = 1$ or

$$T_C = \frac{Jz}{k_B}. \tag{13.57}$$

which shows that the transition temperature is determined by local parameters which is why the critical temperature itself is not a critical variable.

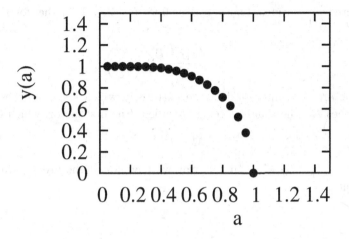

Fig. 13.8 The graphic solution as a function of a.

Now let us find out the precise manner in which the magnetization grows as the temperature passes below T_C in the ferromagnetic phase, and also how the susceptibility grows as the temperature approaches T_C in the paramagnetic phase. To prepare for these, rewrite the equation in terms of T and T_C,

$$<s> = \tanh\left(\frac{mH}{k_B T} + \frac{T_C}{T}<s>\right). \tag{13.58}$$

For small H and small $<s>$, the hyperbolic tangent function may be approximated using its Taylor series expansion,

$$\tanh(x) = x - \frac{x^3}{3} + \frac{2x^5}{15} - \cdots. \tag{13.59}$$

The equation to be solved is then

$$<s> = \left[\frac{mH}{k_B T} + \frac{T_C}{T}<s>\right] - \frac{1}{3}\left[\frac{mH}{k_B T} + \frac{T_C}{T}<s>\right]^3. \tag{13.60}$$

First, to solve it for $<s>$, set $H = 0$, which gives

$$<s>\left[1 - \frac{T_C}{T} + \frac{1}{3}\left(\frac{T_C}{T}\right)^3 <s>^2\right] = 0. \tag{13.61}$$

There are two solutions. One is $<s> = 0$ for $T > T_C$. The other for $T < T_C$ is:

$$<s>^2 = 3 \left(\frac{T_C - T}{T_C} \right) \left(\frac{T}{T_C} \right)^2 . \qquad (13.62)$$

Since we are only interested in the pattern in which $<s>$ vanishes as T approaches T_C, the sought-after pattern is in the first factor, which gives

$$M \sim t^{1/2} \qquad (13.63)$$

where $t = |T - T_C|/T_C$. In this way we have found the order parameter exponent to be

$$\beta = 1/2 . \qquad (13.64)$$

Next, calculate the susceptibility from

$$\chi = \frac{1}{N} \frac{\partial M}{\partial H} = m \frac{\partial <s>}{\partial H} . \qquad (13.65)$$

For a leading order calculation, it is sufficient to differentiate just the first order term in Eq. (13.60) with respect to H, which gives

$$\frac{\partial <s>}{\partial H} = \frac{m}{k_B T} + \frac{T_C}{T} \frac{\partial <s>}{\partial H} . \qquad (13.66)$$

The solution for $\partial <s>/\partial H$ gives

$$\chi \sim (T - T_C)^{-1}, \qquad (13.67)$$

and hence

$$\gamma = 1. \qquad (13.68)$$

Next let us calculate the heat capacity.

$$C_H = \left(\frac{\partial U}{\partial T} \right)_{H=0} \qquad (13.69)$$

where U is the average energy. From Eqs. (13.51) and (13.52) with the external field H set to zero, the average energy is

$$U = <E> = -NJz<s>^2 . \qquad (13.70)$$

For $T > T_C$, since $<s> = 0$, $U = 0$, it follows that

$$C_H = \partial U/\partial T = 0. \qquad (13.71)$$

For $T < T_C$, on the other hand, substituting Eq. (13.62) into Eq. (13.70) and then differentiating it with respect to T, we obtain

$$C_H = NJz3\frac{T^2}{T_C^3} - NJz\left(\frac{T_C - T}{T_C}\right)\frac{2T}{T_C^2}. \tag{13.72}$$

As T approaches T_C, it approaches a non-zero constant. Thus the specific heat exhibits a discontinuity at T_C. According to our definition of critical exponents, the heat capacity exponent is

$$\alpha = 0. \tag{13.73}$$

In order to calculate the rest of the exponents, we have to put the mean field idea in a more sophisticated form which we will do in the next chapter. We started the calculation with three dimensions in mind, but no where during the calculation did we have a chance to do anything to take the dimensionality into account. The only quantity that we put in which has anything to do with the lattice is the coordination number z, but z cannot represent the dimensionality. For example, in two dimensions, $z = 3$ in a triangular lattices while it is 4 in a simple square lattice. In spite of this murky situation, the calculated exponents are not very far away from the known values for three dimensional Ising model and those for two dimensional model. Judging from how little we had to work to get these results, this is a remarkable success. The embarrassment is, however, that there is nothing in the same theory that prevents the finite temperature phase transition in a one-dimensional Ising model. So, we have to clarify the dimensionality issue.

With this in mind, notice that the critical temperature T_C is dependent on the atomic details such as J and z, but the critical exponents are independent of such minute details. Although z is not a direct measure of dimensionality, it does tend to increase with increasing dimensionality. So, let us increase z gradually larger and larger. The transition temperature increases with z, but the exponents are blind of z. Increase it as large as $z = N - 1$. The transition would take place at a very high temperature, but as far as the criticality goes, the exponents remain unchanged. How could this be? Replacing the fluctuating local fields with the global mean field was tantamount to assuming that for every spin all the other spins contribute equally to its local field, or that every spin is every other spins neighbor. This may be justified only if the dimensionality of the lattice is infinite. It is interesting therefore to notice that the mean field values for β and γ are closer to those of three dimensional Ising model than those of the two dimensional model. Actually it is possible that we may not have to go as far as $d = \infty$ to justify the mean field idea. As we will learn in the next chapter, that turns out to be the case. Finally, notice that the mean field values of α, β, and γ satisfy the first scaling law in Eq. (13.22).

13.8 Mean Field Theory: Liquid-Gas Systems

We studied earlier the van de Waals mean field theory for liquid-gas systems. Let us refresh it so that we may use it to study the critical phenomena. When the particles of an ideal gas are allowed to interact with each other, the Helmholtz free energy changes by

$$\Delta F = F_{vdw} - F_{idl} = -Nk_BT \ln\left(\frac{V - Nb}{V}\right) - \frac{aN^2}{V}, \qquad (13.74)$$

where a and b are two constants designed to capture the essential effect of the interactions. The resultant equation of state is

$$P = \frac{Nk_BT}{V - Nb} - \frac{N^2a}{V^2}. \qquad (13.75)$$

At the critical point, the pressure, volume and temperature are given by

$$P_C = \frac{a}{27b^2}; \quad V_C = 3bN; \quad T_C = \frac{8a}{27bk_B}. \qquad (13.76)$$

When the pressure, volume and temperature are scaled with their respective critical value, the equation of state takes the form of

$$\left(\hat{P} + \frac{3}{\hat{V}^2}\right)\left(\hat{V} - 1\right) = 8\hat{T}. \qquad (13.77)$$

Before we start, let me explain why this is a mean field theory. Assume that there is an atom at the coordinate origin. Its neighbor atoms interact with the atom at the origin directly, or indirectly via other particles in between, and constitute the local environment for the atom. When we replaced the local density $n(\vec{r})$ with the global average density, the fluctuation of the local environment was ignored. There is nothing special about being at the coordinate origin. Rarely, if at all, particles would be gathered in such a way that the instantaneous local density is equal to the global average density everywhere. If that happened, all particles would stay absolutely motionless at all times. In real liquids, some particles would be pushed to the left by their neighbors, some to the right, etc. All these site-to-site fluctuation are ignored, whence the theory is a mean field theory.

Let us start with the specific heat

$$C_V = T \left(\frac{\partial S}{\partial T} \right)_V . \tag{13.78}$$

The change in entropy due to the interactions is

$$\Delta S = -\frac{\partial \Delta F}{\partial T} = -N k_B \ln \left(\frac{V - Nb}{V} \right) . \tag{13.79}$$

Since there is no temperature dependence, the specific heat of the interacting molecules is the same as that of their ideal counterpart,

$$C_V = 3 N k_B / 2 , \tag{13.80}$$

hence the specific heat goes like

$$C_V \sim (T - T_C)^{-\alpha}, \quad \alpha = 0 . \tag{13.81}$$

Let us calculate the isothermal compressibility

$$\chi_T = -\frac{1}{V} \frac{\partial V}{\partial P}$$

$$= -\frac{1}{V} \frac{1}{-N k_B T / (V - Nb)^2 + 2N^2 a / V^3}$$

$$= \frac{1}{V} \frac{(V - Nb)^2 / N k_B}{T - 2N^2 a (V - Nb)^2 / V^3 N k_B} . \tag{13.82}$$

Substituting $V = V_C = 3Nb$, we find

$$\chi_T = \frac{4b/3k_B}{T - T_C}, \quad \chi_T \sim (T - T_C)^{-\gamma}, \quad \gamma = 1 . \tag{13.83}$$

Next turn to the isotherm of $T = T_C$ that passes through the critical point. We wish to find out how the volume changes with pressure. Start from Eq. (13.77). It is convenient to rewrite it in terms of $p = \hat{P} - 1 = (P - P_C)/P_C$, $v = \hat{V} - 1$, and $\epsilon = \hat{T} - 1$; these variables represent the degree of deviation from the critical point, and therefore they are all small. It follows that

$$\left[p + 1 + \frac{3}{(v+1)^2} \right] [3(v+1) - 1] = 8(\epsilon + 1) . \tag{13.84}$$

Multiply the equation with$(v+1)^2$, and then expand both sides and combine terms to obtain

$$2p(1 + 7v/2 + 4v^2 + 3v^3/2) = -3v^3 + 8\epsilon(1 + 2v + v^2). \qquad (13.85)$$

Set $\epsilon = 0$, i.e. $T = T_C$, and find that

$$p = -\frac{3}{2}v^3(1 + 7v/2 + 4v^2 + 3v^3/2)^{-1}$$

$$= -\frac{3}{2}v^3(1 - 7v/2 + \cdots) \qquad (13.86)$$

where the expansion is permissible because v is small. The leading order of the isotherm is v^3, hence we may write

$$P \sim (V - V_C)^\delta, \quad \delta = 3. \qquad (13.87)$$

The order parameter must reflect the extent to which the particles are gathered, and for that purpose, the coexistence volumes V_l on the liquid side and V_g on the gas side may serve the role. These values can be obtained by performing the Maxwell construction using the equation of state. Start with Eq. (13.77) and proceed as before,

$$\hat{P} = \frac{8\hat{T}}{3\hat{V} - 1} - \frac{3}{\hat{V}^2}$$

$$= \frac{8(\epsilon + 1)}{3(v + 1) - 1} - \frac{3}{(v + 1)^2}$$

$$= \frac{8(\epsilon + 1)/2}{1 + 3v/2} - \frac{3}{(1 + v)^2}$$

$$= 4(\epsilon + 1)(1 - 3v/2 + 9v^2/4 - 27v^3/8 + \cdots) - 3(1 - v + v^2 - v^3 + \cdots)$$

$$= 1 + 4\epsilon - 6\epsilon v - 3v^3/2 + \cdots \qquad (13.88)$$

which is an expansion up to the third power of two small variables ϵ and v. With this, we now perform the Maxwell construction in the form of

$$\oint v d\hat{P} = 0. \qquad (13.89)$$

Since $d\hat{P} = -6\epsilon dv - 9v^2/2v^2 dv$, the integral is, up to the second order,

$$\int_{v_l}^{v_g} v d\hat{P} = -\frac{6}{2}\epsilon(v_g^2 - v_l^2) = 0 \qquad (13.90)$$

which shows that $v_l = v_g$ up to the shown order of approximation. The corresponding reduced pressures are

$$\hat{P}_g = 1 + 4\epsilon - 6\epsilon v_g - 3v_g^3/2\,,$$
$$\hat{P}_l = 1 + 4\epsilon + 6\epsilon v_g + 3v_g^3/2\,. \tag{13.91}$$

Here $\hat{P}_g = \hat{P}_l$. Subtracting the two,

$$12\epsilon v_g + 3v_g^3 = 0, \quad v_g \sim (T_C - T)^\beta, \quad \beta = 1/2\,. \tag{13.92}$$

All other exponents require more sophisticated version of mean field theory.

Notice that the results for β, γ and α are the same as those for Ising ferromagnets. This shows that the single-component liquid gas systems and the Ising ferromagnets are in the same universality class. The spins are in a solid state while the atoms are in a mobile liquid state, but they share the same set of exponents. It is the fact that for both systems the order parameter has only one component that puts the two to the same class. Finally, it is delightful to notice that the mean field values of the three exponents that we have calculated satisfy the first two scaling laws in Eqs. (13.22) and (13.23).

13.9 Landau Mean Field Theory

Landau constructs a mean field theory in a different and remarkably simple way. Landau's mean field theory is applicable to all second order phase transitions, but we will be speaking below in the language of ferromagnetism. This is only to have something concrete to think of while learning this important subject rather than getting fuzzy for the sake of generality. I said earlier that if some insightful inspiration led you to guess a free energy in a certain parametric form, the parameters may be determined by using the free energy principle. I said so with Landau's mean field theory in mind.

Landau proceeds as follows. How much free energy would it cost if the total magnetization were to be M at temperature T? From Eq. (11.9), the answer is

$$F(M,T) = -k_B T \ln \sum_i \Delta(M - M_i) \exp(-E_i/k_B T), \tag{13.93}$$

where $\Delta(M - M_i)$ is 1 if $M_i = M$ and zero if $M_i \neq M$. It selects among all the microstates only those in which the magnetization M_i is M.

So with M, we first declare a macrostate. The sum then gives the partial partition function limited to the microstates belonging to the declared macrostate and excluding all the rest, and then the logarithm gives the corresponding free energy. It has two independent variables, but M and T are not on the same footing. The equilibrium value of M is a function of T, and therefore it cannot be an independent variable for an equilibrium state. It is there along with T to specify the macrostate in question. If the chosen macrostate is small, $F(M, T)$ would be unduly large. If the macrostate is large, $F(M, T)$ would be small, reaching its minimum for the largest macrostate representing the equilibrium state. So, if we had this function, we would know the nature of the equilibrium state. We would also be able to study how a system thrown into a non-equilibrium state moves toward the equilibrium state, which we will do in the next chapter.

Unfortunately we do not know how to carry out the summation that Eq. (13.93) requires. It was written there only to show what $F(M, T)$ means. We shall guess. We have several clues. First, at the critical point, nothing changes abruptly. Therefore the free energy should be a smooth analytical function of both M and T. Second, the order parameter M is zero in the high temperature phase and starts to grow smoothly when the temperature enters into the narrow critical region below T_C. So, if we pretend that we have the exact function $F_L(M, T)$, we should be able to Taylor expand it in terms of M. That series expansion is our guess for $F_L(M, T)$:

$$F_L(M, T) = \frac{1}{2}g_2(T)M^2 + \frac{1}{4}g_4(T)M^4 - HM, \qquad (13.94)$$

where the first three terms are the expansion of the field-independent part of the free energy. The expansion contains only terms of even powers, which reflects the fact that in the absence of the external field H the free energy should be invariant to the spin inversion $s_i \to -s_i$, i.e. it should be an even function of M. The series ends at the M^4 term for convenience. For the field-dependent part, we have added the $-HM$ term. Think of it as the leading term of the expansion for the field-dependent part of the free energy. The coefficients should be smooth functions of T. We shall choose them so that when the system is at equilibrium near $T = T_C$, F_L can mimic the second order phase transition as best as it can within the limitation set by the functional form given by Eq. (13.94). It is amazing that a singular behavior can emerge from such a smooth analytic function. Since $F_L(M, T)$ is written entirely in terms of the constant global total spin M ignoring local fluctuations, it is a mean field theory.

Now we are ready to find out the value of M corresponding to the largest macrostate, i.e. the equilibrium value of the order parameter. To that end, we demand $\partial F_L / \partial M = 0$, which gives

$$g_2 M + g_4 M^3 - H = 0 \,. \tag{13.95}$$

We shall use this equation to figure out how the order parameter grows, how the susceptibility changes with temperature, and finally to figure out the equation of state at $T = T_C$.

First consider the order parameter itself in the absence of the field. So, set $H = 0$ and solve the remaining part of the equation for M. Obviously $M = 0$ is one solution, regardless of the temperature. It should remain as the lone solution for $T > T_C$. There should be a nontrivial solution for $T < T_C$. Let us seek it in

$$g_2 + g_4 M^2 = 0 \,, \tag{13.96}$$

which gives

$$M = \pm \left(\frac{-g2}{g_4} \right)^{1/2}, \quad \text{for} \quad T < T_C \,. \tag{13.97}$$

We want this to be the solution only for $T < T_C$, which can be secured by choosing

$$g_2 = a(T - T_c), \quad g_4 = b \tag{13.98}$$

where a and b are positive constants. The temperature dependence is very smooth as we required earlier. With both coefficients chosen in this way, let us take a close look at the model.

$$F_L(M, T) = \frac{a}{2}(T - T_C)M^2 + \frac{b}{4}M^4 - MH \,. \tag{13.99}$$

Figure 13.9 depicts F_L plotted as a function of M at various temperatures in the absence of H. For $T > T_C$, the curve has one stable minimum at $M = 0$. For $T < T_C$, the minimum at $M = 0$ becomes an unstable maximum while two stable minima emerge that favor nonzero M and hence a ferromagnetic phase. The minimum position starts from $M = 0$ at $T = T_C$ and increases continuously as T further decreases, as it should. It is a picturesque exposition of a second order phase transition. Moreover, this form of free energy shows an important feature in the shape of the minima. For T away from T_C, the shape of the minima is that of the square term

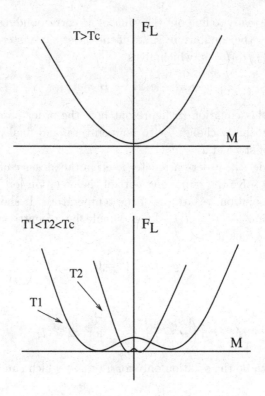

Fig. 13.9   The Landau mean-field free energy model.

M^2. But at $T = T_C$, the minimum is determined by the quartic term M^4, which means that the basin is indeed flatter and wider at T_C than away from T_C.

The order parameter changes with temperature following

$$M = (a/b)^{1/2}(T_C - T)^{1/2} ,   \tag{13.100}$$

which gives the order parameter exponent

$$\beta = 1/2   \tag{13.101}$$

in agreement with the previous mean field theory.

Return back to Eq. (13.95) for the susceptibility. In the presence of the external field H, the same equation now tells us how the field affects the value of M corresponding to the largest macrostate. Rather than solving the cubic equation for M and then differentiating it with respect to H, we may exchange the order; differentiate the equation with respect to H and then solve for the susceptibility. Thus we find

$$a(T - T_C)\frac{\partial M}{\partial H} + 3bM^2\frac{\partial M}{\partial H} - 1 = 0. \tag{13.102}$$

In the second term, M is zero for $T > T_C$ and as given by Eq. (13.100) for $T < T_C$. Thus we find

$$\chi = \frac{\partial M}{\partial H} = \begin{cases} \dfrac{1}{a(T - T_C)}, & \gamma = 1 \quad \text{if} \quad T > T_C \\ \dfrac{1}{2a(T_C - T)}, & \gamma = 1 \quad \text{if} \quad T < T_C \end{cases}. \tag{13.103}$$

The susceptibility does diverge $T = T_C$. From the fluctuation-dissipation theorem, this means that there is a diverging global fluctuation in the magnetization, which we anticipated from the flatter basin of the free energy at $T = T_C$. From the static susceptibility sum rule, this also means that the model supports a long-ranged correlation, but the present form is not suitable to bring out this feature to light.

Return back to Eq. (13.95) again to find out how the order parameter changes with the external field at $T = T_C$. We find

$$M = \left(\frac{H}{b}\right)^{1/3}, \qquad \delta = 3. \tag{13.104}$$

Now let us calculate the heat capacity in the absence of the external field. To this end, substitute Eqs. (13.98) and (13.100) into Eq. (13.94). It is clear that $F_L = 0$ for $T > T_C$ and $-(1/4)(T - T_C)^2 a^2/b$ for $T < T_C$. Thus we find

For $T > T_C$,

$$C_H = -T\frac{\partial^2 F_L}{\partial T^2} = 0. \tag{13.105}$$

For $T < T_C$,

$$C_H = T\frac{a^2}{2b^2}. \tag{13.106}$$

Thus there is a finite discontinuity, hence

$$\alpha = 0. \tag{13.107}$$

Just with what we have gathered so far, it is quite clear that the Landau mean field theory yields the same exponents as those we saw in the two previous mean field theories. Its virtue is that it directly involves the order parameter and therefor it is adaptable to liquid-gas systems, and in fact to many other systems as well. Just say what the order parameter is and what the corresponding field is.

The mean field results are not very far away from the experimental results or the exact model results when such results are available. This is remarkable in view of the simplicity of the mean field theories, but it has proven very difficult to improve beyond the mean field theories. The situation needed some new ideas. Emerging at this point is the idea of the static scaling hypothesis.

Exercise. *Figure out what the following free energy model attempts to represent,*

$$F_L^2(\eta, T) = F_L^1 + \frac{1}{2}g_2\eta^2 - \frac{1}{4}g_4\eta^4 + \frac{1}{6}\eta^6 , \qquad (13.108)$$

where η stands for the order parameter. $F_L^2(\eta, T)$ is the macrostate free energy of phase 2 and $F_L^1(\eta, T)$ is that of phase 1; $g_2 = a(T - T_0)$, $a > 0$, $g_4 > 0$, and $g_6 > 0$. Notice that the g_4 term and the g_6 term compete against each other, the former in favor of phase 2 and the latter in favor of phase 1. The g_2 term sides with the g_6 term at $T > T_0$ and with the g_4 term at $T < T_0$. When the competition changes from favoring one to another, does the order parameter change smoothly or abruptly?

Another version which shows the same type of abrupt change is

$$F_L^2(\eta, T) = F_L^1 + \frac{1}{2}g_2\eta^2 + c\eta^3 + b\eta^4 . \qquad (13.109)$$

13.10   The Static Scaling Hypothesis

It is easy to say 'role up your sleeve and try hard, etc.', but no one can endure for long without an awarding and enjoyable moment. We can certainly use one now, and we have a good one here.

Someone once said that it would be impossible for anyone to walk through the halls of his physics department building without hearing the word "scaling" coming out of a room. This points out how popular the idea of scaling is. We have already observed what might be called a primitive version while studying the van der Waals theory.

The static scaling hypothesis may be presented in several ways. We will first follow a version due to Widom. The hypothesis says that free energy is a homogeneous function of its proper variables. What does this mean? If a function $F(x, y)$ is homogeneous with degree 1, the function is associated with two numbers p and q such that the following holds true, no matter what value is chosen for λ:

$$F(\lambda^p x, \lambda^q y) = \lambda F(x, y). \tag{13.110}$$

This means that if we know $F(x, y)$ for a limited range of x and y, say, $x < L$ and $y < L$, we can tell what the function is outside the range by merely changing the scales (units) of x, y, and F. Suppose that we wish to know $F(x, y)$ for $x < L$ and $y < L$. Then choose λ such that

$$\lambda^p x = X > L, \quad \text{and} \quad \lambda^q y = Y > L. \tag{13.111}$$

Then

$$F(X, Y) = \lambda F(x, y). \tag{13.112}$$

We do not even have to bother going outside the $x < L$ and $y < L$ range. We may stay inside the range and just change the unit of the x axis by λ^p, the y axis by λ^q, and the z axis (for F) by λ and then read the vertical reading! It is a lazy man's dream function!

Let us rephrase this fact in a more practical way. There are two independent variables x and y. Suppose that we plot F versus x with y fixed at various values. The plot is depicted in Fig. 13.10. The relationship between F and x seems to require an infinite number of curves. But notice the following. Since we may choose any value for λ, let us try $\lambda^q y = 1$ or $\lambda = y^{-1/q}$. Substituting this into Eq. (13.110), we find

$$F(y^{-p/q} x, 1) = y^{-1/q} F(x, y), \tag{13.113}$$

where the left-hand side is a function of one variable $z = y^{-p/q} x$. Therefore if we plot $F/y^{1/q}$ as a function of $x/y^{p/q}$, all of those scattered curves collapse into one, as depicted in Fig. 9.3(b). The implication is that x itself and F itself are meaningless. To be meaningful, x has to be measured in units of $y^{p/q}$, and likewise F has to be measured in units of $y^{1/q}$. When F and x are scaled in this way, they have a unique one-to-one relationship. Of course, we may say likewise for F and y in which case x will determine the proper scales.

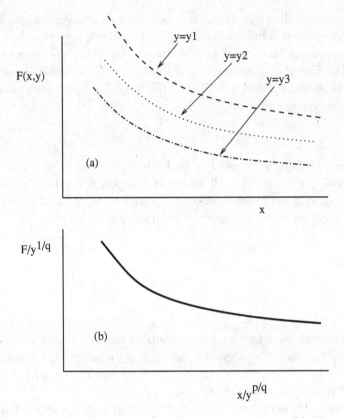

Fig. 13.10 (a) A homogeneous function $F(x, y)$ plotted versus x. (b) When F and x are properly scaled, all of the curves collapse into one.

There are many situations in life where different variables are interrelated in a similar fashion. When a person is stung by bees, a doctor may say something similar. It is not the amount of poison itself that determines how critical the poison is going to be. It is the amount of poison that went into the blood stream divided by the total amount of blood in the body. The reader can add many more similar examples.

Now let us apply the static scaling hypothesis to a ferromagnet in a magnetic field. The appropriate free energy is the Gibbs free energy, namely, $G(t, H)$, where $t = |T - T_C|/T_C$. Assume that $G(t, H)$ is a homogeneous function of t and H, namely,

$$G(\lambda^p t, \lambda^q H) = \lambda G(t, H). \tag{13.114}$$

What does the homogeneity assumption say about the magnetization, susceptibility, specific heat, and so on? Recall that M is to spins what V is to molecules, and mH is to spins what $-P$ is to molecules. According to Eq. (7.6), M may be obtained from G by differentiating it with respect to H. By differentiating M with respect to H, we may then obtain the susceptibility χ. The specific heat C may be obtained from G by differentiating it twice with respect to t.

Differentiating Eq. (13.114) with respect to H, we obtain

$$\lambda^q M(\lambda^p t, \lambda^q H) = \lambda M(t, H).\qquad(13.115)$$

Let $H = 0$ and, since we may choose anything for λ, let $\lambda^p t = 1$, i.e. $\lambda = t^{-1/p}$. Then Eq. (13.115) reduces to

$$M(t, H) = t^{-(q-1)/p} M(1, 0),\qquad(13.116)$$

which gives

$$\beta = \frac{1-q}{p}.\qquad(13.117)$$

To obtain the susceptibility exponent γ, differentiate Eq. (13.115) once more with respect to H to obtain

$$\lambda^{2q} \chi(\lambda^p t, \lambda^q H) = \lambda \chi(t, H).\qquad(13.118)$$

Now let $H = 0$ and $\lambda^p t = 1$ to obtain

$$\chi(t, 0) = t^{-(2q-1)/p} \chi(1, 0),\qquad(13.119)$$

which gives

$$\gamma = \frac{2q-1}{p}.\qquad(13.120)$$

To obtain the specific heat exponent α, return to Eq. (13.114) and differentiate it twice with respect to t to obtain

$$\lambda^{2p} C(\lambda^p t, \lambda^q H) = \lambda C(t, H).\qquad(13.121)$$

Let $H = 0$ and $\lambda^p t = 1$, which gives

$$C(t, 0) = t^{-(2p-1)/p} C(1, 0),\qquad(13.122)$$

and thus

$$\alpha = \frac{2p-1}{p}.\qquad(13.123)$$

Since we do not know what p and q are, we have not really calculated these exponents. However, substitute the three exponents that we have obtained above into Eq. (13.22). All p's and q's are canceled out, leaving us with the result 2 as the scaling law predicts!

The static scaling hypothesis may be understood in the following way. By means of this hypothesis, thermodynamic quantities effectively say to us: "In the critical region, we do not know anything but the correlation length ξ. Everything that we do in the critical region is based on the correlation length. If you doubt, then dissect us. You will find ξ inscribed in our body." Consider a thermodynamic quantity A which comes in the unit of length(L) raised to the power of D_A. According to the last sentence, if we dissect it, we should find

$$[A] = L^{D_A}, \qquad \text{or} \qquad A \sim \xi^{D_A}, \tag{13.124}$$

where the vertical bracket signs mean that we are only specifying the unit of A. Since we know how ξ changes with temperature in the critical region, we now know A's critical behavior. We expect all thermodynamic quantities involved in describing a critical behavior to act according to the above statement to the LETTER. Under this circumstance, the thermal behavior of any critical quantity is totally dictated by its spatial dimensionality and thermal physics becomes just a dimensional analysis.

Let us first find the dimensionality of the free energy G. Well, free energies have composite units and do not come in a single length-related unit. So, what we have to do is to multiply or divide it with some other quantities so that we can convert it into some thing with a well-defined spatial unit. The only physically meaningful such a quantity related to G is $A = G/(k_B T V)$: G has been divided by $k_B T$ to make it unitless and then divided again by V so that we can say "free energy per volume" which is a physically meaningful thermodynamic quantity. We could have multiplied G by V, but then the resultant quantity would have no physical meaning and cannot be regarded as a thermodynamic quantity. Of course we made this construction because we know that the critical property of thus constructed A has to be solely due to G while $k_B T$ and V are there only to give G a length-related unit. This clearly sounds like a cheating, and it is in that sense. That was how I felt, I remember, when I first saw this in a book. A popular question in my class has been: does volume V somehow change with temperature like ξ^3 in the critical region? Of course, V itself does not. Either V is fixed or may vary if the system is tied to a pressure reservoir. In the latter case, volume can fluctuate, but that fluctuation does not go like ξ^3. It is g that the hypothesis claims to behave like ξ^{-d}, and to be in the true spirit of the hypothesis, it is not correct to say that that is because of the volume. In any event, what is important is that it is not illegal because we strictly followed the hypothesis. So, let us go ahead and try it out. For $g = G/(k_B T V)$, we have

$$[g] = \frac{1}{L^d}, \qquad \text{or} \qquad g \sim \frac{1}{\xi^d}. \qquad (13.125)$$

Now we have to deal with the field H and the magnetization M. Following the same strategy, we shall write

$$\left[\frac{HM}{k_B T V}\right] = \frac{1}{L^d} \qquad (13.126)$$

which in turn allows us to write

$$\left[\frac{M}{V}\right] = \frac{1}{L^{D_M}}, \qquad \text{and} \qquad \left[\frac{H}{k_B T}\right] = \frac{1}{L^{D_H}}, \qquad (13.127)$$

under the condition

$$D_M + D_H = -d. \qquad (13.128)$$

To determine D_M, let me rewrite Eq. (13.18) as

$$G(r) = <\delta m(0)\delta m(\vec{r})>, \qquad (13.129)$$

where m is the local magnetization density M/V. We may then read D_M off Eq. (13.20):

$$D_M = -(d - 2 + \eta)/2, \qquad (13.130)$$

which then gives

$$D_H = -d - D_M = -(d + 2 + \eta)/2. \qquad (13.131)$$

This is another spot which the reader may find hard to swallow. The field is just what we apply and is not by itself a critical quantity. Why should it behave like ξ^{D_H}? By itself, it should not. But Eq. (13.127) is a legitimate thing to do as is the first part of Eq. (13.127). These forced us to the second part in Eq. (13.127). That is just how the hypothesis works.

We are now ready to find out all the exponents. For the order parameter exponent β, using Eq. (13.127),

$$\left[\frac{M}{V}\right] = L^{D_M}, \qquad \frac{M}{V} \sim \xi^{(d-2+\eta)/2} = t^{\nu(d-2+\eta)/2}, \qquad (13.132)$$

whence

$$\beta = \nu(d - 2 + \eta)/2 \, . \tag{13.133}$$

Next, for the susceptibility exponent γ,

$$[\chi] = \left[\frac{\partial M}{\partial H}\right] = L^{D_M - D_H} = L^{2 - \eta}, \quad \chi \sim t^{-\nu(2 - \eta)} \, , \tag{13.134}$$

whence

$$\gamma = \nu(2 - \eta) \, . \tag{13.135}$$

For the specific heat exponent α, recall

$$[g] = L^{-d}, \quad g \sim t^{\nu d} \, , \tag{13.136}$$

whence

$$C_H \sim \frac{\partial^2 g}{\partial t^2} \sim t^{\nu d - 2} \, . \tag{13.137}$$

and therefore

$$\alpha = 2 - \nu d \, . \tag{13.138}$$

Finally doing the dimensional analysis on

$$M \sim H^{1/\delta} \tag{13.139}$$

we write

$$L^{d - 2 + \eta)/2} \sim L^{-(d + 2 - \eta)/2\delta} \tag{13.140}$$

which gives

$$\delta = \frac{d + 2 - \eta}{d - 2 + \eta} \, . \tag{13.141}$$

We cannot say anything about ν because everything is based on the correlation length; we simply do not have anything else for the correlation length to rely on. We cannot say anything about η either because it determines the dimension of the order parameter and likewise for everything related to the order parameter. We have determined four exponents. Four is just the right number because there are four scaling laws for them to obey. Now comes the real fun part! Do they obey the scaling laws? Yes, they obey every single one! It is particularly worth to notice Fisher's and Josephson's scaling laws. Since we have nothing to substitute for ν and η, the susceptibility exponent γ which involves both ν and η had better come out as Fisher's law says, and likewise the specific heat exponent α which involves ν had better come out as predicted by Josephson's law, and Oh My! They sure do!

The hypothesis does not give any specific value for any exponent, but it works remarkably well in relating different exponents to each other as required by the scaling law, and by the way, you can tell why those relationships are called scaling laws. This suggests that a successful theory of critical phenomena may have to involve some sort of scaling argument, and we should remember that scaling can make seemingly different things look alike.

Having said all these, here is a caveat. The correlation length is the longest length scale, and therefore it is understandable that it plays such an important role, but it would be wrong to say that it is the only length scale that matters and that all shorter length scales have no role to play. They all play roles in the critical opalescence. Samples in critical region scatter all components of light in all directions to make themselves look milkishy white. If ξ is the only pertinent length scale, the samples should look red.

Chapter 14

Landau-Ginzburg Free Energy Functional and Applications

14.1 Introduction

In the previous chapter we studied in some depth the Landau mean field theory, but that form of mean field theory is not adequate to discuss the correlation function. Emerging to overcome this difficulty is the Landau–Ginzburg theory, which will be the subject but only briefly in Sec. 14.2. Rather remarkably, the correlation function calculated in this version of mean field theory sheds some light on when mean field theories are reliable. According to what is known as the Ginzburg criterion, mean field theories are valid (a) when the temperature is away from the critical temperature or (b) when the dimensionality of the system is equal to, or greater than, four. Following the guide (a), we will apply the Landau–Ginzburg form of free energy to issues related to the first order phase transitions in Sec. 14.3.

According to (b) the mean field theories fail when the dimensionality is less than four. This is due to the inadequate way mean field theories handle fluctuations. Emerging to overcome this shortcomings is the Landau–Ginzburg free energy functional. In Sec. 14.4, we will introduce it and then briefly introduce the idea of the renormalization group theory of Wilson. This is the end of our long adventure on critical point phenomena.

14.2 Landau–Ginzburg Free Energy

In the earlier version of Landau theory, the order parameter M represents the entire sample volume. To include local fluctuations in the order parameter, we divide the volume into small cells. Each cell is small enough to be represented by a differential volume element d^3r but large enough to include many spins in it. Let $s(\vec{r})$ represent the average magnetization of the cell located at $\vec{r}$ and call it the order parameter function. Because $s(\vec{r})$ is the average of many spins, it is no longer limited to $+1$ and -1; it is a continuous function of $\vec{r}$. So, we are now declaring our macrostate with a particular shape of the function $s(\vec{r})$. Let us build the macrostate free energy per volume at $\vec{r}$. We may start by extending the previous Landau free energy into the current version,

$$f_L[s(\vec{r})] = \frac{1}{2}g_2 s(\vec{r})^2 + \frac{1}{4}g_4 s(\vec{r})^4 , \qquad (14.1)$$

where g_2 and g_4 are specified by Eq. (13.98). If necessary, terms of higher order may be added. This is just the density; for the total macrostate free energy, it will have to be summed over the entire space of the volume.

Next we introduce a similarly inhomogeneous $H(\vec{r})$ and add the interaction term $H(\vec{r})s(\vec{r})$ to the free energy function. Here H is the external magnetic field for magnets and an external chemical potential for gas-liquid systems. The order parameter for the latter is related to the density N/V, and conjugate to the particle number N is the chemical potential. Finally, we may add a term to mimic particle interactions. In an Ising model, if a spin agrees with its neighbor environment making the spin distribution smooth, it saves energy and thus free energy. But if it goes against its environment to make the spin distribution to change abruptly, it costs free energy. We can bring that effect here with $(\nabla s(\vec{r}))^2$. In a macroscopic length scale, the gradient term charges hefty free energy cost when there is an interface. Putting all these together and integrating over the entire volume, the total macrostate free energy is given by

$$F_L\{s(\vec{r})\} = \int d^d r \left[\frac{K}{2}\left(\nabla s(\vec{r})\right)^2 + \frac{a}{2}t s(\vec{r})^2 + \frac{b}{4}s(\vec{r})^4 - H(\vec{r})s(\vec{r}) \right] . \quad (14.2)$$

where $t = (T - T_C)/T_C$ and K is another parameter. This is called the Landau–Ginzburg free energy functional or simply Landau free energy and is without a doubt the most frequently used free energy model in condensed matter physics and materials science.

The virtue of the present form of Landau theory is that we may calculate directly the correlation function, which in turn makes it possible to find out the extent of fluctuations. In the previous form, we only had the global total order parameter M, and we could not address a certain spin or the local magnetization at a certain place. Now we can. The actual calculation involves, however, a heavy dosage of Fourier transform and functional derivatives beyond the scope of this book, and so let me just quote the result. The correlation function $G(0, \vec{r}) = <s(0)s(\vec{r})>$ for large r turns out to be:

$$G(r) = \frac{\pi^{(1-d)/2}}{2^{(1+d)/2}} \frac{k_B T}{K/2} \frac{e^{-r/\xi}}{r^{(d-2)/2}} \tag{14.3}$$

which, when substituted into the static susceptibility sum rule, gives $\gamma = 1$, the same mean field result. More significantly, it is now possible to estimate the importance of fluctuations. Remember that the fluctuation-dissipation theorem relates the order parameter fluctuations to the susceptibility and that the static susceptibility sum rule relates the susceptibility to the correlation function. Thus, it is possible to estimate the importance of the fluctuations from

$$Q \equiv \frac{|\int d^d r\, G(r)|}{\int d^d r\, s(r)^2} \tag{14.4}$$

where the fluctuations of the order parameter over the correlation volume $V = \xi^d$ in the numerator is compared with its mean in the denominator. If $Q \ll 1$, fluctuations are minimal and mean field theories are vindicated. On the other hand, if $Q \gg 1$, then fluctuations are substantial and mean field theories fail. How do you calculate Q? It is entirely possible that as long as both the numerator and the denominator are calculated consistently within the same theoretical scheme Q may be far more reliable than either one alone. Then, why not calculate Q using the present mean field theory? The result turns out to be

$$Q \sim t^{d/2-2} . \tag{14.5}$$

According to this, mean field theories would be valid if (a) $d \geq 4$, or (b) $d = 3$, but the temperature is far away from T_C, i.e. outside the critical region. This is called the Ginzburg criterion. It has been applied to many different systems and all indications are that the mean field result of Q is correct. In other words, mean field theories predict their own demise, an interesting story by itself.

Unless the temperature is far away from T_C, mean field theories fail in three dimensions. Critical phenomena need a new idea, an idea drastically different from mean field theories.

14.3   The Renormalization Group Theory

The critical point behavior was discovered during the latter part of the 19th century, but it took nearly a century to understand the essential physics involved in it. The renormalization group (RG) theory not only makes it possible to calculate the critical exponents but more importantly, it explains in a rather picturesque manner the universality of critical behavior. Our purpose here is not to go through the details of the RG theory which is beyond the scope of this book, but to highlight the essential elements of the theory so that the reader can get a feeling for the theory. Advanced and interested readers can find a couple of excellent books in the reference section.

We have painfully learned that for critical phenomena the issue is no longer in finding the largest macrostate. It is how to take into account the fluctuations between all macrostates, the largest, the smallest, and everything in between. How can we do that? We need to let each macrostate have its due chance regardless of its macrostate. Recall what we found in Chapter 10. The partition function is the sum of the Boltzmann factors if we wish to work with microstates. If we wish to work with macrostates, the energy in the Boltzmann factor should be replaced with the macrostate free energy. Thus, the probability of the system visiting the macrostate defined by $s(\vec{r})$ is given by the partial partition function

$$Z_{s(\vec{r})} = \exp(-F_{\mathrm{LG}}\{s(\vec{r})\}/k_BT)\,, \qquad (14.6)$$

where the Landau–Ginzburg free energy functional F_{LG} is now in the most generalized form

$$\frac{F_{\mathrm{LG}}\{s(\vec{r})\}}{k_BT} = \int d^dr \left\{ g_0 + g_2 s(r)^2 + g_4 s(r)^4 + \cdots + \frac{1}{2}K[\nabla s(\vec{r})]^2 + \cdots \right\}. \qquad (14.7)$$

What is inside the large parenthesis represents the local free energy per volume. It includes terms beyond the fourth order and more interface-related terms such as $[\nabla s(\vec{r})]^4$; it will become clear later why we have to be willing to include all of them. The coefficients, g_0, g_2, etc. are the main focus of the renormalization group theory because they characterize the sample under study. Regarding them as a vector in the parameter space,

$$g = (g_0, g_2, g_4, \ldots K, \ldots). \qquad (14.8)$$

we will refer to them simply as g.

The probability given by Eq. (14.6) has to be **summed over all possible macrostates** for the full partition function,

$$Z = \int \mathcal{D}s(\vec{r}) \exp(-F_{\mathrm{LG}}\{s(\vec{r})\}/k_B T). \qquad (14.9)$$

This is the continuum counterpart of Eq. (11.4). The functional integration $\int \mathcal{D}s(\vec{r})$ is an integral over all possible values of $s(\vec{r})$ at all positions $\vec{r}$. It adds the Boltzmann factor of all possible functional forms that the order parameter function can take. It should read 'sum over all possible macrostates'. We are trying to take the fluctuations to the fullest extent.

How do you carry out such an integral? It is indeed a very complex matter, but it can be put in a more manageable form, at least conceptionally, if we switch from the position space to the wave-vector space by expressing the order parameter distribution $s(\vec{r})$ in terms of its Fourier components,

$$s(r) = L^{-d/2} \sum_{k<\Lambda} s(k) \exp(-i\vec{k} \cdot \vec{r}), \qquad (14.10)$$

where L is the linear dimension of the space. We may now express the free energy as a functional of $s(k)$ which turns out to look like

$$\frac{\mathcal{F}_{\mathrm{LG}}\{s(\vec{k})\}}{k_B T} = \sum_{k}(g_2 + Kk^2)|s(k)|^2$$

$$+ \frac{g_4}{4} L^{-d} \sum_{k,k',k''} s(k)s(k')s(-k-k'-k'') + \cdots. \qquad (14.11)$$

Since we are not going to carry out any actual calculation, we will not write down its explicit form. The partition function is now given by

$$Z = \int \mathcal{D}s(\vec{k}) \exp(-\mathcal{F}_{\mathrm{LG}}\{s(\vec{k})\}/k_B T), \qquad (14.12)$$

where

$$\mathcal{D}s(\vec{k}) = \prod_{k} ds(k). \qquad (14.13)$$

All will be in vain if we do not understand what $s(k)$ means. Let me repeat here what has been said in Chapter 1. If $s(k)$ is very large for a small wave vector k, then it means that the order parameter distribution $s(\vec{r})$ has a prominent pattern which tends to repeat itself with wavelength given by $\lambda = 2\pi/k$. In other words, there are many structures of size given by $\lambda = 2\pi/k$. So, the repeating pattern is big. Thus $s(k)$ reveals how the order parameter function is distributed in all length scales. Remember the relationship between the size of the repeating pattern and the corresponding k. All the features of long length scales are in the small k regime while all those of short length scales are in the large k regime. The statement $k < \Lambda$ in Eq. (14.10) declares that the shortest length scale that we are willing to worry about is $2\pi/\Lambda$. And remember how rich the structure of $s(\vec{r})$ is near the critical point. It has repeating patterns of all sizes, which is why we called it a dirty diffraction grating.

To prepare for the next step, let us break up the integral into two parts, one for large wave vector part and one for short wave vector part:

$$\mathcal{D}s(\vec{k}) = \mathcal{D}^> s(k) \times \mathcal{D}^< s(k) \equiv \prod_{\Lambda/l < k < \Lambda} ds(k) \times \prod_{0 < k < \Lambda/l} ds(k) \qquad (14.14)$$

which we will use shortly.

The diverging response functions are a result of long range correlations. Many different small local parts over a long range are dancing together coherently. Critical opalescence, on the other hand, shows that the system does not support just one large dancing group but also many smaller ones of all sizes. This must be a result of the massive fluctuations. Exactly at $T = T_C$, the groups come in all sizes, from the infinitely large ones down to very small ones and everything in between. The renormalization group theory is an attempt to find out the symmetry of these dancing groups. The symmetry operation is called the renormalization group transformation (RGT). When the system is at the critical point, the system should be invariant to the symmetry operation. How would you construct such an operation?

We have some clues. The operation should involve some sort of scaling. Another clue was provided by Kadanoff who has been widely hailed for his idea of coarse graining. We have seen that atomic and molecular minute details in short length scales do not contribute to the critical behavior. If we get rid of all such small details by integrating out the corresponding degrees of freedom, how would the sample look like? This important question was originally raised by Kadanoff and was later picked up by K. Wilson who constructed the RGT using it. To repeat the question, if we get rid of the group of the smallest patches from the sample, what would be the result? Let us assume that it will give another sample. Would that sample be the same as the first one? It cannot because it now does not have the group consisting of the smallest patches that the original sample had. So we anticipate that the coarse graining must be followed by a rescaling so that the group consisting of the second smallest patches may now replace the smallest patches in the previous sample. But that would then reduce the group consisting of the biggest patches to the second biggest. That would be no problem if the initial sample was at $T = T_C$ where the correlation length is infinity because any finite fraction of infinity is still infinity. That may take care of the multi length scale structure. But we still have more to worry about. The current group of the smallest patches has been formed by squeezing the group of the second smallest group, which means that the current group of the smallest patches contain a larger number of spins than the original group that it has replaced. In order to have a self similarity, $s(k)$ itself may have to be rescaled.

The heart of the renormalization group theory is the renormalization group transformation (RGT). Given a sample, namely, $g = (g_0, g_1, g_2, \dots)$, using RGT we generate another sample, $g' = (g'_0, g'_1, g'_2, \dots)$. We will explain how this is done shortly. We manipulate the transformation so that these samples mimic the way a laboratory sample changes its correlation length when we change its temperature near T_C. This manipulation is done in three steps.

(1) We start with a coarse-graining. We carry out the summation in Eq. (14.14) only for $\mathcal{D}^> s(k)$,

$$Z^< = \int \mathcal{D}^> s(\vec{k}) \exp(-\mathcal{F}_{\text{LG}}\{s(\vec{k})\}/k_B T), \qquad (14.15)$$

which gives the partial partition function $Z^<$ involving only the small k components. Should we sum up over this for the remaining small k components, we know that

$$\int \mathcal{D}^< s(\vec{k}) Z^< = Z. \qquad (14.16)$$

In view of this, it is reasonable to assume that $Z^<$ has the same structure that Z has but involving only small wave vector components, i.e.

$$\int \mathcal{D}^< s(\vec{k}) \exp\left[-\mathcal{F}'_{\text{LG}}\{s(k)\}\right] = Z, \qquad (14.17)$$

where

$$\exp[-\mathcal{F}'_{\text{LG}}\{s(k)\}] = \int \mathcal{D}^> s(k) \exp\left[-\mathcal{F}_{\text{LG}}\{s(k)\}/k_B T\right]. \qquad (14.18)$$

This is a very important equation. Let us examine it thoroughly. The integrations on the right hand side remove all the patterns in the short-length regime $\Lambda/l < k < \Lambda$ and therefore the remaining length scales extend only to $2\pi/(\Lambda/l) = 2\pi l/\Lambda$. By performing the integrations, we say that **we do not want to see any short-length details beyond the length scale of $2\pi l/\Lambda$**. So, when the short-length scale structures are removed from the sample, how does the sample now look like? The prime sign on the left hand side says that it looks like the same sample except that the coefficients g are now different. We can guess how this happens. Look at the fourth order term involving g_4 in Eq. (14.11). Depending on the wave vectors, some will remain unaffected by the coarse-graining integration while some others will. In the latter case, the integration can reduce the fourth order term to a second order term involving only two $s(k)$s. Similarly a 6th order term may end up generating a new fourth-order term by the coarse graining. This is why F_{LG} has to be in the most general form.

(2) Shrink the sample by a factor of l. This can be done by changing k to lk, that is,

$$k \to lk \qquad (14.19)$$

wherever we see k. All length scales are reduced by a factor of l, and therefore the shortest length scale is back to $2\pi/\Lambda$.

(3) All the remaining Fourier components corresponding to $k < \Lambda/l$ are now rescaled with $l^{1-\eta}$,

$$s(k) \to l^{1-\eta} s(k). \qquad (14.20)$$

With this rescaling, we decrease the "contrast" of what we see now. Assuming that we are looking at the sample under a light, this rescaling makes the light dimmer.

When RGT is performed on a properly chosen sample, it results in another sample which we may write symbolically as

$$R_l\, g = g'. \qquad (14.21)$$

Exactly at the critical point, there are patterns of all length scales, from zero to infinity, and therefore the new sample should be indistinguishable from the old one. In other words, the parameter space should have fixed points g^* which remain invariant under RGT, namely,

$$R_l\, g^* = g^* \qquad (14.22)$$

and actual calculations bear out this expectation.

Away from the critical point, on the other hand, since the longest length scale is not infinity, RGT should shorten the longest length scale. In other words, RGT should push the sample further away from the critical point. The result should then reveal how the correlation length becomes shorter and shorter as $t = (T - T_C)/T_C$ increases, namely, the correlation length exponent ν. These two expectations are borne out if we choose η properly.

The story goes like this. Consider a point $g^* + \delta g$ where δg is the vector drawn from the fixed point to g. Representing component α of g with g_α, we may linearize R_l and write

$$\delta g'_\alpha = \sum_\beta \left(R_l^L\right)_{\alpha\beta} \delta g_\beta, \tag{14.23}$$

where

$$\left(R_l^L\right)_{\alpha\beta} = \left(\frac{\partial g_\alpha}{\partial g_\beta}\right)_{g=g^*}, \tag{14.24}$$

where the differentiation is taken at $g = g^*$. The presence of the off-diagonal elements suggests that the parameters are tangled up after the action of RGT. The best way to handle this type of messy situation is to represent the vectors (points) in the parameter space in terms of the eigenvectors and represent the results of RGT in terms of the eigenvalues. The eigenvectors and eigenvalues are defined by

$$R_l^L e_j = \rho_j(l) e_j. \tag{14.25}$$

We demand that the operation of RGT is associative, i.e. $R_l R_{l'} = R_{ll'}$, from which it follows that

$$\rho_j(l) = l^{y_j}. \tag{14.26}$$

Using the fact that the set of eigenvectors span the parameter space, we may write

$$\delta g = \sum_j t_j e_j, \tag{14.27}$$

and

$$\delta g' = \sum_j t'_j e_j. \tag{14.28}$$

It then follows that

$$t'_j = t_j l^{y_j}. \tag{14.29}$$

The effect of RGT on a point near a fixed point is different depending on the direction of δg. Along the direction of those eigenvectors with $y_j < 0$, Eq. (14.29) says that RGT drives the point towards the fixed point. Think of the points in this path as the basin of the fixed point. All the points in the basin represent different samples but they differ from each other only in some irrelevant short-length atomic and molecular details which are washed away by RGT. Since RGT drives all of them into the same fixed point, these samples share the same critical behavior. This explains the universality of critical point phenomena. The eigenvectors corresponding to $y_i < 0$ represent properties irrelevant to critical point phenomena.

Now let δg be directed along the direction of an eigenvector with $y_j > 0$. Equation (14.29) now says that RGT pushes the point further away from the fixed point along the direction of that eigenvector. The eigenvectors corresponding to $y_i > 0$ represent relevant properties. Among the eigenvalues with $y_i > 0$, find the largest one and call it y_1. Since the temperature cannot be irrelevant, let us assume that it is represented by e_1. Concentrating on the temperature dependence,

$$R_l^L g(T) \approx g^* + t_1(T) l^{y_1} e_1 + \cdots . \tag{14.30}$$

At the critical point $T = T_C$, $t_1(T) = 0$; if it were not, RGT would drive it away from the fixed point. This means that we can expand $t_1(T)$ near T_C to first order and write

$$t_1(T) = A(T - T_C). \tag{14.31}$$

The sample is now characterized by $(t_1, t_2, t_3 \cdots)$. These parameters contain highly coarse-grained yet local information, and therefore $t_1, t_2, t_3 \cdots$ must be a smooth function of temperature, as Eq. (14.31) claims for t_1. Substituting Eq. (14.31) into Eq. (14.30), we find

$$R_l^L g(T) \approx g^* \pm (l/\xi)^{1/\nu} e_1 + \cdots \tag{14.32}$$

where $\nu = 1/y_1$ and $\xi = A|T - T_C|^{-\nu}$ is identified as the correlation length. The way t_1 changes by RGT reveals how the correlation length changes with the changing temperature!

There are many parallels between RGT and vision. Look at a picture which has objects of many different sizes. If you move further away from the picture, you lose small details; you get less light and thus lose contrast, and see the objects on a reduced scale. With this in mind, look at the portrait in Fig. 14.1. The portrait consists of blocks, and therefore it has

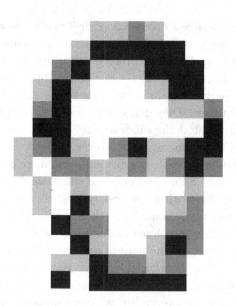

Fig. 14.1   A highly coarse-grained portrait of Lincoln.

no feature in length scales shorter than the block size. Walk away from it until you can no longer see the individual blocks. At this distance, even if there were short-length details in each block, you would not see them. You are now seeing only the features which come in length scales larger than the block size, and apparently there are enough of them to identify the famous face. Do you see Lincoln? If you walk further away, this crucial feature is renormalized away and nothing is left. In the language of RGT, the portrait represents systems away from fixed points.

It is difficult to find a picture resembling a fixed point because the picture has to have features of all length scales including an infinite one. Straight railroad tracks as viewed from the back of a train come close, but only capture finite length scales. In order to have an infinite one as well, just imagine that the width of the railroad track is very very wide. Then one would see very long length scales right in front, the nearby railroad ties, and infinitely small ones far away on the horizon. As the train moves, the longest one turns into the next longest, and what was previously the next longest turns into the next-next longest, and *ad infinitum*. If the longest one were infinitely long, since half of infinity is still infinite, we would find the scene invariant against the RGT. My graphic skills are not good enough to draw this scene.

Finally, let me add a few comments. Our journey was to study the way thermal systems change their behavior as the temperature is lowered, from a very disorderly behavior to an orderly behavior. This transition takes place at a particular well-defined temperature. But the transition comes with a big dance party. It is interesting that they do not dance like they do at high temperatures. They dance together in an orderly fashion rather than different individuals doing whatever they want to do as they do at high temperatures. As the temperature is further reduced, the party gradually fades away and the constituents become more and more orderly. Liquid-gas molecules just gather. Spins just point to the same direction. They still dance together but only with a few nearby neighbors.

If they dance together, should that not make the problem easier rather than harder? If we take a volume as small as 1 cm × 1 cm × 1 cm and put in it a gas or liquid, the number of gas or liquid molecules is as large as 10^{19} and more. The coordinate of each is a degree of freedom. We have a huge number of degrees of freedom, which is why it is hard to calculate everything on a paper. But we can simulate what the molecules and spins do as accurately as needed on a computer and run it to calculate the susceptibility and isothermal compressibility. If the temperature is not near T_C, it is foolish to bother with as many as 10^{22}. Just a few hundreds will do. But if the temperature is near T_C, a few hundreds Will NOT do. You can calculate the susceptibility, but it will remain finite and will not diverge. You have to cover at least ξ^3 for the volume and choose the number accordingly, which calls for a very large number of molecules, i.e. a very large number of degrees of freedom. Each $s(k)$ may be regarded as a degree of freedom. When the problem is stated in this way, I hear the alert reader saying 'thin out some of the degrees of freedom'. That is exactly what RGT does. RGT may be regarded as thinning the degrees of freedom. At $T = T_C$, the number of degrees of freedom is infinity. No matter how many times RGT is performed, we cannot thin them out to a few. Away from $T = T_C$, we can. Think about it. Is it not amazing that 10^{22} coupled degrees of freedom are reduced to a few coupled degrees of freedom? That is what coarse-graining does.

14.4   Phase Separation Dynamics

So far our topic has been static equilibrium properties of matter. For a change, we shall study in this section non-equilibrium phenomena. Take a thermal system in equilibrium in a certain phase in its phase diagram. Suddenly change the condition so that it will be thrown into the middle of another phase region far away in the phase diagram. The system suddenly finds itself in a hopelessly displaced non-equilibrium state and searches for its new equilibrium state. Spinodal decomposition is an example of such processes.

Spinodal decomposition is not only interesting on its own right but it is also very useful from a technological point of view. It provides a very finely dispersed microstructure which can be manipulated to enhance physical properties of matter. The temperature at the non-equilibrium state is far away from T_C and the fluctuation can play an important role but not a dominant role, and therefore the Landau–Ginzburg free energy is well suited for this purpose. The issue simply is: if a system is given an unduly large free energy as it is thrown into an unduly small macrostate, how should the order parameter function change to reduce the free energy, i.e. to reach the largest possible macrostate? So, the issue is the way a small macrostate evolving into a larger and larger macrostate, and the evolving macrostates are represented by the Landau–Ginzburg free energy.

Let me reintroduce the Landau–Ginzburg free energy functional for binary systems. Consider a binary system consisting of two species A and B. Divide the space into small cells. The cells are so small that the chance of more than one particle being in the same cell at the same time is negligible. If cell i is occupied by a particle of species A, the cell is associated with an Ising spin $s_i = 1$. If the cell is occupied by a particle of species B, then it is associated with $s_i = -1$. In this way, the distribution of molecules (atoms) may be described with Ising spins. Combine the cells to form blocks. The blocks are large enough to include many cells but still small enough to be represented by d^3r. Denote the average spin in each block with $c(\vec{r})$ which represents the local concentration of species A or the lack of species B. The order parameter function $c(\vec{r})$ is a continuous function of $\vec{r}$, and $c(\vec{r}) > 0$ means that species A is dominant at $\vec{r}$ while $c(\vec{r}) < 0$ means that species B is dominant. In this way, we again arrive at the Landau–Ginzburg free energy functional,

$$F\left\{c(\vec{r})\right\} = \int d^3r \left\{ f_L[c(\vec{r})] + \frac{1}{2}K[\nabla c(\vec{r})]^2 \right\}. \tag{14.33}$$

Because the kinetic degrees of freedom are not included, the only role that $k_B T$ plays is through g_2 which changes sign when T passes through T_C.

14.4.1   The Time Evolution of the Order Parameter Function

Return to Fig. 12.15 which is a phase diagram for binary mixtures. The system is initially in the high temperature single phase in which the two species are mixed uniformly. Now suddenly quench the temperature well below T_C to displace the system into the two phase region. Here the system is either metastable or unstable against phase separation, but because they are thrown here in a mixed state it takes time for them to phase separate; they will have to go through some tortuous path in the configuration space to find a configuration in which the two species are separated from each other.

The question to answer is: given an initial order parameter function $c(\vec{r}, 0)$, how does it evolve as time progresses? The easy and quick answer would be: it should evolve so as to decrease the free energy. The order parameter function initially given represents a small (not probable) macrostate with an unduly large amount of free energy. We have to find an equation which can guide the order parameter function to change in a way that helps reduce the free energy. The atoms need to do some highly complex migration, some this way and some that way. And this traffic, to use the metaphor we introduced in Sec. 3.8, is controlled by the chemical potential.

The question of whether $c(\vec{r})$ should increase (for more A) or decrease (for more B) at $\vec{r}$ depends on the chemical potential at $\vec{r}$.

$$\mu(\vec{r}') = \frac{\delta F}{\delta c(\vec{r}')}, \tag{14.34}$$

which says that the local chemical potential $\mu(\vec{r})$ is the rate of increase of the free energy per each unit of increase in $c(\vec{r})$. So, we may still use the metaphor we used in Chapter 3 and regard it as a measure of 'unlivability'. But the order parameter function $c(\vec{r})$ is a conserved quantity. If a particle moves out of a block, it has to go somewhere; if $c(\vec{r})$ decreases at $\vec{r}$, the order parameter has to increase somewhere in the neighborhood. So, the desired direction of migration has to be determined by comparing the local 'unlivability' with those of the surrounding blocks.

Let us take the functional derivative in Eq. (14.34). The functional derivative asks: if the concentration changes at $\vec{r}'$ while remaining the same everywhere else, how much does the free energy change? Let $c(\vec{r})$ change by $\delta c(\vec{r})$ at $\vec{r}'$. To take the functional derivative, replace $c(\vec{r})$ in Eq. (14.33) with $\delta c(\vec{r})\delta(\vec{r} - \vec{r}')$. The delta function ensures that the integral picks up only what comes out of the block at $\vec{r}'$ and nothing from all other blocks. The interface term $[\nabla c(\vec{r})]^2$ requires an integration by parts, after which we find

$$\mu(\vec{r}) = g_2 c(\vec{r}) + g_4 c(\vec{r})^3 - K\nabla^2 c(\vec{r})\,. \tag{14.35}$$

Since $\nabla^2 c(\vec{r})$ involves not only $c(\vec{r})$ at $\vec{r}$ but also $c(\vec{r})$ at the surrounding blocks, the interface term provides a coupling between nearby blocks, as it should. Particles should move in the direction in which the "unlivability" decreases most rapidly, namely along the direction of $-\nabla\mu(\vec{r})$. That is the sought-after answer: the traffic controller tells A particles to move along this direction (B particles in the opposite direction) to reduce their free energy. So, we may regard $-\nabla\mu(\vec{r})$ as the driving force trying to rearrange the order parameter distribution. It will then cause an order parameter current which we will denote with $\vec{j}$. The relationship between the driving force and the resultant current should be just like Ohm's law which relates the electric force to the resultant electric current in conductors. We may similarly relate the two on a phenomenological base with

$$\vec{j}(\vec{r}) = -M\nabla\mu(\vec{r})\,, \tag{14.36}$$

where M is the measure of mobility. It is to the order parameter current what the conductivity is to the electric current.

If there is a current from point X to point Y, the current should reduce the concentration at X and increase it at Y. The order parameter conservation may be enforced with the continuity equation,

$$\frac{\partial c(\vec{r}, t)}{\partial t} = \nabla\vec{j}(\vec{r}, t)\,, \tag{14.37}$$

where t represents time. Substituting Eq. (14.36) into Eq. (14.37), we finally obtain the Cahn–Hilliard equation

$$\frac{\partial c(\vec{r}, t)}{\partial t} = M\nabla^2[g_2 c(\vec{r}, t) + g_4 c(\vec{r}, t)^3 - K\nabla^2 c(\vec{r}, t)]\,, \tag{14.38}$$

which instructs $c(\vec{r}, t)$ how to change in order to reduce the free energy. We have just sung the last verse of our free energy song!

The Cahn–Hilliard equation is highly non-linear and cannot be solved analytically. That is not a problem. We can solve it on a computer by writing a program. We may start from an initial order parameter function and update it using the $\partial c(\vec{r})/\partial t$ that the Cahn-Hilliard provides. To that end, some preparations are necessary. First, notice that we need to specify three parameters, a, b and K to precisely define the system to be studied. In addition, we need to specify the temperature $T - T_C$. No one can possibly solve the equation for all possible sets of these parameters and the temperature. It would be nice if we could solve the equation just once in such a way that from that solution we could deduce the solution for any given set.

This is indeed possible. Change the three main variables, the local concentration c, the time t, and the position coordinate r, in the following way:

$$c = \left(\frac{|g_2|}{g_4} \right)^{1/2} \phi , \qquad (14.39)$$

$$t = \frac{K}{2Mg_2^2} \tau , \qquad (14.40)$$

$$r = \left(\frac{K}{|g_2|} \right)^{1/2} R . \qquad (14.41)$$

I will leave it as an exercise for the reader to prove that the rescaled variables, ϕ, τ, and R are all unitless. In terms of these unitless variables, the Cahn–Hilliard equation now reads

$$2\frac{\partial\phi(\vec{R},\tau)}{\partial\tau} = \nabla^2[\pm\phi(\vec{R},\tau) + \phi(\vec{R},\tau)^3 - \nabla^2\phi(\vec{R},\tau)] , \qquad (14.42)$$

where the first term takes a plus sign for $T > T_C$ and a minus sign for $T < T_C$. All parameters have disappeared. After solving the equation for ϕ as a function of $\vec{R}$ and τ, the system-specific information can be put back, if you so desire, by merely scaling the three variables following Eqs. (14.39), (14.40) and (14.41).

For the purposes of learning, and for some more serious purposes as well, we can get a good sense of what the equation does by solving it in two dimensions. Make a square lattice with a small lattice spacing, and indicate the order parameter at each lattice site by $\phi(i,j)$. For the initial configuration at $T > T_C$, choose random values for $\phi(i,j)$ in the range between $+1$ and -1 so that $\sum_{i,j} \phi(i,j)$ is zero, which means a critical (50–50) mixture, or with a bias for plus (or minus) so that the sum is not zero, which means an off-critical mixture. Because the Cahn–Hilliard equation enforces the order parameter conservation, the mixing ratio remains constant throughout the ensuing time evolution.

Next, approximate the differentiation operators using the first order finite difference method. For example, $\partial\phi(x,y)/\partial x$ at the site (i,j) may be approximated as $\phi(i+1,j) - \phi(i,j)$. In this way, one can compute all the partial derivatives appearing in the right-hand side of Eq. (14.42), and then update $\phi(i,j)$ for each time increment $\delta\tau$. This completes our preparations.

We are now ready to implement the equation to relocate a thermal system in equilibrium to a totally different part of the phase diagram, and that very suddenly prohibiting them to follow a reversible equilibrium path. If we translate this in human terms, it would mean that we suddenly displace a group of happy content people to a totally new environment where they have to find a new way to become happy and content again. That would be very mean. That would be mean just as well if we did to animals, but there is one exception.

I live in a house next to a large vacant lot full of trees. This is home to a nation of squirrels. They seem to like all the fruits that I manage to grow in my backyard, even more than we do. Over the past five or six years, we have hardly tasted any; the squirrels wanted them all. If I could sit down with them and cut a deal, I would be happy for a mere one percent, but I know that they will not settle for anything less than one hundred percent. Thoughts of a violent solution, a BB gun, did flash through my mind but I could not do it. A more peaceful solution was suggested by a colleague of mine who is a master squirrel trapper. From a nearby farming and home supply store, I bought a nice trap which is a wire cage of about 30 cm × 30 cm × 70 cm. It has a door and a seat. I invite them to enjoy some nuts on the seat. When they jump on the seat, the door closes. Not all members of the Lee family were thrilled with this idea of getting rid of the squirrels, and so we had some serious discussions and investigations on where to relocate them. Our first choice was the university campus because the campus is full of live oak trees, but we thought that each tree was owned by resident squirrels who may not be very kind to a newcomer. So we decided to take them to a nearby forest about 20 minutes away. We thought that the forest should be large enough to accommodate plenty of newcomers and provide a more continuous supply of food throughout the year.

Whenever I find a trapped squirrel waiting for me, I take the cage to the forest, leave it on the ground, and then open the door. The way the squirrels run away from the cage is a truly heart-rending scene. There are two types of squirrels. The first type will run away from the cage at lightning speed and climb up a tree. The second type starts running just as fast, but seem to fear that I am watching them, and so they suddenly stop when they find something to hide under, wait a while, and then run again. I always wish them a happy new life in their new home. I would love to learn how they adjust to the new environment, but I will never know. I do know one thing for sure. They were very happy when they recovered their freedom, and I hope they will appreciate the wide open forest where they can roam about. I may have acted against some law of ecology, but I certainly did not act against the law that we treasure in this book.

Back to the thermal world! How do the atomic squirrels handle the displacement and relocation? As you have seen, the Cahn–Hilliard equation makes it sure that the order parameter function is updated each time so as to decrease the free energy. But it is a local decision rather than a global decision. If we were to update at $\vec{r}$, we are only taking into account the block at $\vec{r}$ and those surrounding it. Then what is being done at one block can easily be against what is done at a nearby block; it is utterly unlikely that all A particles will march to one side while all B particles to the opposite side. That is why it takes time for them to phase separate. That is why they form some interesting and useful patterns of phase separation on the way.

Since the two species are initially mixed, we start out with an order parameter function which changes randomly with $\vec{r}$; it can be generated easily using a random number generator. Using the Cahn–Hilliard equation with the minus sign in front, we then update the order parameter function. The solution turns out to be different depending on the mixing ratio.

In the metastable region in which, say, A is the minority, the initial phase-separated domains are small droplets of A-dominated phase in a sea of of B-dominated majority phase, as we discussed in Sec. 12.8. Those droplets that overcome the barrier will grow for a while because there is enough imbalance in the chemical potential which pulls A atoms into the region surrounding the droplets, but the process cannot continue indefinitely because the available A atoms are soon depleted. Thereafter the droplets begin to merge to reduce the interface cost. The net effect is that big droplets grow at the expense of small droplets. This phase separation mechanism is called nucleation and growth.

In the unstable region where the mixing ratio of the two species is closer to 50/50, since neither of the two species are dominant, the phase-separated domains tend to grow into a shape resembling seaweeds, or a mesh of tunnels which seem to turn, merge, or close in an unpredictable manner.

Fig. 14.2   A sketch of a very typical spinodal pattern.

Immediately after the quench, all the initial thermal fluctuations (in local compositions) that existed before the quench stay and begin to grow; with no barrier to overcome, the phase separation begins immediately. In the end, one long wavelength mode dominates to form phase-separated domains which look like those shown in Fig. 14.2. As time further progresses, the pattern coarsens and the seaweeds grow bigger and bigger to reduce the interface cost. This process is called spinodal decomposition.

The morphology of the evolving phase-separated domains is a matter of great interest to materials scientists. Most useful materials are composite materials consisting of more than one component. The physical properties of the composites depend heavily on the morphology of the phase-separated domains. One wishes to learn how to manipulate the phase separating process to obtain the optimum morphology. In order to do so, one must first understand the morphology-forming process, and the Cahn–Hilliard equation serves that purpose very well.

Let us add one more feature to the equation. Suppose that the system has frozen impurities that prefer one of the two components more than the other. These impurity atoms are just sitting at certain randomly chosen locations and stay without compromising with the thermal fluctuations. Figure 14.3 shows the evolving pattern where the small dots represent the impurity atoms. The impurity atoms prefer to to be surrounded by one

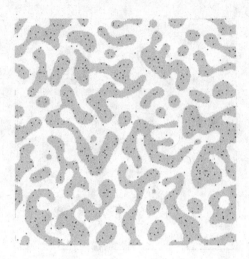

Fig. 14.3   A spinodal pattern in the presence of frozen impurity atoms which prefer the gray component. The small dots represents these impurity atoms.

species over the other, which is an expensive proposition because granting this preference for every impurity atom would vastly increase the interface cost. The two competing factors, the need to please the impurity atoms and the need to minimize the interface, must be optimally balanced. At the time the picture was taken, most of the dots were indeed surrounded by their preferred component (the gray), but notice that several dots have been abandoned in hostile territory.[14] Do the resultant domains look like various animals to you?

To conclude the section, if the issue is how fast one can convert the disorderly mixed state in the high temperature phase into an orderly unmixed final state in the low temperature phase, one should gradually lower the temperature so that the system can make the first order phase transition through a reversible path which would consist entirely of equilibrium states on the way. Instead, we quenched the temperature and made it hard for the displaced system to find the final equilibrium state. I made this remark partly for its own reason and partly to remind you that our underlying theme is the order and disorder. What I want to make clear is that we are using the terms 'order' and 'disorder' here strictly in the sense we declared in an earlier chapter: an orderly state can only be realized in a smaller number of ways than a disorderly state can be. Without doing any actual calculation, can you convince yourself on this point? Take a two

---

[14]For details, see J. C. Lee, *Physica A* **20**, 127 (1990).

dimensional square lattice to make it simple. Half of the sites are to be occupied by A particles and the other half by B particles. How many ways are there to separate A and B particles? Draw a contiguous line so that it will divide the lattice into two halves. Yes, there are quite a few ways. That is the number of ways of realizing the ordered state. Now using a random number generator, pick randomly half of the sites. These sites are going to be occupied by A particles and the rest by B particles. This can be done in a much larger number of ways. (it would be very rare, but if it happened that all the selected sites are connected, it should be thrown out.) That is the number of ways the mixed state can be realized. Do you not sense that the latter number is much larger than the former?

Exercise. *This exercise is for all computer and high-tech wizards which I am not. Suppose that we mix ink and water at a ratio of, say, 1 to 100. What would be the probability of finding the ink molecules spontaneously aggregating to write "THERMAL PHYSICS" in the middle of the water? I would suggest the following two-dimensional experiment. Write the letters on a piece of paper with a thick marker pen, and then make the letters bigger with a lighter pen to gradually mesh the letters into the white background which represents the water component. Digitize it and carry out the integration in Eq. (14.33). Repeat the experiment with "COMPUTER SCIENCE", and then "GREEK PHILOSOPHY". Which is more probable?*

Chapter 15

Quantum Fluid

15.1 Introduction

We started out the book with the theme of entropy and disorder. Thermal systems have unlimited appetite for entropy and disorder at high temperatures. However, the inter-particle interactions begin to play a role as the temperature is lowered and limit thermal systems to a much lower entropy and thus a higher degree of order. One cannot claim to have seen how orderly a thermal system can be until one studies quantum fluids. Unfortunately, quantum fluids require very sophisticated physics. So, just like the previous chapter, read this chapter as a story. The story covers only what I consider the key elements that causes the orderly behavior. It is not meant to be comprehensive. We will introduce the notion of elementary excitation to show how atoms dance together in an orderly fashion and discuss the Landau theory of superfluidity. This will be followed by a discussion on the Bose-Einstein Condensation in cold atoms. As all atoms turn into one super atom, this is the ultimate order. Unfortunately, the quantum mechanics necessary here is more than we prepared for in Chapter 2, and the beginners should read it just as a story.

Interacting fermions and bosons exhibit remarkable varieties of low temperature behavior. When the temperature is below a certain transition temperature, conducting electrons in metals seem to experience no resistance from the ions of the lattice, and likewise, liquid Helium-4 atoms seem to flow through narrow pores with no viscosity from the pore walls. These are a result of the interactions and the physics is far more complicated than for non-interacting ideal gases. But it turns out that in the low temperature regime, particles do not act individually but act collectively. The energy of these collective motions come in quanta and each of them is therefore called the elementary excitation. Moreover, the elementary excitations possess both momentum and energy, and therefore they are also called quasiparticles. In this way, the interacting particles turn into noninteracting or very weakly interacting gas of quasiparticles. We missed our old friends, the ideal gas particles, but we find here their cousins, the ideal gas of quasiparticles. When physicists enter into a new unfamiliar territory, they cannot help but trying to frame the new problem in the old familiar way. And we often succeed!

When we throw a pebble into a pond, the pebble is trying to transfer its momentum and energy to the water molecules in the pond. We know that the transfer is always carried out, no matter how we throw the pebble. But imagine just for a moment that because the water molecules are in a very peculiar condition, the transfer is impossible. Then what happens? Since the pebble keeps the same momentum and energy, it should continue its motion through the pond without causing any disturbance as if the water molecules did not exist! The lack of interaction between the pebble and water is mutual. Make a wall out of these pebbles and put the wall in the middle of a water stream. The water will pass right through it without slowing down one bit as if the wall did not exist! Of course this does not happen with water, but it does happen with liquid Helium-4. That is the story of superfluidity. The story of superconductivity is similar. The key to understanding these remarkable phenomena is the peculiar way the quasiparticles possess momentum and energy. This requires a very complex physics beyond our means. For the superfluidity of interacting bosons, however, a relatively simple phenomenological theory exists due to Landau, providing us with a chance to see how the transfer can be impossible.

Another important topic of interacting bosons is the Bose-Einstein condensation in cold atoms. This is also a topic beyond the scope of this book, but my motivation is to highlight the remarkable order in the B-E condensate. In every sense of the word, this is the most remarkable ordered state that I know of.

15.2 Superfluidity

The boiling temperature of liquid He^4 is near absolute zero: far, far below the 100 degree temperature of water. When we cool Helium gas below this temperature, the gas atoms liquefy and the liquid boils violently. Once the temperature passes below T_λ, however, the liquid suddenly becomes calm and looks just like any other ordinary liquid. But this is no ordinary liquid. Try to dip some up with a ceramic spoon. The liquid passes through the small pores of the ceramic as if the pore walls did not exist, in much the same way as water passes through a mesh of wire when we try to cool hot spaghetti. The viscosity of the pore in the ceramic spoon appears to have vanished. Indeed Helium-4 liquid behaves as if it consisted of two components, a superfluid component and a normal component; what goes through the ceramic pores is the superfluid component. Where does this strange idea come from?

Just imagine that liquid Helium is at the absolute zero temperature. The liquid is in its ground state, the state where the total energy of the atoms is the lowest at E_0. The exact nature of the ground state is uncertain but just think of it as the analogue of the condensate of the ideal Bose gas and call it the condensate. Now imagine that we raise the temperature slightly. The atoms pick up the thermal energy and will be excited. According to Landau, the excitation energy is given by

$$E_n - E_0 = \sum_k \hat{\epsilon}_k n_k \tag{15.1}$$

where E_0 is the ground state energy and n_k are any positive integers including zero, and $\hat{\epsilon}$ is called the elementary excitation because the total excitation energy can be just one of it, two of it, or any integer number of it. The subscript k warns us that the elementary excitations also carry momentum $p = \hbar k$. Since they carry both momentum and energy, they are taken as quasiparticles. The important difference between these quasiparticles and what we normally call particles is that while the latter particles conserve their numbers the formers do not. It is these quasiparticles that constitute the normal component, and the condensate or the background from which they emerged may be regarded as the superfluid component.

In the above the excitation was made possible by raising the temperature. We may say that by raising the temperature we created those quasiparticles. Is there any other way to create them? That should be possible at least in principle by sending into the liquid a large classical object. Consider the possibility of creating just one quasiparticle. If the proposed process is possible, the momentum and energy of the newly created quasiparticle can only come from the classical object. Then the conservation laws would demand that

$$\frac{1}{2}M_0 V'^2 = \frac{1}{2}M_0 V^2 - \tilde{\epsilon}_k \tag{15.2}$$

and

$$M_0 \vec{V}' = M_0 \vec{V} - \hbar \vec{k} \qquad (15.3)$$

where V' is the speed of the classical object after the encounter with the liquid. Square the momentum equation and multiply the result by $1/2M_0$ and then subtract the energy equation from the result to obtain

$$\hbar \vec{V} \cdot \vec{k} = \hbar V k \cos(\theta) = \frac{\hbar^2 k^2}{2 M_0} + \tilde{\epsilon}_k . \qquad (15.4)$$

Now let M_0 go to ∞ so that we may ignore the first term in the last equation. This leaves us with

$$\hbar \vec{k} \cdot \vec{V} = \hbar k V \cos(\theta) = \tilde{\epsilon}_k . \qquad (15.5)$$

Since $\cos(\theta)$ cannot be greater than than unity, it says that

$$V > \text{minimum of} \left(\frac{\tilde{\epsilon}_k}{\hbar k} \right) . \qquad (15.6)$$

Unless the speed of the classical object is greater than this minimum, the classical object with all its momentum and energy cannot transfer any to the liquid and the proposed creation of quasiparticle cannot happen. Then the classical object moves through the liquid without any resistance. This is how Landau argues for the superfluidity.

According to Landau the energy spectrum of $\tilde{\epsilon}_k$ is as shown in Fig. 15.1. The slope of the dashed-line gives the critical velocity or the minimum speed necessary for the superfluid behavior. You can see that $\tilde{\epsilon}_k / \hbar k$ starts with a finite value. It stays that way throughout the phonon region and then as it enters the roton region it begins to decrease, reaching the minimum shown by the dotted line. The slope of the dotted line gives $V_{min} = \Delta / \hbar k_0 \approx 5 \times 10^3 cm/sec$

The figure is a plot of the energy versus momentum relationship given by

$$\tilde{\epsilon}_k = \begin{cases} c\hbar k & \text{for phonon part} \\ \Delta + \hbar^2 (k - k_0)^2 / 2\sigma & \text{for roton part} \end{cases} , \qquad (15.7)$$

where

$$\begin{cases} c = 239 \text{ m/sec}, & \text{speed of sound} \\ \Delta / k_B = 8.65 \text{ K} \\ k_0 = 1.92 \text{ A}^{-1} \\ \sigma / m = 0.16 \end{cases} \qquad (15.8)$$

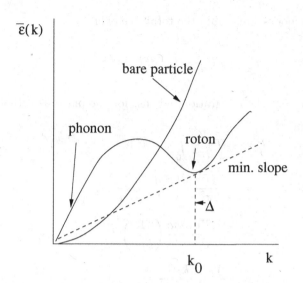

Fig. 15.1 Depicted in the figure is the elementary excitation energy as a function of wave vector k. It is compared with the bare particle energy.

and m in the last line is the mass of a helium atom. The low wave vector portion of $\hat{\epsilon}_k$ consists of phonons and the higher wave vector portion rotons. Such relationships between energy and momentum are called a dispersion relationship, a terminology borrowed from optics. Notice how different the dispersion relationship is compared to that of the classical ideal gas particles or the bare atomic particles. For the latter, when momentum is $p = \hbar k$, energy (or the frequency) is $\hbar^2 k^2 / 2m$.

How did Landau figure out the dispersion relationship? He got the clue from a specific heat data. He viewed the specific data given to him as that of quasiparticles.

So, let us calculate the partition function of the quasiparticles. The partition function is the sum of Boltzmann factors over all micro states and the microstate energies are given by: $0, \tilde{\epsilon}, 2\tilde{\epsilon}, 3\tilde{\epsilon}, \ldots$.

In fact we did this before for black-body radiation. Proceeding the same way for just one mode of wave vector k,

$$Z_k = \sum_{n=0}^{\infty} e^{-\beta n \tilde{\epsilon}} = \frac{1}{1 - e^{-\beta \tilde{\epsilon}}} \tag{15.9}$$

and the corresponding Helmholtz free energy is

$$F_k = k_B T \ln \left(1 - e^{-\beta \tilde{\epsilon}_k}\right) . \tag{15.10}$$

Adding all modes, we obtain the total free energy

$$F = k_B T \frac{V}{2\pi^2} \int_0^\infty dk \, k^2 \ln\left(1 - e^{-\beta \tilde{\epsilon}_k}\right) . \tag{15.11}$$

Let us carry out the integration but just for the phonons with $\tilde{\epsilon}_k = \hbar c k$.

$$
\begin{aligned}
F &= k_B T \frac{V}{2\pi^2} \int_0^\infty dk \, k^2 \ln\left(1 - e^{-\beta \hbar c k}\right) \\
&= -\frac{k_B T V}{2\pi^2} \frac{\beta \hbar c}{3} \int_0^\infty dk \, \frac{k^3}{e^{\beta \hbar c k} - 1} \\
&= -\frac{k_B T V}{2\pi^2} \frac{\beta \hbar c}{3} \left(\frac{1}{\beta \hbar c}\right)^4 \int_0^\infty dx \, \frac{x^3}{e^x - 1} \\
&= -\frac{V}{2\pi^2} \frac{1}{3} \frac{(k_B T)^4}{(\hbar c)^3} \frac{\pi^4}{15} .
\end{aligned}
\tag{15.12}
$$

In the first integral, the upper limit has been extended to ∞ because the extension hardly adds anything to the integral. To see this, consider $k = k_0$ which is outside the phonon region where $\hbar c k_0 / k_B T = 35$ and $\exp(-35)$ may safely be regarded as zero. For the second line, integration was performed by parts and the integrated part vanished at both limits. The change of variable for the third line is obvious. The integral appearing in the last equation is $\Gamma(4)\zeta(4) = \pi^4/15$.

Entropy is then given by

$$S = -\left(\frac{\partial F}{\partial T}\right)_V = \frac{V}{6\pi^2} \frac{4 k_B^4 T^3}{(\hbar c)^3} \frac{\pi^4}{15} \tag{15.13}$$

and the specific heat by

$$C = T\left(\frac{\partial S}{\partial T}\right)_V = \frac{V}{2\pi^2} \frac{4 k_B^4 T^3}{(\hbar c)^3} \frac{\pi^4}{15} . \tag{15.14}$$

The entropy rapidly vanishes to zero as T approaches to zero, but if everything is so organized and orderly, why does it not approach zero faster? Where is the multiplicity? A given amount of excitation energy may be realized in many ways with different phonon numbers in different phonon modes.

The calculation for the roton part proceeds similarly. The free energy comes out to be

$$F = -\frac{V\hbar^2}{3\pi^2\sigma}e^{-\Delta/k_BT}$$

$$\times \left[\frac{3}{8}\sqrt{\pi}\left(\frac{2\sigma}{\hbar^2}\right)^{5/2}(k_BT)^{5/2} + 3k_0^2\frac{\sqrt{\pi}}{4}\left(\frac{2\sigma}{\hbar^2}\right)^{3/2}(k_BT)^{3/2}\right]. \quad (15.15)$$

The roton entropy and specific heat are:

$$S = \frac{V}{2\pi^2}k_0^2\sqrt{\pi}\left(\frac{2\sigma}{\hbar^2}\right)^{1/2}k_Be^{-\Delta/k_BT}\left(\frac{\Delta}{(k_BT)^{1/2}} + \frac{3}{2}(k_BT)^{1/2}\right), \quad (15.16)$$

$$C = \frac{V}{2\pi^2}k_0^2\sqrt{\pi}\left(\frac{2\sigma}{\hbar^2}\right)^{1/2}k_Be^{-\Delta/k_BT}\left(\frac{\Delta^2}{(k_BT)^{3/2}} + \frac{\Delta}{(k_BT)^{1/2}} + \frac{3}{4}(k_BT)^{1/2}\right).$$

$$(15.17)$$

Combining the results for the phonon part and the roton part, we have the total specific heat. Rotons are an important heat absorbing mechanism at temperatures near $k_BT = \Delta$, but as the temperature approaches zero, the mechanism rapidly ceases to be active and the phonon part dominates. It was the temperature dependence of this specific heat that gave the clue to Landau. This is how he determined the energy spectra of the elementary excitations and all the parameters. This was stunning to me when I first learned it many years ago and it still is now.

Return back to Eq. (15.7) and examine what the first line says for a low energy excitation. The excitation mechanism is the sound wave. There is no doubt that this is a well-coordinated collective motion involving every single atom. The excitation energy can be as low as we wish, and notice that for a low energy excitation, the wave vector k has to be very small, which means that the wavelength has to be very long. If we want a really really low energy excitation, the wavelength has to be really really long.

Compare this to the same type of long wavelength sound wave in a solid bar. Every atom is executing a simple harmonic motion, moving back and forth, but there is a tiny systematic time lag between neighboring atoms; if there were not, the whole bar would be moving back and forth, which we know does not happen unless we pull one end back and forth. Indeed if we pull a spoon placed on a table at one end, the whole thing moves without leaving a single atom behind, and this does not excite the atoms. This is due to the rigidity of the solid. In order to excite such tightly held atoms, we have to gently tap it to create a sound wave in it. That is the only way the rigidly held atoms can take a low energy excitation. Now consider the atoms executing a really long wavelength sound wave. If we just look at the atoms in a small finite portion Δx which is much less than the wavelength, the atoms would hardly look like executing a wave motion. Depending on where Δx is taken, the atoms would look like all displaced forward by the same amount or backward by the same amount, or not displaced at all. If we judge by each of these findings, we would say that there is no excitation. That is how the atoms are fooling the rigidity. The atoms want to be free like they were when they were in a gaseous phase at high temperatures. This is the best they can do to free themselves from the rigidity. It is a large scale conspiration involving every single atom.

Viewed in this way, there is a perfect analogy between the superfluid atoms in liquid Helium and these atoms in a spoon. When we pull a spoon, the whole thing moves. That is a superfluidity. The superfluidity of Helium atoms is the same manifestation of rigidity, the rigidity being that of the condensate. In both cases the long-wavelength phonon is called the Goldstone mode.

The background in which Landau did the work is also stunning. Kapitza discovered the superfluidity of liquid Helium, but no one had the slightest idea how to explain it. There was a promising man whom he believed to have the capacity to do the urgent job, but he was in a jail or more likely a labor camp in Siberia for dissidents. Kapitza successfully persuaded his government (then the Soviet Union) to release Landau. Landau did not disappoint Kapitza: within two years he did what we have learned above and received a Nobel award. Then in late fifties, a fatal auto accident struck him. I heard that a mechanical engineer immediately designed a bed to suit his injured body and several airplanes were ready to fly to anywhere to bring medicine or doctors. They tried that hard to save the human treasure and kept him alive for a few years, but he never fully recovered and died in 1962. You can tell how warm and caring teacher he also was by reading the letters he wrote to aspiring beginning physicst students who sought his advice.[15]

---

[15]L. D. Landau's plain talk to students of physics, *American Journal of Physics* **45** (1977) 415.

15.3 Bose–Einstein Condensation

In the above Landau theory, there is no mention of Bose-Einstein condensation, but Fritz London suggested that the superfluidity of liquid helium is a result of BEC. This inspired a great deal of effort to detect BEC in liquid Helium, but the idea is supported only by an indirect evidence, according to which BEC is reduced to less than 10 percent by the strong interactions between Helium atoms. The search was then redirected to find a way to work with weakly interacting atomic bosons in gaseous phase. A gas phase is desirable because we would then be free from all the complications that arise due to the complexities of the strong interatomic interactions, but it comes with a very expensive trade; we have to bring the gas atoms to the low temperature required for BEC, which is not an easy task. BEC is a quantum phenomenon and in order for it to occur it is necessary that

$$\Lambda_{th} > (V/N)^{1/3} \tag{15.18}$$

which says that the thermal de Broglie wavelength must be larger than the average interpaticle spacing of the gas atoms. Since the density is low in gas than in liquid, this is a formidable demand as it requires to lower the temperature to the order of nk or 10^{-9} K. This difficulty was overcome by Cornell and Wieman at Boulder and by Ketterle at MIT in 1995 who successfully created the conditions required for BEC. They trap a gas in a small region of space using a magnetic trap and then cooled the gas by means of a laser cooling technique. They observed a true BEC for the first time since Einstein predicted it nearly 70 years ago, and the evidence was overwhelming. Let me quote what Cornell and Wieman say in their Nobel lecture.

"We were experienced enough to know that when results in experimental physics seem too good to be true, they almost always are! We worried that in our enthusiasm we might confuse the long-desired BEC with some spurious artifact of our imaging system. However, our worries about the possibility of deluding ourselves were quickly and almost entirely alleviated by the anisotropy of the BEC cloud."

The bosons in the condensate constitute a hitherto unseen state of matter. In BEC, all atoms are in the same state sharing the same space. They may be regarded as a super atom of a size large enough to allow a photo picture.

Let me outline the physics. The key is the wave function of the condensate. It can be calculated because the interactions between the gas atoms are relatively weak and allows a relatively simple approximation. Under conditions appropriate for the gas atoms used in the experiments, the interactions may be simplified with

$$V_{int} = U_0 \delta(\vec{r} - \vec{r}') = \frac{4\pi\hbar^2 a}{m} \delta(\vec{r} - \vec{r}') \tag{15.19}$$

where a is called the s-wave scattering length. Due to the delta function, the effect of the interaction on any boson atom from all other bosons may be averaged out, rendering the two-particle interactions into self interactions. Thanks to this remarkable simplification, we can speak of a legitimate single particle wave function (in spite of the correlation) to be called $\phi(\vec{r})$, which is normalized to

$$\int d^3r |\phi(\vec{r})|^2 = 1 \tag{15.20}$$

where $|\phi(\vec{r})|^2 = n(\vec{r})$ is the probability density of finding a boson atom at $\vec{r}$. This is the lowest single particle orbital, but unlike in the ideal boson gas we studied earlier, it is not a plane wave because the bosons are subject to an external potential which confines them into an anisotropic region of space.

Here comes the crucial step. The wave function of the condensate is given by

$$\Psi(\vec{r}) = \sqrt{N}\phi(\vec{r}). \tag{15.21}$$

It is meant to represent the N bosons locked in the single particle state $\phi(\vec{r})$. We can see that it is just that because

$$\int d^3r |\Psi(\vec{r})|^2 = \int n(\vec{r}) d^3r = N \tag{15.22}$$

which says that the 'atomic cloud' contains N atoms as it should.

Now we have to determine $\Psi(\vec{r})$. To that end, we have to express the energy of the condensate in terms of Ψ. The energy operator is

$$H = \sum_{i=1}^{N} \left[\frac{\hbar^2}{2m} \nabla_i^2 + V_{ex}(\vec{r}_i) \right] + \sum_{i<j} U_0 \delta(\vec{r}_i - \vec{r}_j) \tag{15.23}$$

where V_{ext} represents the interaction with the magnetic trap and is an anisotropic harmonic potential,

$$V_{ext} = \frac{1}{2} m\omega_1^2 x^2 + \frac{1}{2} m\omega_2^2 y^2 + \frac{1}{2} m\omega_3^2 z^2. \tag{15.24}$$

The expected value of energy is then given by

$$\bar{E} = \int \Psi^*(\vec{r}_1, \vec{r}_2, \cdots, \vec{r}_N) H \Psi(\vec{r}_1, \vec{r}_2, \cdots, \vec{r}_N) d^3 r_1 d^3 r_2 \cdots d^3 r_N$$

$$= N \int d^3 r \left[\frac{\hbar^2}{2m} |\nabla \phi(\vec{r})|^2 + V(\vec{r}) |\phi(\vec{r})|^2 + \frac{(N-1)}{2} U_0 |\phi(\vec{r})|^4 \right]. \quad (15.25)$$

The third term inside the parenthesis represents boson-boson interaction energy. To see how it came out to be so, notice that the interaction energy between boson i at $\vec{r}$ and boson j at $\vec{r}'$ is

$$E_{ij} = \int \phi^*(\vec{r}) \phi(\vec{r}) U_0 \delta(\vec{r} - \vec{r}') \phi^*(\vec{r}') \phi(\vec{r}') d^3 r d^3 r'$$

$$= U_0 \int \phi^*(\vec{r}) \phi^*(\vec{r}) \phi(\vec{r}) \phi(\vec{r}) d^3 r \quad (15.26)$$

and that there are $N(N-1)/2$ such pairs. Now rewrite the expectation value in terms of the condensate wave function. The result is

$$\bar{E} = \int d^3 r \left[\frac{\hbar^2}{2m} |\nabla \Psi(\vec{r})|^2 + V(\vec{r}) |\Psi(\vec{r})|^2 + \frac{U_0}{2} |\Psi(\vec{r})|^4 \right] \quad (15.27)$$

where we approximated $N(N-1)$ with N^2. This is what we wanted to have.

We have energy as a functional of $\Psi(\vec{r})$ and also N as a functional of $\Psi(\vec{r})$. And our task is to determine $\Psi(\vec{r})$. How? Is there a free energy which we can construct out of energy and particle number so that we can sing the free energy song? It is the grand potential Ω given by

$$\bar{\Omega} = \int d^3 r \left[\frac{\hbar^2}{2m} |\nabla \Psi(\vec{r})|^2 + V_{ext}(\vec{r}) |\Psi(\vec{r})|^2 + \frac{U_0}{2} |\Psi(\vec{r})|^4 \right]$$

$$- \mu \int d^3 r \Psi^*(\vec{r}) \Psi(\vec{r}). \quad (15.28)$$

How fitting it is! We sang the free energy song many times but never for the grand potential. It takes its turn now. To apply the minimum free energy principle, take the functional derivative of $\bar{\Omega}$ with respect to $\Psi(\vec{r})$, and then set it to zero and solve it to determine $\Psi(\vec{r})$. Let us proceed. So we may minimize Ω with respect to $\Psi(\vec{r})$ or $\Psi(\vec{r})^*$. Choosing the latter, we have Then

$$\frac{\delta(E - \mu N)}{\delta \psi^*(\vec{r})} = 0 \quad (15.29)$$

gives

$$-\frac{\hbar^2}{2m}\nabla^2\psi(\vec{r}) + V_{ext}(\vec{r})\psi(\vec{r}) + U_0|\psi(\vec{r})|^2\psi(\vec{r}) - \mu\psi(\vec{r}) = 0. \qquad (15.30)$$

This is called the Gross–Pitaevskii equation. One can solve this equation numerically. We will, however, take a shortcut. If the the inter-atomic interaction is very large so that U_0 is large, or the number N is very large so that $|\Psi(\vec{r})|^2$ is very large, then the first term $\nabla^2\psi(\vec{r})$ may be ignored. The solution to the remaining three terms is then simply

$$n(\vec{r}) = |\Psi(\vec{r})|^2 = \frac{1}{U_0}\left[\mu - V_{ext}(\vec{r})\right]. \qquad (15.31)$$

Since $n(\vec{r})$ cannot be negative, the solution is limited only within the region of space where the above is positive; $n(\vec{r})$ is zero outside the range. The matter wave cloud representing the condensate in this range must add to N atoms, and therefore

$$N = \int n(\vec{r})d^3r = \frac{1}{U_0}\int d^3r\left[\mu - V_{ext}(x,y,z)\right] \qquad (15.32)$$

which, referring to Eq. (15.24), calls for an integration over a volume of space resembling an American football squeezed along one of the two short axes perpendicular to the long axis. This can be done simply by scaling each coordinate by

$$x = \sqrt{\frac{2\mu}{m\omega_1^2}}x', \qquad \text{etc.} \qquad (15.33)$$

The integration is then over a unit sphere, which gives

$$N = \left(\frac{2\mu}{m}\right)^{3/2}\frac{1}{\hat{\omega}^3}\frac{\mu}{U_0}\frac{8\pi}{15} \qquad (15.34)$$

where

$$\hat{\omega} = (\omega_1\omega_2\omega_3)^{1/3}. \qquad (15.35)$$

We have determined μ in terms of N:

$$\mu = \frac{15^{2/5}}{2}\left(\frac{Na}{\hat{a}}\right)^{2/5}\hbar\hat{\omega}, \qquad (15.36)$$

where a comes from $U_0 = 4\pi\hbar^2 a/m$, and

$$\hat{a} = \sqrt{\frac{\hbar}{m\hat{\omega}}}. \tag{15.37}$$

We have done all this to get some idea about the size of the condensate. Now that we have μ in terms of N, we could substitute it into Eq. (15.31) and find out how large the volume of space is where $n(\vec{r})$ is not zero. A more convenient measure of the size of the condensate is

$$\hat{R} = (R_1 R_2 R_3)^{1/3} = \left(\sqrt{\frac{2\mu}{m\hat{\omega}_1^2}} \sqrt{\frac{2\mu}{m\hat{\omega}_2^2}} \sqrt{\frac{2\mu}{m\hat{\omega}_3^2}} \right)^{1/3}, \tag{15.38}$$

where R_1, R_2 and R_3 are the three semi axes of the American football. Using Eqs. 15.36 and 15.37, and doing some rearrangements, we find

$$\hat{R} = 15^{1/5} \left(\frac{Na}{\hat{a}} \right)^{1/5} \hat{a}. \tag{15.39}$$

For the original Boulder experiment where Rb^{87} was used, $\omega_3 = 2\pi 220$ Hz, $\omega_1 = \omega_2 = \omega_z/\sqrt{8}$; $a = 100a_0 = 100 \times 0.529 \times 10^{-8}$ cm, and $\hat{a} = 1.22 \times 10^{-4}$ cm. Substituting these numbers into the above equation, we find

$$\hat{R} = 0.976 \times 10^{-3} \text{ cm}. \tag{15.40}$$

This is small to our daily standard, but it is to atoms what the entire Pacific ocean is to us! They took a picture of this super atom, or rather the super atom on its way of breaking up. So, the proof of BEC is direct, not a typical subtle indirect one. What posed for long as a brutally difficult problem has been conquered, brutally conquered! The bosons lost their individual identity to form a 'super atom', all sharing the same space. From our perspective, this is the ultimate order. What more can you ask?

As it happens when major discoveries are made in physics, it almost always find a way to technological applications. We now have atomic laser. BEC can be used to slow down lights to a speed less than bicycle speed, even to stop it completely and then later to reactivate it! This has the potential to bring lots of technological inventions. Many useful applications are expected in the near future.

Far far away there is a fairy tale land. The land is divided into ten districts and each district elects a Representative. The elected representatives go to their capital city Anthonville. Anthonville is located on top of a very high mountain and it is so cold throughout the year that no ordinary citizen can live there. There is, however, an ever-present master spirit. The master spirit meets the elected officials at the foothill of the mountain and performs a big ceremony which includes a passionate debate on important issues facing the land. At the end of the ceremony, the elected officials recite certain words again and again and drink something that the master spirit has prepared for them. In this way, they are ready to be Anthonized. When they climb up the high mountain following the master spirit, they turn into one giant. You can no longer see each representative. Instead, all you can see is one giant visible miles and miles away from any part of the land. It is a giant, but it is very gentle, walks and talks like any citizen. The land is governed by this giant affectionately called Tony by the citizens. When the term ends, Tony is de-Anthonized by the master spirit. With just about one or two hours before the effect of the potion runs out, Tony is de-Anthonized. Tony maintains the same size and shape, but it suddenly becomes less bright, and one Representative emerges. That representative apparently was not just a part of Tony but was in every part of Tony. This keeps going on until Tony becomes very fuzzy and barely visible, and then the last representative emerges. Tony disappears from the sight, and people cry like babies. Then they elect their representatives again and become happy when they see another Tony.

Chapter 16

Computer Simulations

16.1 Introduction

The last two chapters were devoted to show how strange thermal physics can be when atoms and molecules behave collectively for more order rather than individually. Except the beginning Landau theory, however, there was not much we could calculate or derive. I would like to close this book with something we can actually do. What do you say? I trust that the reader can write a simple computer program and run it on your computer. In this chapter, I will offer a crash course on a popular method of computer simulations: the Monte Carlo method. The computer simulation can only be performed for a finite system which does not support an infinite correlation length and therefore does not support the true critical behavior. This difficulty can be overcome thanks to the finite size scaling.

16.2 Monte Carlo Method I

The title 'Monte Carlo' carries many different hats. It is not an exaggeration to say that the title goes whenever random numbers are used. So among many, we will study two Monte Carlo schemes pertinent to us. Their purpose is to realize a certain probability distribution. A good example is the Maxwell velocity distribution. Suppose that we measure the speed of a molecule over and over again. We may simulate this action and make a list of the outcome of each measurement. The frequency of a certain value of velocity appearing in this list should be proportional to the Maxwell velocity distribution.

Let us start with a much simpler probability distribution. Suppose that the outcome of x can only be x_1, x_2 or x_3, and the probability of x

taking on x_1, x_2, and x_3 are 0.5, 0.1 and 0.4, respectively. To simulate the process, generate a random number r between 0 and 1. The random number generator should give a number between 0 and 1 without favoring any part over the rest but in an unpredictable way. Such a perfect generator is hard to come by, but many approximate ones exist and are reasonably good for our purpose.

To carry out the simulation, generate a random number r. The outcome of x is determined by r in the following way.

$$x = \begin{cases} x_1, & \text{if} \quad 0 < r \le 0.5 \\ x_2, & \text{if} \quad 0.5 < r \le 0.6 \\ x_3, & \text{if} \quad 0.6 < r \le 1 \end{cases} \tag{16.1}$$

Notice that we have stacked up the three probabilities to divide the range from 0 to 1 into three regions. Since the probability of the random number to be in any of the three regions is proportional to its width, this procedure chooses x according to the given probability distribution. Repeat this many times to generate a series of x. No one can tell whether this is an actual data taken by an experiment or a simulated data.

Now apply the same technique to a continuous probability distribution function $f(x)$ where x can be any thing between 0 and infinity. Actually $f(x)$ is the probability density distribution function because $f(x)dx$ gives the probability for x to be any thing in the range between x and $x + dx$. Construct an accumulated probability distribution function $F(x)$:

$$F(x) = \int_0^x f(x')dx' , \tag{16.2}$$

where $F(x)$ gives the probability for x to be any thing between 0 and x.

Now we are ready to simulate the actual outcome of x. Generate a random number r. The outcome is determined by solving

$$F(x) = r . \tag{16.3}$$

Again repeat this many times and generate a series of x. Shall we say it again? "No one can tell whether the set is a simulated data set or an outcome of an actual experimentation."

This practice of Monte Carlo method is not limited to thermal physics. In atomic physics or nuclear physics, interactions between different particles are always given by cross sections. From the given cross sections, one can construct a normalized probability distribution, and then the accumulated probability distribution. In such cases, the accumulated probability distribution function is given by

$$F(x) = \frac{1}{Z} \int_0^x f(x')dx' ,$$ (16.4)

where

$$Z = \int_0^\infty f(x')dx' .$$ (16.5)

16.3 Monte Carlo Method II: The Master Equation

The purpose of this Monte Carlo is to simulate the equilibrium behavior of thermal systems. This is an extremely ambitious attempt and the fact that such a simulation is possible is a significant triumph. Let me tell you why. Suppose that we try to carry out the canonical ensemble average for a system as small as 100 Ising spins on a computer. A total of 2^{100} microstates need to be generated. This is far too many, even for the fastest computers today. In the language of the above Monte Carlo Method I, x is a variable in a space of 100 dimensions. The sheer vastness of the space makes it impossible to generate that many configurations. Even if the fastest computer ran ever since big bang, it would not have done it. Monte Carlo I is helpless for this purpose. The task is just like asking a blind man standing on his bare feet in Mississippi to find out the tallest mountain peak in North America.

Why not then randomly select a limited number of microstates and carry out the average over them, just like pollsters do when they take public opinion polls? Unfortunately random sampling does not work either because the important microstates, namely, those that carry a substantial Boltzmann factor, are *not* distributed randomly in the microstate state space, but rather hidden here and there together. Why does random sampling work so well for pollsters? Take any city for an example. It is true that people living in a section of a city tend to share the same opinions, but their names do not appear together in the telephone book on the same page. So, if you randomly take a number from the telephone directory, there is a fair chance for every body to be selected.

Is there still a way to generate just enough microstates over which the ensemble average can be carried out? The answer is yes. Many of the 2^{100}

microstates, even if we have generated them, make little or no contribution towards the ensemble average because their Boltzmann factors are so small. What then is the purpose of generating them in the first place? Rather than generating all microstates indiscriminately and then later weighing them only to discard most of them, why not weigh the generating process itself so that a microstate corresponding to a large Boltzmann factor has a better chance of being generated than those corresponding to a small Boltzmann factor? In other words, why not generate the microstates according to their importance? A pool of microstates generated in this way (called importance sampling) can be as small as on the order of 10^6, but because they represent all the microstates in proportion to their Boltzmann factors, they give the correct ensemble average. We can generate such a pool of microstates in the following way.

Imagine that our Monte Carlo walker is wandering around in the microstate space following a transition probability function $W(x, x')$ which gives the probability that the walker will move to microstate x' within one unit of time, given that it is now in microstate x. Given $W(x, x')$, we should be able to obtain the probability function $P(x, t)$, the probability of the walker visiting microstate x at time t. How would the probability function $P(x, t)$ change after time dt? It is clear that

$$\frac{\partial P(x, t)}{\partial t} = -\sum_{x'} P(x, t)W(x, x') + \sum_{x'} P(x', t)W(x', x), \qquad (16.6)$$

which is called the master equation. The first term accounts for all the processes in which the walker moves out of x while the second term accounts for all the processes in which the walker moves into x. The equation defines a dynamics by choosing $W(x, x')$, but the dynamics itself is not our purpose which is why I said just "imagine".

Since the transition probability density is independent of time, it is reasonable to expect that the probability function will ultimately reach a stationary state in which it no longer changes. Several questions may arise. First of all, how do we know $P(x, t)$? Why do we have to wait long to reach the stationary state? Suppose that the walker made 1000 visitations until time t, but among all the microstates he visited, x_1 was visited only once. Then $P(x_1, t) = 1/1000$. Microstate x_{500} is in the middle of the largest macrostate, but it has never been visited so far. So $P(x_{500}, t) = 0$. This is because the walker started from microstate x_1 which is in a tiny little macrostate far away from the largest macrostate. In order for the walker to come anywhere near the largest macrostate, it takes a long time.

Until then $P(x,t)$ will keep on changing. Once the walker comes near the largest macrostate, then and only then $P(x,t)$ will no longer change by any significant amount. For example, unless we wait literally an infinite amount of time, it is unlikely that the walker will return to x_1, and therefore in the stationary state $P(x_1, \infty)$ will be negligible.

Now, we know what we want for the stationary probability distribution. We want it to be the Boltzmann distribution. So suppose that $P(x, \infty)$ turned out to be the Boltzmann distribution. Then the question is: what is the transition probability density function $W(x, x')$ that has brought our walker to this stationary state and is keeping it there? In this stationary state, the walker is visiting each microstate according to the Boltzmann factor, i.e. a state corresponding to a large Boltzmann factor very often and a state corresponding to a small Boltzmann factor very rarely. The walker would be sampling for us the microstates according to the Boltzmann factors. Therefore, the pool of the microstates visited by the walker is exactly what we want in order to perform the proposed importance sampling!

Since $dP(x, \infty)/dt = 0$ in the stationary state, we have

$$\sum_{x'} P(x, \infty) W(x, x') = \sum_{x'} P(x', \infty) W(x', x) \,, \tag{16.7}$$

where $P(x, \infty)$ is given by

$$P(x, \infty) \sim \exp(-E(x)/k_B T) \,. \tag{16.8}$$

To solve Eq. (16.7) for $W(x, x')$, the left and right sides of the equation need not match (balance) term by term, but we demand that they do and rewrite the equation in the form

$$P(x, \infty) W(x, x') = P(x', \infty) W(x', x) \,, \tag{16.9}$$

or

$$\frac{W(x, x')}{W(x', x)} = \frac{P(x', \infty)}{P(x, \infty)} = \exp(-\Delta U/k_B T) \,, \tag{16.10}$$

where

$$\Delta U = U(x') - U(x) \,. \tag{16.11}$$

Let me rewrite Eq. (16.10):

$$\frac{W(x, x')}{W(x', x)} = \frac{\exp(-U(x')/k_B T}{\exp(-U(x)/k_B T} \,. \tag{16.12}$$

It guarantees us that **the microstates will be visited according to their Boltzmann factors**, i.e. those of lower energy will be visited more often than those of higher energy exactly in the same manner as the real system does.

Equation (16.9) is called the condition of "detailed balance". We adopted this condition because it gives us an easy solution as we will find below. We are not after the general solution to Eq. (16.7). We are only after a $W(x, x')$ which will guarantee us that the Monte Carlo walker will visit the microstates according to their Boltzmann factors and we are not concerned about the sequence taken by the walker.

The simplest solution to Eq. (16.10) is

$$W(x, x') = \begin{cases} 1 & \text{when} \quad \Delta U \leq 0, \\ \exp(-\Delta U/k_B T) & \text{when} \quad \Delta U > 0. \end{cases} \tag{16.13}$$

where the first line says 'unity if the movement is downhill' while the second line says 'exp(−increase in energy$/k_B T$) if the movement is uphill.'

Notice that $W(x, x')$ is not symmetric for x and x', which is why it is important to read the equation as suggested above. Partly for a practice and partly to show that Eq. (16.13) solves Eq. (16.10), assume that $U(x') < U(x)$, i.e. $\Delta U < 0$. Then, $W(x, x')$ is a downhill movement and therefore $W(x, x') = 1$, but $W(x', x)$ is an uphill movement, and therefore $W(x', x) = \exp\{-[(U(x) - U(x'))]/k_B T\} = \exp(\Delta U/k_B T)$. Substituting both into Eq. (16.10), it is clear that Eq. (16.13) is indeed the solution. Equation (16.13) is called the Metropolis algorithm.

There are also two symmetric solutions,

$$W(x, x') = \frac{1}{2}(1 - \tanh \Delta U/2k_B T), \tag{16.14}$$

and

$$W(x, x') = \frac{\exp(-\Delta U/k_B T)}{1 + \exp(-\Delta U/k_B T)}. \tag{16.15}$$

The dynamics following Eq. (16.15) is known as the Glauber dynamics.

We assumed that the transition probability density $W(x, x')$ only depends on x and x' and *not* on the past history that has brought the system to x. Such a process is called a Markov process. The time series of the visited microstates is said to form a Markov chain.

.The Metropolis algorithm works in the following steps.

(i) Choose an initial microstate x.

(ii) Choose randomly or sequentially another microstate x' which can easily be realized from x.

(iii) Calculate ΔU.

(iv) If ΔU is equal to or less than zero, the walker moves to x'.

(v) If ΔU is greater than zero, the walker moves to x' with probability $p = \exp(-\Delta U/k_B T)$. To perform this step, generate a random number r between zero and one. We let the walker move to x' if $r < p$. When the gamble fails, i,e., $r > p$, the system remains at x once more; the initial microstate x is counted as having been visited once more, i.e. $x' = x$.

Examine how each trial ends. If it ends at step (iv), the selection of x' is biased in favor of larger Boltzmann factor. The same is true when the gambling fails. Under these two scenarios, the walker would be heading towards a local minimum energy state. It should have a chance to get out, and it does. It is the third scenario, namely, a successful gambling that takes the walker out of the minima, and therefore it is actually the most crucial step.

We repeat this process on the order of 10^6 times to prepare a pool of microstates. We can then calculate all equilibrium averages as

$$< A >= \frac{1}{\hat{N}} \sum_x A(x) \,, \qquad (16.16)$$

where $\hat{N}$ is the number of microstates in the pool. Note that the sum in Eq. (16.16) does not have a Boltzmann factor. The transition probability $W(x, x')$ has been chosen such that microstates are sampled according to the Boltzmann factor, and therefore those important microstates corresponding to a large Boltzmann factor appear in the pool many, many times while those corresponding to a small Boltzmann factor appear very few times. When an average is performed on such a pool, no one can tell whether it is an actual measurement or a simulated data. Well, let us not over do this. It is so to the extent that the microstate energies used for the simulation are a faithful representation of the real microstates.

Exercise. *Let us try out this idea using a very simple example. Assume that microstates are distributed according to* $\exp(-x^2/2)$ *where x is no longer a symbol; it is a number which is distributed between $-\infty$ to*

$+\infty$. *Calculate $<x^2>$ using the Metropolis algorithm. We know the correct answer in advance by carrying out the integration,*

$$<x^2> = \int_{-\infty}^{\infty} dx x^2 \exp(-x^2/2)/Z \, , \qquad (16.17)$$

where $Z = \int_{-\infty}^{\infty} dx \exp(-x^2/2)$. Random sampling does not work here because while x is distributed over an infinite range, the x's that contribute significantly to the average are in a relatively small range around $x = 0$. Make sure that you have an adequate random number generator on your computer. You may generate the Markov chain $(x_1, x_2, x_3, \ldots, x_N)$ in the following way. Imagine a walker who walks only on the x-axis. At each step, the walker can either (i) move to the left to decrease its x-coordinate, (ii) move to the right to increase its x-coordinate, or (iii) stay at the same position. Choose any initial position x_1, and a small, but not too small, step size Δx. Generate a pair of random numbers, r_1 and r_2; r_1 is to be used to decide the direction of the next movement and r_2 is to be used in case the walker has to gamble. Try $x_1 - \Delta x$ for x_2 if $r_1 < 0.5$, but try $x_1 + \Delta x$ if $r_1 > 0.5$. In either case, if $x_2^2/2 < x_1^2/2$, then select x_2 to enter the pool. Otherwise, gamble using r_2. This completes one Monte Carlo step. Generate another pair of random numbers and repeat the process for x_3, and so on. Now, take a look at the chain $(x_1, x_2, x_3 \cdots)$. It should look like footsteps taken by a drunken man who moves to the left and to the right in a seemingly unpredictable fashion. He lingers around $x = 0$ most of the time, but notice that once in a while he moves a good distance away. This is a crucial element of importance sampling. If the walker only lingers around $x = 0$ and never ventures far out, the result will not be correct. Plot the frequency at which each value of x is visited. Does it look like a "Boltzmann factor"? Compute $<x^2>$. In this exercise problem, we can see without any effort where the important microstates are. In the real world problems, we cannot see it, but we do not have to because the Monte Carlo walker sees it for us.

Let us now apply the method to study a system of interacting Ising spins. The microstate energies are given by

$$E\{s\} = -J \sum_{<ij>} s_i s_j \, , \qquad (16.18)$$

where $J > 0$ is the coupling constant, and $<ij>$ means that the sum covers all pairs of spins i and j which are the nearest neighbor of each other.

We need two parameters to specify the system, J and $k_B T$, but only one unitless parameter $\epsilon = k_B T / J$ enters into the Monte Carlo computer code. Run the simulation covering a range of ϵ. From the results, we can, if we wish, deduce what the results would be for any combination of $k_B T$ and J.

The exercise problem above should be good enough to lead you to write your own code. Computer simulation can be a great learning tool, but to fully appreciate it one has to write ones own code. To some beginners, that can be a difficult proposition and I am well aware of what causes errors most often for beginners. It is identifying the nearest neighbors correctly. This can be done in several different ways, and some are more efficient than others. For beginners, however, the efficiency is not an issue; it is writing code correctly even if it may be slow. A slow but cautious and correct code is far better than those incorrect ones written by high flying but error-prone wizards. Once one learns how to do it correctly, then and only then the issue of efficiency may be addressed.

Let me suggest a very safe way. Make a file identifying all the neighbors for each spin. Call the file $II(i, j)$. If $II(i, j) = k$, it says that 'the jth neighbor of spin 'i' is 'k'. Let us start with N Ising spins on a closed one-dimensional chain. Because it is closed, all spins have two neighbors to interact with. For $i = 1$, $II(1, 1) = N$, $II(1, 2) = 2$; for $i = N$, $II(N, 1) = N - 1$, and $II(N, 2) = 1$. For all the rest of the spins in the interior, $II(i, 1) = i - 1$, $II(i, 2) = i + 2$.

Now consider an Ising model in two dimensions which is our main goal. For the sake of simplicity, let us start with a small system of 10×10 spins on a two-dimensional square lattice. Label the spins sequentially. Those on the top row carry the labels 1 through 10, and those on the next row 11 through 20, and etc. Those on the edge have only two neighbors to interact with while those in the interior have four. To eliminate this edge effect, close the sheet by folding it from left to right and from bottom to top so that every spin has four neighbors to interact with. By doing so, we have imposed a periodic boundary condition. Unlike the one-dimensional chain, however, this is difficult to do unless the sheet is an elastic rubber sheet, but what matters is the connectivity of the spins and not the Pythagorean geometry. The neighbor file $II(i, j)$ has four entries for j. We need a convention for the order in which we enter the four. Let $j = 1$ be its left neighbor, $j = 2$ its bottom neighbor, $j = 3$ the right neighbor, and $j = 4$ the top neighbor. I went around starting from the left counter clock way. That was how I always suggested to my students. One day, Eddy showed me his code and results. I was so impressed with his results that I wanted to check his code in more detail because I saw in him immediately a promising candidate for an undergraduate research. Well, his neighbor file was different: $J = 1$ is the left neighbor, $j = 2$ the right neighbor, $j = 3$ the top neighbor, and $j = 4$ the bottom neighbor. I asked "Any reason for choosing this particular order?" His hand started moving from the left part of his chest to his right part, then to his forehead and then slowly to the bottom of his chest. He said "This is how I ask for a blessing". I can certainly use as much blessing as Eddy did, but changing directions make me dizzy. So let me continue with my monotonic couterclock way, and let us do it manually. If there is any chance of making an error, we will divide the spins into groups and do it separately for each group. So let us do it first for the four spins on the four corners.

$$\begin{array}{llll}
\text{II}(1,1) = 10, & \text{II}(1,2) = 11, & \text{II}(1,3) = 2, & \text{II}(1,4) = 91. \\
\text{II}(10,1) = 9, & \text{II}(10,2) = 20, & \text{II}(10,3) = 1, & \text{II}(10,4) = 100. \\
\text{II}(91,1) = 100, & \text{II}(91,2) = 1, & \text{II}(91,3) = 92, & \text{II}(91,4) = 81. \\
\text{II}(100,1) = 99, & \text{II}(100,2) = 10, & \text{II}(100,3) = 91, & \text{II}(100,4) = 90.
\end{array}$$

Do similarly for all the spins on the left column. For example, taking one spin from the middle part of the left column, say, $i = 11$:

$$\text{II}(11,1) = 20, \quad \text{II}(11,2) = 21, \quad \text{II}(11,3) = 12, \quad \text{II}(11,4) = 1$$

where you can see a pattern. Do similarly for all the spins on the right columns, on the top raw and the bottom raw.

For those spins deep in the interior, the pattern is more explicit:

$$\text{II}(i,1) = i - 1, \quad \text{II}(i,2) = i + 10, \quad \text{II}(i,3) = i + 1, \quad \text{II}(i,4) = i - 10.$$

Once you have this file made at the beginning of your program, then the rest should be manageable.

- The microstate x is simply a particular spin configuration:

$$x = \{s(1), s(2), s(3), \cdots s(100)\}$$

where all the spins are up, or down, or any combination.

- To try the next microstate x', you can flip any one spin leaving the rest as they are in x. Go around sequentially so that all the spins have a chance to flip. If you would prefer to go around randomly, that will also work.

$$x' = \{s(1), s(2), s(3), \cdots s(i) = -s(i), \cdots s(100)\}$$

where the i-the spin has been flipped while the rest remain the same as in x.

• The key part is where the walker is debating whether to flip spin 'i' to move on to the proposed new microstate x' or leave it as it is to stay at x. That part of the program would look like

```
engy = s(i)*(s(II(i,1)) + s(II(i,2)) + s(II(i,3)) + s(II(i,4)))
delu = -engy-engy
if(delu.lt.0)then
    s(i) = -s(i)
else
call rand(r)
boltzmn = exp(-dele/epsilon)
if(boltzmn.gt.r)then
    s(i) = -s(i)
end if
end if
```

In the first line, engy is the nearest neighbor interaction energies of spin i. In the second line, -engy would be the energy if spin i flipped. Since all we need is the change in energy and since the rest of the spins do not change, we need not worry about the rest of the spins and their interaction energy. This is in Fortran, and a subroutine called 'rand' is called to generate random numbers.

Complete the code without any calculation and make it sure that it is working correctly. Then gradually add one by one calculating parts. For the specific heat and susceptibility, prepare four storage "buckets" for U, U^2, M, and M^2. After each round, calculate these four quantities. Call the current U and M 'unow, emnow', which may be computed after each round:

```
unow = 0.0
emnow = 0.0
do i = 1,100
unow = unow + S(i) * (S(II(i,1)) + S(II(i,2)) + S(II(i,3)) + S(II(i,4)))
emnow = emnow + S(i)
end do
unow = unow/2.0
```

where all the interaction energies have been counted twice and hence the correction in the last line.

Now dump these into their respective storage buckets.

$$U = U + \text{unow}$$

$$U2 = U2 + \text{unow*unow}$$

$$EM = EM + \text{emnow}$$

$$EM2 = EM2 + \text{emnow*emnow}$$

After $\hat{N}$ rounds, each bucket should have $\hat{N}$ entries. Calculate their average for $<U>$, $<U^2>$, $<M>$, and $<M^2>$. Finally, calculate the specific heat and the susceptibility by calculating δU^2 and δM^2.

If you do this, you will be well rewarded; you can get a good sense of what is going on. You can take a snapshot of spin configurations and plot various quantities. Now if you wish to be more ambitious and want to calculate the exponents, move on to the next section.

16.4   Finite Size Scaling

If you actually carry out the Monte Carlo simulation, you will be disappointed because your results will not exhibit the expected singular behavior. You will only see a mild peak for the susceptibility and hardly anything that can be called a peak for the specific heat. This is a finite size effect. The singular behavior of the response functions is a result of the diverging correlation length which a finite system cannot support.

This impasse can be overcome, thanks to an idea known as finite-size scaling. Consider how the correlation length changes with temperature in a finite system of size L. As the temperature approaches T_C (the critical temperature of the infinite system), the correlation length ξ will grow and at some point reach L. Once it reaches this point, the temperature can get closer to T_C, but ξ will no longer grow. This is just like an indoor tree which can only grow as high as the room ceiling. The temperature at which the correlation length reaches its maximum possible value L depends on the system size. If the system's size is very small, the saturation of ξ will happen while the system is still far from T_C, but if the system size is large, it will happen when the temperature is much closer to T_C. Is there a temperature at which we can *absolutely and positively* guarantee that the correlation length has already saturated to the value of L regardless of the system size? Yes, $T = T_C$!

Now, compare a finite system of linear size L at $T = T_C$ with an infinite system at $T \neq T_C$. Suppose that we choose T such that the correlation length of the infinite system is $\xi(t) = L$ where $t = |T - T_C|/T_C$. Since the correlation length of the finite system is the same as that of the infinite system, we may state that the finite system is a faithful representation of the infinite system at that temperature. Therefore, what happens in the infinite system as the temperature T approaches T_C may be represented with a series of finite systems ordered in increasing size. For a given temperature of the infinite system, the system size of the corresponding finite system may be obtained from

$$\xi(t) \sim t^{-\nu} = L, \tag{16.19}$$

or

$$t = L^{-1/\nu}. \tag{16.20}$$

The criticality of χ and C in the infinite system, $\chi \sim t^{-\gamma}$ and $C \sim t^{-\alpha}$, may then be represented with finite systems at T_C in the form of

$$\chi \sim L^{\gamma/\nu}, \tag{16.21}$$

and

$$C \sim L^{\alpha/\nu}. \tag{16.22}$$

To implement this, compute C and χ starting from a small-sized system, such as $L = 5$, and repeat the computations with gradually increasing L. Plot the results for C and χ as a function of L on a log-log graph. The result should be a straight line. The good news is that L does not have to be infinite to reach this power law regime. The slopes may be identified as γ/ν and α/ν. This trick requires T_C to be known. The critical temperature of the two-dimensional Ising model is given by $J/k_B T_C = -(1/2)\ln(\sqrt{2} - 1)$.

The Monte Carlo scheme that we have studied may be applied to mobile molecular systems, but the task of writing a code is much more difficult. The Lennard-Jones inter-molecular interaction is not as long ranged as the Coulomb potential between charged objects, but it is not as short-ranged as the exchange interactions of Ising spins either. Moreover, since the molecules are moving around, we cannot have a prefixed nearest neighbor table. All these make it difficult to write a code. Actually, for liquid systems there is a better way to simulate than MC; it is called the method of molecular dynamics. In this method, one simulate the dynamics of the constituent molecules following Newton's second law of motion. There is no brilliant finessing like we saw in the Monte Carlo. In fact it is the most brutal simulation method relying on the sheer power of computer. But, if one can write a code, it is priceless. Unfortunately one has to deal with the same difficulty mentioned above. For interested readers, I refer to the very readable book by Allen and Tildesdale.

It is late October here in Hattiesburg. While there remains a good deal more to learn, it is too nice to stay inside. There is going to be a festival downtown this weekend. I will stop the book here and join the crowd to roam the downtown blocks. The brown Pontiac still runs, but it will stay in the garage.

That was about 10 years ago for the first edition. Now it is late spring in Hattiesburg. The Brown Pontiac has retired since then. My dispute with the squirrels? After Katrina, the big pear tree in our backyard was so badly damaged that there has not been as much fruit supply as the squirrels became used to. I gave up my dispute with them anyway and they took all the pears as usual. That was not enough for them. So last year, they started stealing my tomatoes to make up. I grew my vegetable garden for a long time over the years. I have never seen this before, but it is true. I could not let that go. I took out the trap and I invited them with plenty of cracked pecans all around. As smart as they are, the cracked nut was apparently irresistible. Three ventured into the trap. They have been relocated to a heavily wooded area. New residents arrived, but they did not discover that they could eat my tomatoes throughout the summer. So we are getting along OK right now. Wish me good luck that the peace will last.

Bibliography

The literature on Thermodynamics and Statistical Mechanics is vast. I do not claim to have seen them all. I can only say that the following books are on my book shelve and that I consulted some of them while writing this book. I consider the first four books to be classics. They provided me with a crash course in thermal physics, and I wish to salute the authors for these excellent books. The rest are either more recent publications or deal with more advanced subjects of statistical mechanics.

States of Matter by David L. Goodstein (Dover, New York)

Fundamentals of Statistical and Thermal Physics by Frederick Reif (McGraw-Hill, 1985)

Thermodynamics and Introduction to Thermostatistics by Herbert B. Callen (John Wiley, 1985)

Thermal Physics by Charles Kittel and Herbert Kroamer (Freeman, New York, 1980)

Statistical Physics: An Introductory Course by Daniel J. Amit and Yosef Verbin (World Scientific, Singapore, 1995)

An Introduction to Thermal Physics by Daniel V. Schroeder (Addison Wiley, 2000)

Thermal Physics by Ralph Baierlein (Cambridge Press, 1999)

Statistical Thermodynamics by Robert P. H. Gasser and W. Graham Richards (Worlds Scientific, 1995)

Statistical Mechanics and Thermodynamics by Claude Garrod (Oxford, 1995)

A Modern Course in Statistical Physics by L. E. Reichel (U. of Texas Press, 1988)

Statistical Mechanics by Shang-Keng Ma (World Scientific, 1985)

Introduction to Modern Statistical Mechanics by David Chandler (Oxford, 1987)

Statistical Mechanics by Donald A. McQuarries (Harper Collins, 1976)

Statistical Mechanics by Kerson Huang (John Wiley, 1987)

Statistical Physics by L. D. Landau and E. M. Lifshitz (Pergamon, 1980)

Statistical Physics: Statics, Dynamics and Renormalization by Leo P. Kadanoff (World Scientific, 2001)

Entropy, Order Parameters, and Complexity by James P. Sethna (Oxford, 2006)

Statistical Physics of Particles by Mehran Karddar (Cambridge, 2007)

Statistical Physics of Fields by Mehran Karddar (Cambridge, 2007)

The book by Kittel and Kroemer has very nice problems, some are solved and some not; we have borrowed some of them. Amit and Verbin and Garrod offer a wonderful collection of exercise problems with solutions.

A very detailed mathematical exposition of free energies can be found in the book by Callen. Callen also offers a more in-depth exposition of Shannon's information theory.

Except at the beginning, we covered little or no thermodynamics for spins; we relied heavily on statistical mechanics for spins. For an introduction to the thermodynamics of magnetism on a more general ground, a good book is **Statistical Physics** by F. Mandl (John Wiley, 1971). The brief section on magnetic work is superb.

We applied the Monte Carlo method only to a two-dimensional Ising model. The Monte Carlo method may be applied to liquid-gas molecular systems as well, but because the molecular interaction is not short-ranged, the simulation is not as straightforward as for the Ising spins. The actual molecular motions may be simulated using the method called Molecular Dynamics. For both Monte Carlo and Molecular Dynamics for liquids, a good reference is **Computer Simulation of Liquids** by M. P. Allen and D. J. Tildesley (Oxford Press, 1994).

For a more intensive introduction to critical phenomena, you will do well to read **Introduction to Phase Transitions and Critical Phenomena** by H. E. Stanley (Oxford, 1971).

For a coverage of non-equilibrium thermodynamics along the line discussed in Sec. 3.9, see **Nonequilibrium Thermodynamics** by Donald D. Fitts (McGraw-Hill, 1962) and **Thermodynamics of Irreversible Processes** by Rolf Haase (Dover, 1990).

If you wish to study how the partition function may be calculated for Ising models in two dimensions, see Stanley's book. Landau and Lifshitz present a slightly different version for the two-dimensional model.

If you wish to study the fluctuation-dissipation theorem in the context of transport phenomena (a subject which we did not cover), Chandler offers an excellent introduction to the vast subject area. McQuarries provides more extensive coverage of the subject.

For a full coverage of the renormalization group theory of second order phase transitions, I have two excellent books: **Modern Theory of Critical Phenomena** by Shang-Keng Ma (W. A. Benjamin, 1976), and **Lectures on Phase Transitions and the Renormalization Group** by Nigel Goldenfeld (Addison Wesley, 1992). The latter book also covers first order phase transitions.

If the finite-size scaling fascinated you and you wish to read more, see **Finite Size Scaling and Numerical Simulation of Statistical Systems** edited by V. Privman (World Scientific, 1990).

Thermal physics is no longer a peripheral subject. It is everywhere, but its role is most prominent in condensed matter physics. When you feel you are ready, you may first start with the book by Goodstein and then **Principles of Condensed Matter Physics** by P. M. Chaikin and T. C. Lubensky (Cambridge University Press, 1997).

There are those who wish to construct thermal physics based on the Shannon information. **A Farewell to Entropy: Statistical Thermodynamics Based on Information Theory** by Arieh Ben-Namim (World Scientific, 2008); **Statistical Thermodynamics** by H. S. Robertson (Prentice Hall, 1993).

Finally, if you have difficulty writing code for the two-dimensional Ising model, Chandler has one in his book. Similar programs may also be found in **Thermal and Statistical Physics Simulations** by H. Gould, L. Spornick, J. Tobochnik (Wiley, 1995), **Computational Physics** by N. Giordano (Prentice Hall, 1997), and **Computer Simulation Methods** by D. W. Heermann (Springer-Verlag, 1990).

Index